Nearshore Sediment Dynamics and Sedimentation

GREAT BRITAIN: General location map.

Nearshore Sediment Dynamics and Sedimentation

An Interdisciplinary Review

Edited by

J. Hails and A. Carr

Natural Environment Research Council,
Institute of Oceanographic Sciences, Taunton

A Wiley–Interscience Publication

JOHN WILEY & SONS
London · New York · Sydney · Toronto

Library of Congress Cataloging in Publication Data:

Main entry under title:

Nearshore sediment dynamics and sedimentation.

Chiefly papers read at a symposium organized by the editors on behalf of the British Geomorphological Research Group, held in the Dept. of Geography, University of Southampton, 27 October 1973.

'A Wiley–Interscience publication.'
1. Coast changes—Congresses. 2. Marine sediments—Congresses. 3. Oceanography—Congresses. I. Hails, J. II. Carr, Alan, 1930– III. British Geomorphological Research Group.

GB450.N42 551.3′6 75–6950

ISBN 0 471 33946 6

Typesetting by William Clowes & Sons Ltd., London, Beccles and Colchester and printed by photolithography and bound in Great Britain at The Pitman Press, Bath.

Contributing Authors

A. J. BOWEN	*Department of Oceanography, Dalhousie University, Halifax, Nova Scotia, Canada*
A. T. BULLER	*Tay Estuary Research Centre, Dundee University, Newport-on-Tay, Fife DD6 8EX, Scotland*
M. B. COLLINS	*Department of Geology, Imperial College, London.*
R. G. D. DAVIDSON-ARNOTT	*Geography Division, Scarborough College, University of Toronto, West Hill, Ontario.*
G. EVANS	*Department of Geology, Imperial College, London.*
C. D. GREEN	*Tay Estuary Research Centre, Dundee University, Newport-on-Tay, Fife DD6 8EX, Scotland.*
J. T. GREENSMITH	*Department of Geology, Queen Mary College, London.*
B. GREENWOOD	*Geography Division, Scarborough College, University of Toronto, West Hill, Ontario, Canada.*
P. HOLMES	*Department of Civil Engineering, University of Liverpool.*
D. A. HUNTLEY	*Department of Oceanography, Dalhousie University, Halifax, Nova Scotia, Canada.*
P. H. KEMP	*Department of Civil and Municipal Engineering, University College, London.*
P. D. KOMAR	*School of Oceanography, Oregon State University, Corvallis, Oregon 97331, USA.*
J. MCMANUS	*Tay Estuary Research Centre, Dundee University, Newport-on-Tay, Fife DD6 8EX, Scotland.*
W. R. PARKER	*Institute of Oceanographic Science, Taunton, Somerset, England.*
W. A. PRICE	*Hydraulics Research Station, Wallingford, Berkshire, England.*
A. H. W. ROBINSON	*Department of Geography, The University, Leicester LE1 7RH, England.*
E. V. TUCKER	*Department of Geology, Queen Mary College, London.*
D. H. WILLIS	*Hydraulics Research Station, Wallingford, Berkshire, England.*

Preface

This volume contains invited papers, mostly those that were read at a symposium organized by the editors on behalf of the British Geomorphological Research Group and held in the Department of Geography, University of Southampton, on 27 October 1973. The purpose of the meeting was to bring together engineers, mathematicians, physicists, sedimentologists, physical oceanographers and geomorphologists, and to review some of the recent developments in nearshore zone research, particularly studies into sediment and water mass movement, the parameters controlling wave dissipation along the shoreline, the relationship between offshore sediment supply and shoreline equilibrium, and the problems associated with coastal engineering projects.

Although many of the papers refer specifically to research around the coast of the British Isles (see frontispiece) the authors have compared their findings and conclusions with those of other workers studying similar problems overseas. Some of the invited papers concentrate mainly on North American examples.

The editors have decided to publish in its original form the discussion on each paper read at the symposium and it is hoped that further dialogue will ensue between reader and contributor once the book is published. Thus every attempt has been made to meet the interests of a large audience, whatever the discipline of the reader may be.

The general arrangement of topics represented by the sequence of chapters does not strictly follow the symposium proceedings. Chapters 1 to 4 are basically quantitative and are concerned with nearshore currents, the mechanics of wave motion, including wave energy dissipation at the shoreline, and a comparison of the hydrodynamics of steep and shallow beaches. Chapter 5 considers the latest trends in the application of research to resolve coastal engineering problems within the United Kingdom; beach nourishment schemes are considered in the light of the limitations of more conventional methods used to arrest coastal erosion.

Chapters 6 and 7 evaluate sediment transfer in the nearshore zone. The former deals with subaqueous bars in Kouchibouguac Bay, New Brunswick, Canada, while the latter refers to sediment mobility and erosion on a multibarred foreshore in southwest Lancashire, United Kingdom. Chapters 8 and 9 concentrate on estuarial examples, the latter in particular in a broader context.

Chapter 10 looks into some controls on sedimentation in the northern part of the Outer Thames Estuary and compares the formation and reworking of a chenier plain in this part of England with the more classic examples of Louisiana and elsewhere. Finally, Chapter 11 reports the preliminary findings of a comprehensive study of suspended sediment transport over the intertidal flats of the Wash. This is the largest embayment in England and its hinterland is, in many ways, analogous to the polder-lands of northwest Europe.

Although each chapter is self-contained it should not be read in isolation from the others because of the interrelated processes operating in the nearshore zone. For example, the interrelationship between sediment dynamics and ultimate nearshore bedforms is emphasized. We have briefly summarized in the Introduction a few of the more important problems that await solution and that are not covered in any great detail in the individual chapters.

The bibliography at the end of each chapter includes not only source material but what is considered to be useful suggested reading for students. Most of the illustrations are original but where authors have reproduced plates and diagrams from earlier publications full references are given.

Every attempt has been made to use metric units, in either the c.g.s. or SI systems, in the text, but since some of the contributors have referred to, or quoted from, published material citing Imperial units, it has been necessary in a few instances to retain these. Notations are listed at the end of each of those chapters which have distinct mathematical bias, because the symbols and definitions vary slightly between one author and another.

We are extremely grateful not only to the authors but especially to colleagues and friends who have assisted us by their encouragement, in particular to those who have spent many hours, cumulatively, critically reading the manuscripts. These include P. Ackers, R. G. C. Bathurst, E. Derbyshire, K. R. Dyer, J. R. D. Francis, P. F. Friend, A. R. Halliwell, N. Heaps, F. J. T. Kestner, C. A. M. King, C. Kidson, I. N. McCave, R. Miller, K. V. H. Smith, N. B. Webber and D. J. A. Williams. Our sincere thanks also go to M. J. Clark and R. J. Small, Department of Geography, University of Southampton, for their assistance with the organization of the meeting. We would also like to thank the staff of John Wiley & Sons Limited for their valuable cooperation in producing this book.

28 August 1974

J.R.H.
A.P.C.

Contents

Introduction

The purpose of this volume is to focus attention on some of the more recent developments in research into the nearshore zone. Although several of the papers refer to studies undertaken within the United Kingdom, the problems they discuss and evaluate are common to most coastlines and therefore they may be regarded as a variation on a general theme. Nearshore sediment dynamics and sedimentation is a subject that embraces a broad spectrum of disciplines and has aroused the interest of coastal engineers, mathematicians, physicists, sedimentologists, physical oceanographers and geomorphologists for some considerable time. However, since the majority of research programmes have hitherto been unrelated to each other, it is difficult to present a coherent, interdisciplinary review of previous work in this Introduction. As one of us (Hails, 1974) has discovered recently from an extensive review of the literature, it is also somewhat difficult to ascertain what, if indeed any, particular trend is emerging in current research because many problems, particularly those confronting the coastal engineer, are practical rather than academic and consequently demand a rapid solution. Everyone is aware of the increasing pressure on the coastal zone with the advent of more recreational amenities, such as marinas, and the demand for land reclamation, particularly for industrial use (see, for example, Inman and Brush, 1973). Many of these, and allied coastal engineering problems, have provided a major impetus for research into nearshore sediment dynamics.

Until recently research interests have been directed mainly towards theoretical studies and laboratory wave tank investigations of sand movement by oscillatory waves. The constraints of laboratory work are well known, especially such limiting factors as scale effects and the fact that experiments are two-dimensional. The question how well waves generated in wave tanks and models assume the behaviour of their natural prototypes has been frequently raised. Certainly, the distinctive differences both in the mode of breaking and in the after-effects of the wave once it has broken have not been satisfactorily reproduced so far in models (Inman and Bagnold, 1963). However, one must also bear in mind the present limitations of mathematical theory, which is invariably based on simplified and sometimes invalid assumptions. Unfortunately, natural limiting physical factors such as heavy swell and wave turbulence in the breaker zone have restricted *in situ* field measurements, whilst the increasing recreational pressure on formerly less-frequented beaches tends to limit the type of instrumentation that can be installed and left safely for long-term data collection. There is also, of course, the extreme difficulty in the field of being able to vary one parameter at a time and to distinguish between sub-environments; complications such as measuring porosity are particularly evident in the beach zone, for example.

Those people actively engaged in nearshore zone research are readily aware of the existing gap in our knowledge. Perhaps one of the basic, yet most important, questions is how far experiments, both in the field and the laboratory, are representative, particularly since field data are usually obtained either for a short period only or at infrequent intervals over long periods. In neither case do they necessarily relate to the most significant natural events that would help to complement laboratory investigations. There is also the question of gaps in direct sediment measurement which is, at present, done vicariously by measuring hydraulic parameters. These problems are slowly being resolved now that a few researchers, with more adequate financial and manpower resources, are becoming more able to obtain continuous measurements by means of relatively sophisticated instruments in the nearshore zone. This fact is exemplified by the scale and continuity of studies undertaken by research institutes as opposed to those conducted by individuals in universities.

At the time of writing, many important questions remain unanswered about the mechanics of wave motion and how wave energy is transformed and dissipated in and near the surf zone. For example, to what degree does water particle motion keep material in suspension and thus render it available for longshore transport by currents. Similarly, it is necessary to account for the effect of this motion on the size-sorting of sediment by differential velocities in an onshore–offshore direction. As recently as 1972, Longuet-Higgins observed that there was no satisfactory theory to explain breaking waves, either in deep or shallow water, But a number of semi-quantitative theories for predicting both longshore currents and sediment transport are now available. It is hoped that in the near future quantitative and predictive data will be generally available on the movement of material suspended by wave and current action and material moved along the seabed by saltation or as bed load by shear stress in the boundary layer, Perhaps, ultimately, more work will be done on the deposition of sediment in those coastal environments that are characterized by exceptionally strong tidal currents, about which little is known at present.

Evaluation is also required into the effects of wave height, wave period, breaking waves and the differences between storm waves and swell, for example, in relation to the grading of beach material. Such an evaluation must be considered, both along and down the beach face, by using combined field measurements and computational/theoretical wave refraction programmes. The validity of computed wave refraction programmes, both one with another and with 'real sea' observations, is still seriously questioned by some workers. It is also necessary to examine the relationship between the parameters just mentioned and the dynamics of offshore banks and the seabed in the neritic zone. As far as possible, offshore areas and adjacent sectors of the coastline to be studied in order to establish general principles should have a reasonably simple geometric form so that observations in the field can be more readily compared with computed theoretical models. The applicability of mathematical models has been challenged in one of the chapters of this book where the author has found it difficult to prove mobility and continuity of transport from the supposed source area to the point of deposition especially in foreshore areas characterized by complex sand transport patterns.

Although the papers in this volume do not purport to answer all the questions posed here, they do at least shed some light on the complexity of the nearshore zone and into contemporary thought and lines of research.

References

Inman, D. L., and R. A. Bagnold, 1963, Beach and nearshore processes, Part II: Littoral processes. In M. N. Hill (Ed.), *The Sea*, Vol. 3, *The Earth Beneath the Sea History*, Interscience, New York, pp. 529–553.

Inman, D. L., and B. M. Brush, 1973, The coastal challenge, *Science*, **181**, 20–32.

Longuet-Higgins, M. S., 1972, Recent progress in the study of longshore currents. In R. E. Meyer (Ed.), *Waves on Beaches and Resulting Sediment Transport*, Academic Press, New York, pp. 203–248.

Hails, J. R., 1974, A review of some current trends in nearshore research, *Earth-Sci. Rev.*, **10** (in press).

J.R.H.
A.P.C.

□ □

CHAPTER ONE

Wave Conditions in Coastal Areas

P. Holmes

Abstract

The paper discusses the determination of wave conditions appropriate to the computation of sediment movement in coastal areas. A standard method of specification of wave conditions in deep water is presented. The phenomena which occur as waves propagate into shallow water are discussed, with particular emphasis on wave refraction processes. A means of computing onshore wave conditions from offshore wave data is described, illustrated by the results of some computations which have been carried out.

Introduction

Littoral drift may be defined as the transportation of beach material resulting from a number of causative factors of which wave action is particularly important. The rate and direction of littoral movement depend on the alongshore component of wave energy, the physical characteristics and availability of beach material, and other features such as tidal range and beach slope. The detailed mechanism of sediment transportation through wave action is not clearly understood and reliance has to be placed on empirical relationships between wave conditions on the beach and the resulting littoral movement.

This paper attempts to briefly outline methods of evaluating wave characteristics appropriate to the computation of littoral transport.

Wave conditions in deep water

Classical water wave theories are based on deterministic concepts in that they deal with specific values of the important wave characteristics of height, H, length, L, period, T, and velocity, C. Linear wave theory, which strictly applies to waves of low heights, provides a solution for the free surface in the form of a sinusoid, as shown in Figure 1.1.

However, before discussing the evaluation of wave characteristics, it needs to be noted that waves in deep water are neither sinusoidal nor deterministic. The

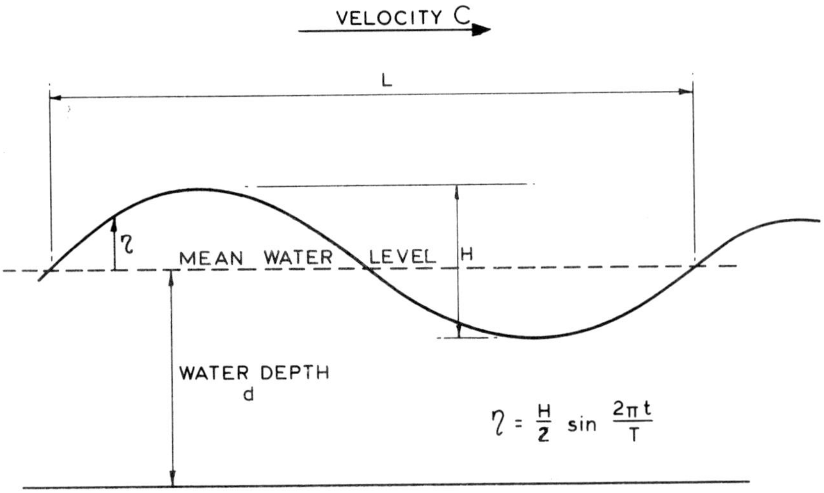

Figure 1.1. Definition of wave parameters.

fluctuations of water-surface elevation may be considered as a random process, described by an energy density spectrum which indicates the distribution of wave energy as a function of wave period, or its inverse, wave frequency, and direction. At the present time only limited information is available on directional spectra. There are also difficulties in using spectral concepts in situations where the response of a system, in this case the littoral drift, is a non-linear function of wave height. For these reasons the wave conditions need to be specified in a somewhat simpler manner.

Figure 1.2 shows a typical section of a wave record on which the definitions of wave height and 'zero-crossing' intervals, T_Z, are given. Thus, for a particular sea-state, there exists a large number of wave heights and zero-crossing intervals and the parameters normally used to represent particular wave conditions are the 'significant wave height', $H_{1/3}$,—defined as the average height of the highest

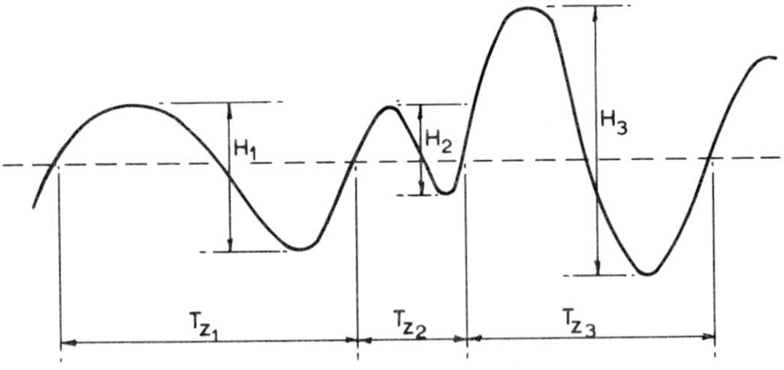

Figure 1.2. Definition of wave height and zero-crossing period.

one-third of the waves—and the mean zero-crossing interval, T_Z. Draper (1963) describes a convenient method for the evaluation of these parameters from a wave record.

If $H_{1/3}$ and T_Z are determined from short wave records of 10 or 15 minutes duration taken at three-hourly intervals over a full year, the results can be presented as a bivariate histogram as shown in Figure 1.3. For littoral drift computations, histograms would be prepared for each month of the year. Such a representation of one year's wave climate is deficient in an important respect, no

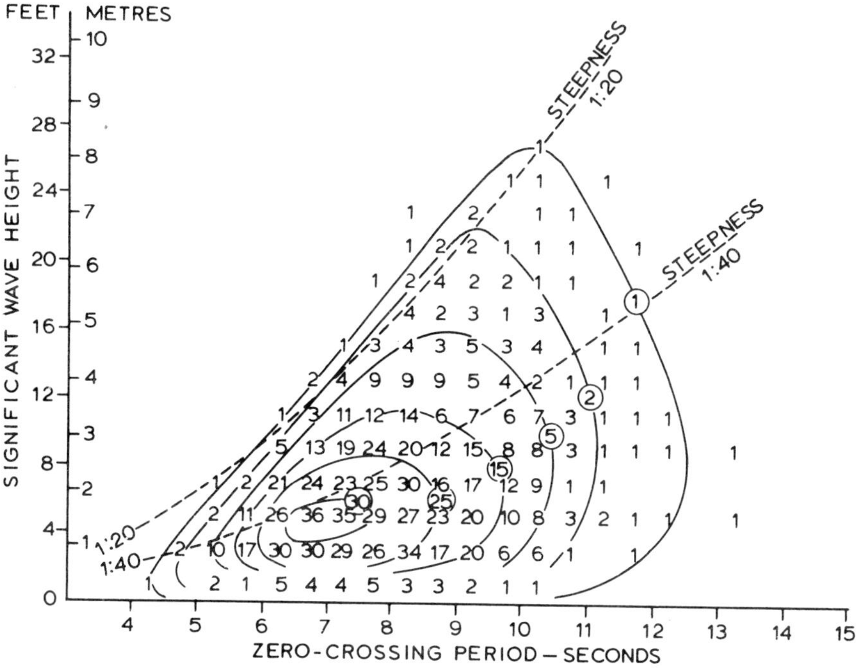

Figure 1.3. Bivariate histogram relating significant wave height to zero-crossing period (after Draper, 1963).

information being given on wave direction. At the present time very few routine wave recording systems are available which provide directional information about the wave conditions. It is therefore necessary to obtain such information by implication from the direction of the winds which generated the waves—on a month by month basis. This latter information is obtainable from meteorological charts which, when combined with measured wave data, provides a trivariate definition of wave conditions in terms of wave height, $H_{1/3}$, wave period, T_Z, and wave direction, θ.

In the absence of recorded wave data the alternative procedure of hindcasting wave conditions has to be used. This process requires the evaluation of wind velocity, directions, duration and fetch from meteorological charts. Hence, by use of semi-empirical relationships between these parameters, a complete month by month wave climate can be obtained (Pierson, Neumann and James, 1955). The computational effort involved in hindcasting is considerable and the results are not reliable in that information on wind velocities over large oceanic areas is subject to considerable error. It is often necessary to 'calibrate' hindcast solutions against limited amounts of wave data which might be available on a particular coastline.

Nearshore wave conditions

The water wave phenomena relevant to the prediction of nearshore wave conditions are those of shoaling, refraction, breaking and diffraction.

As a wave travels into shallow water a change in wave height occurs because of changes in the group velocity, C_g, of the wave. This velocity, which is related to the wave propagation velocity, C, is that at which energy is transmitted in a wave system. The process is termed wave shoaling, a shoaling coefficient, D_d, being used to quantify the change in wave height such that

$$H = D_d . H_0 \tag{1.1}$$

where H_0 is the deep-water wave height and the shoaling coefficient is given by

$$D_d = \left[\tanh \frac{2\pi d}{L}\left(1 + \frac{4\pi d/L}{\sinh 4\pi d/L}\right)\right]^{-1/2} \tag{1.2}$$

It should be noted that the shoaling process is linear in terms of wave height. In deep water D_d equals unity, its value reduces to a minimum of 0·91 at $d/L_0 = 0{\cdot}15$ and thereafter increases without limit. In practice a limit is reached at the point of wave breaking.

The wave propagation velocity, C, is given by

$$C = \sqrt{\frac{gL}{2\pi} \tanh \frac{2\pi d}{L}} \tag{1.3}$$

from which it can be seen that in a decreasing depth of water, $\tanh 2\pi d/L$ reduces, non-linearly, and the wave decelerates. For a particular wave front approaching a uniformly sloping beach at an angle, any particular portion of the wave front will decelerate at a different rate to other parts of the front, resulting in a change in the direction of wave propagation. This process is termed wave refraction and is illustrated in Figure 1.4, where rays are plotted at right-angles to the wave crests. It is assumed that the wave energy between any two rays or orthogonals is preserved, in which case, convergence of orthogonals would result in wave energy being concentrated over a smaller area of ocean surface, leading to an increase in wave height. Conversely, diverging orthogonals would result in reduced wave heights. If b_0 and b are the distances between orthogonals in

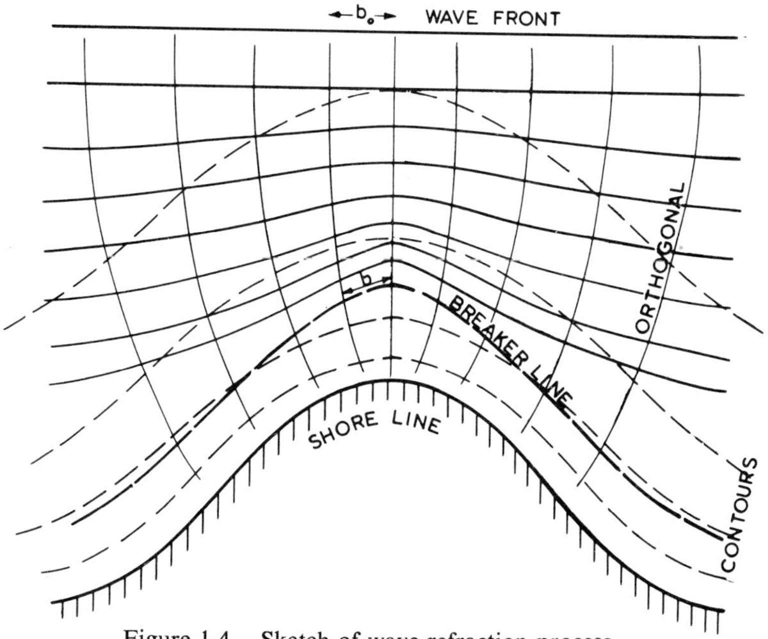

Figure 1.4. Sketch of wave refraction process.

deep and shallow water respectively, then by equating wave energies at two locations

$$wH_0^2b_0C_{g_0} = wH^2bC_g \tag{1.4}$$

where subscript 0 refers to deep-water conditions. Thus

$$H = \sqrt{\frac{b_0}{b}}\,.\,D_d\,.\,H_0 = K_rD_d\,.\,H_0 \tag{1.5}$$

where K_r is the refraction coefficient. Again it may be noted that refraction is a linear process in wave height.

The presence of a current will also result in wave refraction. Wavelength, velocity and height change, as does the direction of wave propagation if the current is at an angle to the wave direction.

The order of magnitude of these changes can be seen by considering a wave travelling in still, deep water meeting a current running with or against the wave. It can be shown (Wiegel, 1964; Johnson and Eagleson, 1966) that the wavelength changes from L_0 to $\bar{L}$ according to the equation

$$\frac{\bar{L}}{L_0} = \left(\frac{1+a}{2}\right)^2 \tag{1.6}$$

where

$$a = \sqrt{1 + 4\left(\frac{U}{C_0}\right)} \tag{1.7}$$

U is the current velocity and C_0 is the wave velocity in deep water. Similarly the wave height will change from H_0 to $\overline{H}$ given by

$$\frac{\overline{H}}{H_0} = \sqrt{\frac{2}{a(1 + a)}} \tag{1.8}$$

Thus, for a following current running at 10% of the wave velocity, the wavelength increases by a factor of 1·19 and the wave height decreases by a factor of 0·88. For an opposing current of the same relative magnitude the corresponding values are 0·79 and 1·21 respectively.

Appropriate solutions are also available for waves and currents intersecting at finite angles. A common consequence of this intersection is that of a possible reduction in wavelength; an associated increase in wave height results in increasing wave steepness, defined as the ratio of wave height to wavelength, eventually leading to wave breaking. In deep water a criterion for wave breaking is often given as $H/L = 1/7$ but for random waves the limit is in the region of 1/16 to 1/20.

The familiar phenomenon of wave breaking in shoaling water is extremely complex in hydrodynamic terms but for purposes of predicting wave conditions at the point of breaking the semi-empirical relationship $H = 0{\cdot}78d$, where d is the water depth, is normally used.

Absorption is the loss of wave energy due to bottom friction and percolation. Such energy losses are often very small unless the waves travel for long distances in very shallow water. A percolation coefficient is given by Bretschneider and Reid (1954) in the form

$$K_p = \left[\tanh\left(\frac{2\pi d}{L}\right)\right]^B \tag{1.9}$$

where $B = 8\pi p/\nu mt$, m being the bed slope, g the acceleration due to gravity, ν the kinematic viscosity of seawater and p a permeability coefficient for the porous seabed. The porous layer, the flow through which results in percolation losses, is assumed to extend a distance of $0{\cdot}3L$ below the seabed. Thus, percolation losses would result in a reduction in wave height, given by

$$H = K_p H_0 \tag{1.10}$$

where K_p is always less than unity.

The corresponding coefficient for friction losses as given by Bretschneider and Reid (1954) is more complex in form,

$$K_f = \left[\left(A\int_{\infty}^{x} \phi_f . \mathrm{d}x\right) + 1\right]^{-1} \tag{1.11}$$

where $x = d/T^2$, $A = fH_0/mT^2$ and ϕ_f is a function of d/T^2 given by

$$\phi_f = \frac{64\pi^2}{3g^2}\left[\frac{K_s}{\sinh(2\pi d/L)}\right]^3 \tag{1.12}$$

f is a dimensionless parameter representing a friction coefficient for the seabed, m is the slope of the seabed directed along an orthogonal and K_s is the shoaling coefficient defined previously. In assessing friction losses over an offshore topography the equation for the friction factor can be simplified by assuming that the seabed is locally horizontal between two calculation points along an orthogonal. Under this assumption, the relative wave heights between two points will be given by

$$H_2 = K_f H_1 \tag{1.13}$$

where

$$K_f = \left[\frac{f.H_1}{T^4}.\frac{\phi_f.\Delta_s}{K_s} + 1\right]^{-1} \tag{1.14}$$

and Δ_s is the distance travelled by the wave between points 1 and 2.

This procedure can be incorporated into refraction computations at some penalty in calculation time. Most importantly, K_f is itself a function of the local wave height and therefore the 'unit wave height' assumption which applies to shoaling and refraction process can no longer be used. This means that calculations for a particular wave period must be repeated over a wide range of wave heights if friction losses are to be included. The assignment of a value to the friction factor, f, is also difficult in view of the very limited amount of data available on such factors.

Diffraction is the term given to the process by which wave energy is transmitted into the lee of an obstacle such as a breakwater. Diffraction can, therefore, be interpreted as resulting in a lateral spreading of wave energy and although such a process might not appear relevant to nearshore wave predictions, it does arise from the refraction process.

If adjacent orthogonals intersect, then the refraction coefficient defined previously goes to infinity. Such a solution is inadmissible on physical grounds and in this situation there is a lateral energy transfer which could reasonably be termed diffused diffraction. Battjes (1963) proposed a solution for this phenomenon in mathematical terms and routines can be included within refraction computations to solve for the redistribution of wave heights. At the present time the solution is lengthy and the effort involved is not justified in relation to other errors inherent to the prediction of sediment motion (Heaf, 1974).

Numerical solutions of wave shoaling and refraction

The refraction coefficients applicable to a particular nearshore region can be determined from plots of the orthogonal paths originating in deep water. Graphical methods of obtaining such solutions are given by Wiegel (1964), but they are laborious and time consuming. Computer solutions provide an alternative, several programs being available (e.g. Dobson (1967) and Heaf (1974)) and the technique is regularly used in the design of coastal engineering works. The

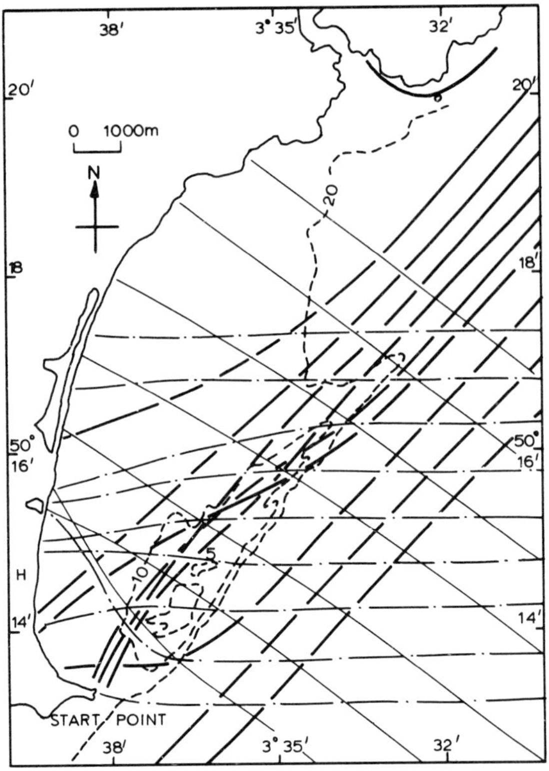

Figure 1.5. Wave refraction solution for Start Bay, Devon, England, showing orthogonal paths. Contours in fathoms; H = Hallsands.

computation is based on the principle that the orthogonal delineates a path of minimum travel time which give rise to a 'ray equation' and Munk and Arthur (1951) give the details of the theory.

Figures 1.5, 1.6 and 1.7 illustrate refraction solutions for three different coastal topographies, based on computations carried out by the author and Heaf.

The basic input data consist of an adequate numerical representation of the nearshore topography, extending to a depth of 0·5 times the deep-water wavelength. Typically the data will consist of a rectangular grid of several thousand values of water depth, the grid size being dictated by the availability of reliable topographical details in the area of concern. Grid sizes might be from 100 to 500 metres square.

The variable input parameters are the wave characteristics of initial direction and period and tide-level relative to the chart datum. Wave height is not generally required as an input since refraction and shoaling processes are linear in wave height. However, this does not apply if friction and diffused diffraction are to be included since the latter are non-linear processes.

The program computes the orthogonal paths from deep water into the shoreline, refraction and shoaling coefficients are evaluated at preselected intervals along the orthogonals and at each such point a note is made of the maximum wave height which could exist there, based on the simple breaking criterion noted previously.

Figure 1.5 illustrates the pronounced effect on wave conditions of the Skerries Bank, Start Bay, Devon, which is indicated by the 10 fathom contour. For these computations a grid of 80 × 60 points was used, the top left-hand corner being located at 50° 20′ 25″ N, 3° 30′ 6″ W, with the longer side running from north to south. The distance between adjacent grid points was 190·5 m, the total area covered by the grid being 174 km².

Figure 1.5 shows the orthogonal paths for waves of 12 second period with deep-water headings of southwest, west and northwest (i.e. approaching from northeast, east and southeast, respectively). In the former two cases the offshore shoal results in a focusing of wave energy onto particular sections of the coastline. It is known that for waves heading in a southwesterly direction the wave attack

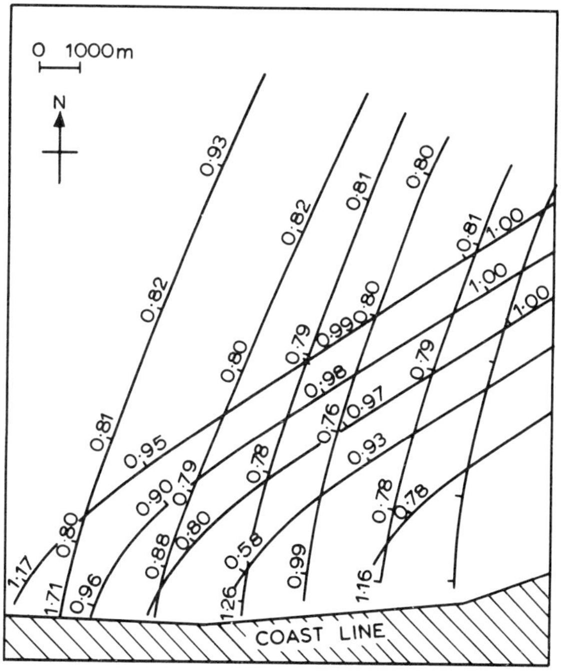

Figure 1.6. Typical computer output for a gently sloping topography, Zuara, eastern Mediterranean. Numbers shown are the product of refraction and shoaling coefficients.

on the southerly section of the bay is extremely severe under storm conditions. This has resulted in the destruction of the village of Hallsands and in significant changes in the sediment characteristics in this area. The degree of focusing for waves heading in a westerly direction is somewhat less, whilst a northwesterly heading results in a reasonably equitable distribution of wave-energy along the shoreline.

Figure 1.6 demonstrates the relatively simple solution obtained for an almost uniformly sloping offshore topography in the Eastern Mediterranean. The product of refraction and shoaling coefficients is plotted on the orthogonals and these numerical values are interpreted as the height of a wave of unit height in deep water. In this case the orthogonals are almost parallel and changes in wave height result largely from the shoaling coefficient.

Figure 1.7 shows a more complex refraction pattern for an area off the north-east coast of the United Kingdom. The interpretation of this solution as a preliminary to sediment computations is made difficult by an almost complete lack of field data against which theoretical solutions may be checked. In Figure 1.7 there is a considerable degree of intersection of orthogonals and the numerical

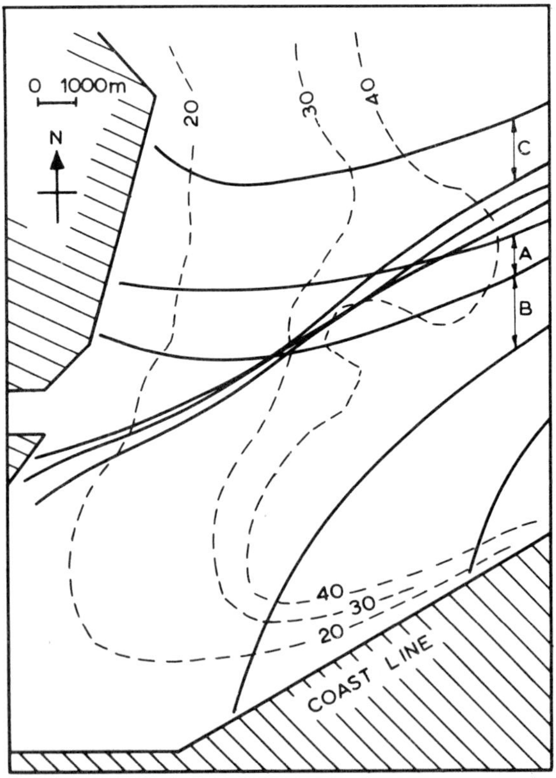

Figure 1.7. Refraction solution near a coastal inlet, Moray Firth, Scotland. Contours are in metres.

solutions for refraction coefficients reflect this in having very large values. Pierson (1951) gave a justification for an approximate limit of $K_r^2 \leqslant 2{\cdot}0$ in this situation but some small-scale laboratory experiments (Sharp, 1965) have indicated that K_r^2 could attain a value of 4·0. It is therefore prudent to base sediment motion calculations on both upper limits. Battjes (1963) proposed that following an orthogonal intersection point it would be reasonable to continue the evaluation of refraction coefficients on the basis of orthogonal separation as discussed above. In Figure 1.7, therefore, there are areas of coastline which will experience wave action originating from separate offshore regions. For example, north of the inlet the wave energy from sector C will be augmented by that from sector A and the separate activities will independently contribute to the sediment motion in that area. Similarly, energy in sector B will coincide with that from sector C immediately off the inlet, although, considering the relative spacing of the orthogonals bounding sectors B and C, the resultant wave heights will be low.

Influence of wave conditions on littoral drift

It is not the purpose of this paper to discuss the evaluation of sediment motion in detail but a brief comment is included to demonstrate the link between such calculations and the wave conditions. Detailed information and discussions of littoral drift may be found in Ippen (1966) and Bijker (1968).

The potential littoral transport, Q, past a point on a shoreline resulting from a wave of a given period, deep-water height and direction is given by

$$Q = K_1 . E \text{ (volume per month)} \tag{1.15}$$

where K_1 is a constant (dimensional) and E represents the resolved wave energy along the shoreline. This equation and those following are taken from *C.E.R.C. Technical Report No. 4* (1966) and are empirical.

The longshore wave energy is given by

$$E = K_2 . E_1 F(i, j, k) \tag{1.16}$$

where $E_1 = K_3 H_b^2 T \sin \alpha_c \cos \alpha_c$ and $F(i, j, k)$ is the fraction of a month for which waves of a given height, period and direction occur. $F(i, j, k)$ will vary from month to month and is obtained from the data discussed previously. H_b is the wave height at the breaking point and α_c is the angle, in degrees, between the orthogonal and the gradient of the beach, also at the breaking point. T is the wave period and K_2 and K_3 are dimensional constants. H_b and α_c are the two important parameters which are evaluated from the wave refraction, shoaling and breaking computations.

Thus the month by month definition of wave conditions in deep water, coupled with refraction computations for a particular coastal area, provide the essential data for inclusion in sediment motion predictions based on the above equations or others similar thereto. It may be noted that the properties of the beach material are not included in the above equations.

The errors involved in predicting littoral transport may be considerable, especially in view of the fact that the net annual transport may be small compared with individual monthly values. Table 1.1, given by Street, Mogel and Perry (1969) illustrates this point and also the differences which result from use of $K_r^2 \leqslant 2{\cdot}0$ or $K_r^2 \leqslant 4{\cdot}0$ referred to above. The transport figures quoted are for a typical coastal section of San Francisco County, California.

Table 1.1. Potential littoral transport (m^3/month)

Month	$K_r^2 \leqslant 4{\cdot}0$	$K_r^2 \leqslant 2{\cdot}0$	Month	$K_r^2 \leqslant 4{\cdot}0$	$K_r^2 \leqslant 2{\cdot}0$
January	+639	+308	July	−111	−111
February	+752	+505	August	−99	−99
March	+186	+84	September	−47	−47
April	+135	−1	October	−4	−32
May	−76	−91	November	−29	−44
June	−164	−163	December	+88	+22
			TOTAL	+1270	+331

Conclusions

A proper definition of deep-water wave conditions in terms of wave height, period, direction and percentage occurrence, together with an evaluation of the predominant shallow-water wave phenomena of refraction, shoaling and breaking, provide the necessary wave data for prediction of sediment movement in coastal areas. There is a serious lack of wave data in shallow water which would allow a more precise interpretation of theoretical predictions. Similarly, there is a need for the monitoring of beach changes associated with specific wave conditions which would lead to a reduction of the errors inherent in existing methods of predicting littoral transport.

Notation

$A = \dfrac{fH_0}{mT^2}$

C = wave velocity

C_g = group wave velocity

d = water depth

D_d = shoaling coefficient

E = resolved wave energy along the shoreline

f = dimensionless parameter representing a friction coefficient for the sea-bed

g = acceleration due to gravity

H = wave height

H_b = wave height at breaking point
H_0 = deep-water wave height
$H_{1/3}$ = significant wave height (i.e. highest $\frac{1}{3}$ of the waves)
K_f = frictional coefficient
K_p = permeability coefficient
K_r = refraction coefficient
K_s = shoaling coefficient
$K_{1,2,3}$ = dimensional coefficients
L = wavelength
m = slope of the seabed along an orthogonal
p = permeability coefficient for the porous seabed
Q = potential littoral transport
T = wave period
T_Z = zero crossing period
$x = d/T^2$

α_c = angle in degrees between the orthogonal and the beach gradient at the breaking point
θ = wave direction
ν = kinematic viscosity of seawater
ϕ_f = a function of d/T^2 (see text)

References

Battjes, 1963, Refraction of water waves, *Am. Soc. civ. Engrs, J. Wat. Ways Harb. Div.*, **94**, 437–452.
Bijker, E. W., 1968, Littoral drift as a function of waves and currents, *Proc. 11th Conf. cst. Engng., London*, Vol. 1, pp. 415–435.
Bretschneider, C. L., and R. O. Reid, 1954, Modification of wave height due to bottom friction, percolation and refraction, Beach Erosion Board, *Tech. Memo. No. 45.*
Dobson, R. S., 1967, Some applications of a digital computer to hydraulic engineering problems, Stanford University *Tech. Report No. 80.*
Draper, L., 1963, Derivation of a 'Design Wave' from instrumental records of sea waves, *Proc. Instn. civ. Engrs*, **26**, 291–305.
Heaf, N. J., 1974, *Wave Refraction*, Ph.D. Thesis, University of Liverpool.
Ippen, A. T. (Ed.), 1966, *Estuary and coastline hydrodynamics*, McGraw-Hill, New York.
Johnson, J. W., and P. S. Eagleson, 1966, Coastal processes, Chap. 9 in Ippen, 1966.
Munk, W. H., and R. S. Arthur, 1951, Wave intensity along a refracted ray, U.S. Dept. Commerce, Nat. Bur. Standards, *Circular 521.*
Pierson, W. J., 1951, The interpretation of crossed orthogonals in wave refraction phenomena, Beach Erosion Board, *Tech. Memo. No. 21.*
Pierson, W. J., G. Neumann, and R. W. James, 1955, Practical methods for observing and forecasting ocean waves by means of wave spectra and statistics, U.S. Navy Hydrographic Office, *Publication No. 603.*
Sharp, W., 1965, *Water Wave Refraction over a Submerged Shoal*, M. Eng. Thesis, University of Liverpool.

Street, R. L., T. Mogel and B. Perry, 1969, Computation of the littoral regime of the shore of San Francisco County, California, by automatic data processing methods, U.S. Army District Corps of Engineers, *Final Report, Contract No. DACW 07-68-0054.*

U.S. Army, Coastal Engineering Research Center, 1966, Shore protection, planning and design, *Tech. Report No. 4*, 3rd ed.

Wiegel, R. L., 1964, *Oceanographical Engineering*, Prentice-Hall, Englewood Cliffs, N.J.

Discussion

J. R. Hardy, Department of Geography, University of Reading. How do you overcome problems of instability in the computation of the orthogonals as they approach the shore and the gradient of the seabed steepens? My own experience, with a much coarser grid than your 200 m interval, gave instability problems and the procedure frequently broke down. This made it very difficult to make any prediction of likely conditions close inshore.

We find that using 12 data points for the topography and a second order curve you generally get a reasonable fit. But if we run into a difficult instability problem we 'cheat' and halve the step length. If this still proves unstable we halve the step length again. We have not run into any stability problems with our offshore predictions for the Dogger Bank.

J. R. Hardy. I understand that one object of the work is to provide a measure of coastal energy but I am not clear what measure of energy is provided by your refraction program. Are the figures printed on your output the height as obtained from the D/L_0 ratio, are they the refraction coefficients derived by Pierson, or some other figure? Do you take any account of the convergence or divergence of the orthogonals, bearing in mind that, in theory at least, energy is conserved between orthogonals?

Use whichever wave height you prefer. In the empirical approach adopted by civil engineers, different people have used different criteria. The refraction process does not itself make any decision about what value of wave height or period go into its computing.

W. A. Price, Hydraulics Research Station, Wallingford. The refraction process conserves wave energy between orthogonals. The intensity of wave energy can, therefore, vary from one compartment to another. Shouldn't an attempt be made to spill wave energy from one compartment to another in a wave refraction programme?

We have, in fact, started to spill energy in this way and have compared it with laboratory experimental work although the latter has been restricted to small waves. However, this refinement increases the computer run time and a program would have to be written to take it into account. While, from a research standpoint, it would be desirable, it is difficult to justify from a practical application point of view.

Graham Evans, Geology Department, Imperial College, London. As a sedimentologist, interested in regional sedimentation of coastlines and the general wave climate of these coastlines (for example, to predict long-term directions of littoral drift and concentration of heavy mineral, placer deposits), one is particularly interested in availability of data. Do any published sources exist, which are readily available, on a worldwide scale?

There is *Ocean Wave Statistics* by Hogben and Lumb, published by HMSO in 1967. That, so far as I am aware, is the best for worldwide statistics, but it has its limitations because it is based on visual observations from ocean-going vessels. The problem from an engineer's point of view is that observations are unreliable when waves become large. Hogben and Lumb have done their best to clarify the situation by comparing the data with those obtained from ocean weather ships. They derived wave statistics for 50 areas covering the world's shipping routes based on Marsden squares. The atlas classifies the wave data into heights, periods and directions for various sea states and seasons of the year. Nevertheless, the only real detail is internally within particular countries such as Canada, the United States, and the U.K. There may be some general data on the west coast of Africa, but the most detailed source for such cases consists of papers on the analysis of waves for a specific harbour project.

CHAPTER TWO

Nearshore Currents: Generation by Obliquely Incident Waves and Longshore Variations in Breaker Height

PAUL D. KOMAR

Abstract

Within the nearshore zone wave-induced longshore currents may be generated either by an oblique wave approach to the shoreline, by longshore variations in the wave breaker height, or by combinations thereof. The purpose of this paper is to present equations that combine the two generating mechanisms and to investigate conditions where the two mechanisms might oppose one another. The velocity $\bar{v}_l$ at the mid-surf position is found to be given by

$$\bar{v}_l = 2{\cdot}7 u_m \sin \alpha_b \cos \alpha_b - \frac{\pi\sqrt{2}}{c_f \gamma_b^3}\left(1 + \frac{3\gamma_b^2}{8} - \frac{\gamma_b^2}{4}\cos^2 \alpha_b\right) u_m \frac{\partial H_b}{\partial y}$$

where α_b is the breaker angle, $\partial H_b/\partial y$ is the longshore variation in wave height, and u_m is the orbital velocity evaluated at the breaker zone. The drag coefficient c_f is 0·008 to 0·018 under normal field conditions and γ_b is the ratio of the wave breaker height to water depth with a value between 0·8 to 1·2. A comparison between theory and all existing field and laboratory data confirms the term due to the oblique wave approach; there is presently no data to test the contribution of the $\partial H_b/\partial y$ term.

Under certain circumstances the two driving forces may oppose and balance such that $\bar{v}_l = 0$. One such case, found in wave basin experiments, is a cuspate shoreline wherein rip currents generated cusps but then ceased to exist once the cusps had reached an equilibrium development and the forces balanced to give zero longshore current velocities and, therefore, no further littoral sand transport. Rip currents can therefore generate a cuspate shoreline but may no longer be present at the time of cusp observation.

A complete solution is also obtained for the distribution of the longshore current velocity across the surf zone, resulting both from an oblique wave approach and from a longshore variation in wave height. It is found that a near balance can be achieved with $\bar{v} \simeq 0$ across the entire surf zone as well as at the mid-surf position.

Introduction

Two wave-induced current systems are generally recognized in the nearshore zone which dominate the water movements in addition to the to-and-fro motions produced by the wave orbits directly. These are: (a) a cell circulation system of rip currents and feeding longshore currents and (b) longshore currents produced by an oblique wave approach to the shoreline.

The cell circulation, Figure 2.1, was described by Shepard and Inman (1950a, 1950b); their study obtained the first comprehensive series of field measurements. Bowen (1969a) and Bowen and Inman (1969) demonstrated both theoretically and experimentally that variations in the wave breaker height along the length of the shore produce variations in the wave set-up, the rise in water level above the still-water level, as well as a longshore directed component of the wave momentum flux (radiation stress). The set-up shoreward of the higher breakers is greater so it raises the water level above that shoreward of the small breakers. This provides the necessary longshore head of water to drive the feeder longshore current and produce the rip currents, the water in the surf zone flowing alongshore from positions of high breaking waves to positions of low breakers where the current turns seaward as a rip current. The original variations in wave breaker height can be produced either by wave refraction or, in the absence of this, as Bowen and Inman (1969) demonstrated, by interactions between the incoming waves and edge waves trapped within the nearshore region, so producing a systematic variation in the wave height along the shore.

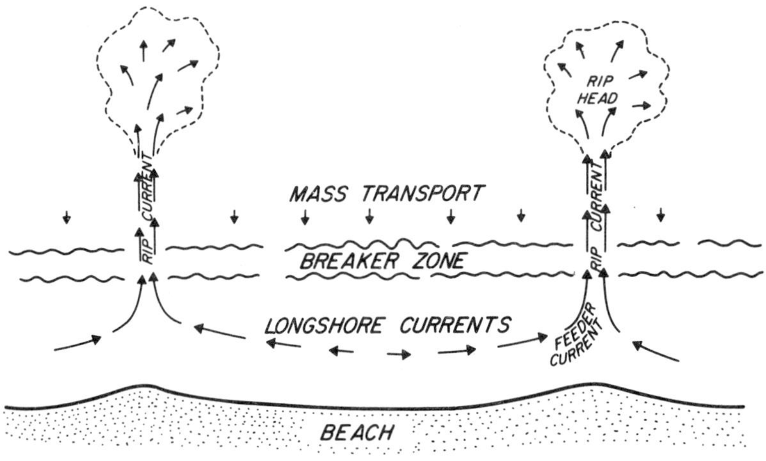

Figure 2.1. The nearshore cell circulation consisting of (1) feeder longshore currents, (2) rip currents and (3) a slow mass transport of water shoreward (after Shepard and Inman, 1951).

Longshore currents can also be generated in the nearshore zone by waves striking the beach at an angle to the shoreline trend. This current, especially, is responsible for the longshore drift of sediment on the beach so that there has been considerable interest in the mechanism of generation and a number of theories have been proposed. Galvin (1967) reviewed the early theories and arrived at the justifiable conclusion that a satisfactorily proven prediction was not available at that time. Since then, there has been considerable progress through the studies of Bowen (1969b), Longuet-Higgins (1970a, 1970b) and Thornton (1970) who attributed the generation of the current to the longshore component of the radiation stress, the excess momentum flux due to the presence of the waves (Longuet-Higgins and Stewart, 1964).

Longshore currents can therefore be driven by longshore variations in the wave breaker height, by an oblique wave approach to the shoreline, or by combinations thereof. The purpose of this paper is to present an analytical equation that combines the two generating mechanisms and to investigate conditions where the two mechanisms might oppose one another. Emphasis is placed on the presentation of usable predictive equations for engineering and other practice. This derivation was presented earlier in an unpublished technical report (Komar, 1972) which accompanied a paper on the formation of giant cusps along shorelines (Komar, 1971). In an experimental study it was found that an equilibrium cuspate shoreline could develop in which the two generating mechanisms for longshore currents exactly oppose and balance one another. This paper will therefore end with a general review of the relationship between the nearshore current and the rhythmic cuspate shorelines that may be produced by the currents.

Longshore current generation

Bowen (1969a, 1969b), Bowen and Inman (1969), Longuet-Higgins (1970a, 1970b) and Thornton (1970) have made considerable progress in our knowledge and understanding of the nearshore currents through applications of the concepts of the radiation stress in water waves developed by Longuet-Higgins and Stewart (1960, 1964). Radiation stress is defined as 'the excess flow of momentum due to the presence of the waves'. Momentum flux is a tensor quantity with four components; the x and y fluxes of x momentum and y momentum (the x coordinate is normal to the shoreline, positive onshore, the direction of wave advance; the y coordinate is parallel to the shoreline and wave crests). The radiation stress (momentum flux) across a plane x = constant, in the direction of wave advance (the x direction), is shown by Longuet-Higgins and Stewart to be given by

$$S_{xx} = E\left[\frac{2kh}{\sinh(2kh)} + \frac{1}{2}\right] \tag{2.1}$$

where E is the wave energy density related to the wave height H by $E = \rho g H^2/8$ [ρ is the water density], h is the still-water depth, and $k = 2\pi/(\text{wavelength})$. Because the pressure departs from the hydrostatic when the waves are present, there is also a flux of momentum parallel to the wave crests. This flux, a flux of y momentum across the plane y = constant, is shown to be given by

$$S_{yy} = E\left[\frac{kh}{\sinh(2kh)}\right] \tag{2.2}$$

Longuet-Higgins and Stewart (1963) have shown theoretically, and Bowen, Inman and Simmons (1968) have demonstrated experimentally, that the radiation stress produces negative and positive changes in the mean water level in the nearshore region, known respectively as wave set-down and set-up. There is a water set-down just outside and within the breaker zone, followed by a set-up within the surf zone. Balancing the pressure gradient of the sloping water surface within the surf zone against the incoming momentum, they obtain

$$\frac{d\bar{\eta}}{dx} = \left[\frac{1}{1 + 3\gamma^2/8}\right] \tan \beta \tag{2.3}$$

for the gradient of the set-up $\bar{\eta}$, the difference between the still-water level and the mean water level in the presence of the waves. The beach slope $\tan \beta$ has been substituted for the gradient in the still-water depth, $-dh/dx$. The ratio

$$\gamma = \frac{H}{\bar{\eta} + h} \tag{2.4}$$

where H is the wave height, is found to be nearly constant so that Equation 2.3 indicates that in the surf zone the mean sea level slope, $d\bar{\eta}/dx$, will be constant and proportional to the beach slope. Equation 2.4 assumes a constant dissipation of the wave energy across the surf zone, a situation more probable with spilling breakers than under plunging breakers. Sonu (1972a), for example, noted differences in the wave set-up pattern for the two breaker types under field conditions.

Bowen (1969a) and Bowen and Inman (1969) demonstrated that variations in the wave set-up in the longshore direction may provide the necessary longshore head of water that contributes to the formation of longshore currents which feed rip currents. The variations in the wave set-up result from longshore variations in the incoming wave height. The larger waves break in deeper water than the smaller waves so that the wave set-up begins further seaward where the larger waves occur. Because of this, the actual water level rise associated with the high waves is considerably greater than the rise resulting from the lower waves. This dependence upon the wave height is shown in Figure 2.2, taken from Bowen (1969a). Inside the surf zone the mean water level is higher shoreward of the larger breakers than shoreward of the small waves. Therefore, a longshore pressure gradient or head of water exists which will drive a longshore current from positions of high waves to adjacent positions of low waves.

It would appear from Figure 2.2 that, since the set-down outside the larger waves is greater, there would be a similar tendency for water to flow from positions of low breakers to positions of high breakers just outside the break point, opposite in direction to the current generated within the surf zone. It turns out, however, that the S_{yy} component of the radiation stress prevents this latter current from developing and at the same time enhances the current within the surf zone. Since the wave height H is taken to vary alongshore, S_{yy} will similarly vary and there will exist a longshore gradient

$$\frac{\partial S_{yy}}{\partial y} = \frac{1}{4} \rho g H \frac{\partial H}{\partial y} \left[\frac{2kh}{\sinh (2kh)}\right] \tag{2.5}$$

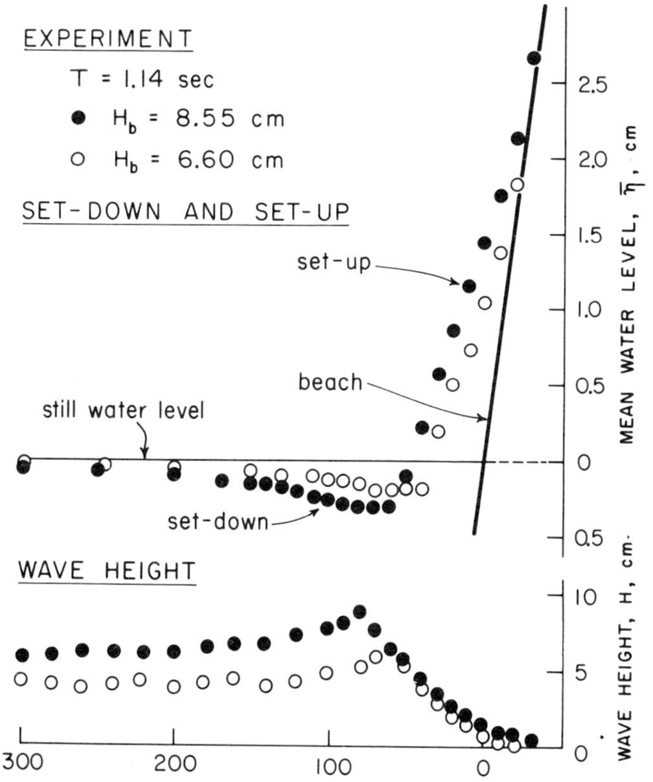

Figure 2.2. Wave set-down and set-up for two different initial wave heights. Note that the larger wave height causes a greater set-down and that, although the water slope of the set-up is approximately the same for both wave heights, the water level is higher shoreward of the larger breakers since the set-up begins in somewhat deeper water (after Bowen, 1969a).

which in shallow water reduces to

$$\frac{\partial S_{yy}}{\partial y} = \tfrac{1}{8}\rho g H \frac{\partial H}{\partial y} \tag{2.6}$$

As determined by Bowen (1969a), the longshore variation in the wave set-down outside the breaker zone is balanced by the gradient $\partial S_{yy}/\partial y$ so that no net force exists outside the breaker zone to produce water circulation. Within the surf zone however the forces combine to give the total driving force

$$F_{\text{cell}} = \rho g(\bar{\eta} + h)\frac{\partial \bar{\eta}}{\partial y} + \frac{\partial S_{yy}}{\partial y} \tag{2.7a}$$

$$= \rho g(\bar{\eta} + h)\frac{\partial \bar{\eta}}{\partial y} + \tfrac{1}{8}\rho g H \frac{\partial H}{\partial y} \tag{2.7b}$$

for the cell circulation. This force produces the observed flow of water within the surf zone away from the region of high waves towards the position of low waves where the flow then turns seaward as a rip current.

It is well known that when waves approach a straight beach at an angle, a longshore current is established flowing parallel to the shoreline in the nearshore zone. The velocity of the current decreases quickly to zero outside the breaker zone so it is clearly wave induced and cannot be attributed to ocean currents or tides. Bowen (1969b), Longuet-Higgins (1970a, 1970b) and Thornton (1970) have attributed the formation of such currents to the longshore component of the radiation stress given by

$$S_{xy} = En \sin \alpha \cos \alpha \tag{2.8}$$

where α is the angle the wave crest makes with the shoreline and n = (wave group velocity)/(wave phase velocity). S_{xy} is the y directed (longshore directed) momentum flux component passing through the x = constant plane (a plane parallel to the shoreline). The longshore current is shown to be produced by a local thrust exerted in the surf zone, given by

$$\tau_y = -\frac{\partial S_{xy}}{\partial x} \tag{2.9a}$$

$$= -\frac{5}{16}\rho g(\bar{\eta} + h)\gamma^2 \sin \alpha \cos \alpha \frac{\partial(\bar{\eta} + h)}{\partial x} \tag{2.9b}$$

$$= -\frac{5}{16}\rho g(\bar{\eta} + h)\gamma^2\left[\frac{1}{1 + 3\gamma^2/8}\right] \tan \beta \sin \alpha \cos \alpha \tag{2.9c}$$

In formulating a relationship for the generated longshore current due to an oblique wave approach, $\bar{v}$, Longuet-Higgins (1970a) balances this thrust against the bottom drag of the generated longshore current given by

$$\mathrm{Drag}_y = \frac{2}{\pi} c_f \rho u_m \bar{v} \tag{2.10}$$

where c_f is a constant drag coefficient and u_m is the maximum value of the wave orbital velocity, assumed to be sinusoidal. By neglecting any horizontal momentum exchange, there would then be a balance between the stress of Equation 2.10 and the total driving force of the current given by Equation 2.7b and Equation 2.9c. Equation 2.7b must be modified since the wave crests now make an angle α with the y axis; Equation 2.7b is for $\alpha = 0$. This modification is in the $\partial S_{yy}/\partial y$ term since now in shallow water

$$S_{yy} = E(\tfrac{3}{2} - \cos^2 \alpha)$$

The total driving force attributed to longshore variations in wave height and set-up is then

$$F_{\text{cell}} = \rho g(\bar{\eta} + h)\frac{\partial \bar{\eta}}{\partial y} + \tfrac{1}{4} g H \frac{\partial H}{\partial y}(\tfrac{3}{2} - \cos^2 \alpha) \tag{2.11}$$

Combining this with the thrust due to the oblique wave approach (2.9c) and balancing it against the drag (2.10), gives

$$\bar{v} = \frac{\pi}{2c_f} u_m \left[\frac{5}{4} \frac{1}{1 + 3\gamma^2/8} \tan \beta \sin \alpha \cos \alpha - \frac{4}{\gamma^3} \left(1 + \frac{3\gamma^2}{8} - \frac{\gamma^2}{4} \cos^2 \alpha \right) \frac{\partial H}{\partial y} \right] \tag{2.12}$$

for the longshore current, having used $u_m = 0{\cdot}5\gamma(gh)^{1/2}$ for the orbital velocity. This relationship for the longshore current depends on both the longshore variation in wave height and on an oblique wave approach, the equation to be derived. If $\partial H/\partial y = 0$, there is no longshore variation in wave height and Equation 2.12 then reduces to the relationship derived by Longuet-Higgins (1970a, Equation 53). If $\partial H/\partial y$ is positive then the driving forces are opposed to one another and the current is reduced. It is apparent that there might be a balance between the two mechanisms such that $\bar{v} = 0$. This was the case for the currents on the flanks of the cuspate shoreline mentioned in the Introduction, and will be examined again in a later section. If $\partial H/\partial y$ is negative then the driving forces act together in the same direction and the current $\bar{v}$ generated would be larger than produced by the two driving forces acting individually.

A comment is required on the variability of α across the surf zone due to wave refraction. Longuet-Higgins (1970a) assumed α small so that $\cos \alpha \simeq 1$, and so was not retained. Wave refraction effects on $\sin \alpha$ were considered but since $\sin \alpha/C$ is constant, independent of x, it was not differentiated. Bowen (1969b) on the other hand immediately assigned α its value at the breaker zone so that it became a constant. A middle path may be taken by giving $\cos \alpha$ its value at the breaker zone and, at the same time, allowing $\sin \alpha$ to vary. This is most important in the ultimate solution of Equation 2.22 which is evaluated later in the paper. Equation 2.12 differs from that given by Longuet-Higgins (1970a) only in the retention of the $\cos \alpha$ factor.

Equation 2.12 has some inherent shortcomings because of the assumptions involved in its derivation. First, horizontal momentum exchange across the surf zone was not included. Because of this, Equation 2.12 predicts a linear increase in the velocity from zero at the shoreline to a maximum at the breaker zone where there is also a discontinuity, the velocity dropping to zero in still deeper water. This of course is not very realistic. The one advantage of Equation 2.12 is that it supplies a relationship that is straightforward to apply. Later in the paper a complete solution is derived, including the horizontal momentum exchange.

Also Equation 2.12 considers only the longshore component of the nearshore velocity. Any offshore component is not included and as a result the possible presence of rip currents is ignored. This is the severest shortcoming of the equation, making it inapplicable close to rips where the offshore component becomes large compared to the longshore component of the velocity. Finally, longshore variations in the surf zone width, resulting from changes in the breaker height, are not included. This limitation in most cases is not severe, requiring only that the y derivatives be small compared to the x derivatives (for example, $\partial H/\partial y \ll$

$\partial H/\partial x$). Although a more complete derivation might be desired in some cases, it would considerably increase the complexity and so is not included here.

Comparison with data

Unfortunately, simultaneous measurements of the wave parameters, the angle of wave approach, the longshore variation in the wave height, and the resulting longshore current do not exist. All of the available data examine the contribution of the oblique wave approach separately, ignoring any contribution resulting from longshore variations in the wave height. Therefore the full Equation 2.12 cannot be tested at the moment. Setting $\partial H/\partial y = 0$ in Equation 2.12, the relationship reduces to that given by Longuet-Higgins (1970a, Equation 53)

$$\bar{v}_1 = \frac{5\pi}{16}\left[\frac{1}{1 + 3\gamma^2/8}\right]\frac{\tan\beta}{c_f}\, u_m \sin\alpha_b \cos\alpha_b \tag{2.13}$$

where the current $\bar{v}_1$ midway between the breaker zone and the shoreline is related to $u_m = [2E_b/\rho h_b]^{1/2}$ now evaluated only at the breaker zone and related to the wave breaker angle α_b. The equation has been written in this form since in most data applications the mid-surf longshore current is related to the breaking wave parameters.

The earlier sand transport studies by Komar and Inman (1970) had suggested that

$$\bar{v}_1 = 2{\cdot}7 u_m \sin\alpha_b \tag{2.14}$$

This was prompted by the agreement between two independent estimates of the littoral sand transport, the agreement only being possible if the longshore current attributed to an oblique wave approach is given by the relationship of Equation 2.14. Their study found no dependence on the beach slope $\tan\beta$, as would be suggested by Equation 2.13, and this was substantiated by direct measurements of the longshore current. Figures 2.3 and 2.4 contain all the available measurements of longshore currents; Figure 2.3 examines Equation 2.13 and Figure 2.4 tests Equation 2.14. A comparison suggests that since γ remains nearly constant, then

$$\frac{\tan\beta\cos\alpha_b}{c_f} \simeq \text{constant} \tag{2.15}$$

that is, the drag coefficient c_f increases proportionally with an increase in beach slope. The constancy of this ratio has also been suggested on theoretical grounds by Komar (1971). However, Longuet-Higgins (1972) has indicated that Equation 2.15 may be more apparent than real since with increasing beach slope, the increasing horizontal mixing by eddies within the surf zone would be more important than an increase in c_f to keep the ratio constant.

A problem not generally recognized in obtaining most of the field data utilized in Figures 2.3 and 2.4 is the presence of the cell circulation in addition to the currents resulting from an oblique wave approach. Commonly the two systems

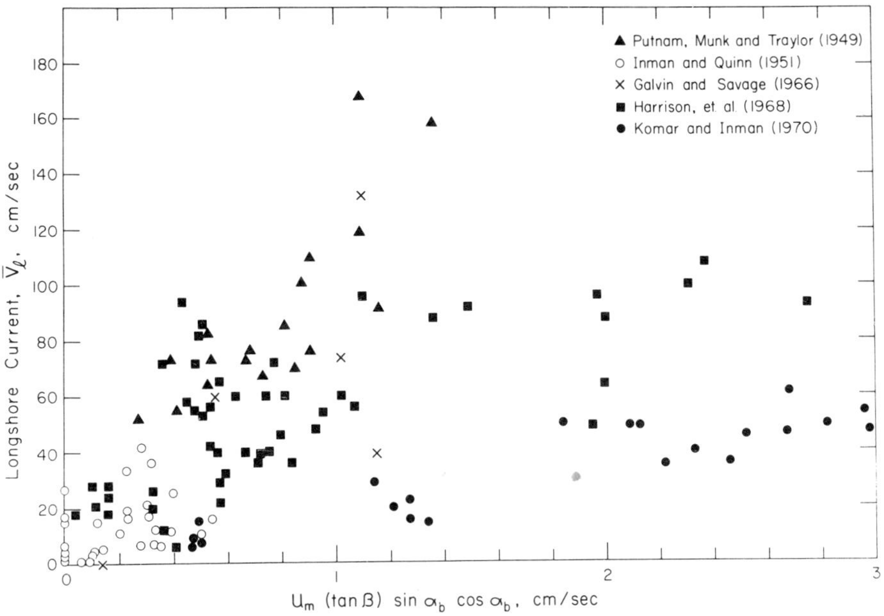

Figure 2.3. Examination of the longshore current relationship (Equation 2.13) deduced theoretically by Longuet-Higgins (1970a) utilizing all available field data. Compare with Figure 2.4; (after Komar and Inman, 1970).

occur simultaneously so that the observed currents are the sums of the two systems. If care is not taken, the measurements obtained would be the current resulting from the combined systems, not just the current resulting from the oblique wave approach. The measured longshore current $\bar{v}_1$ would in fact be caused by both an oblique wave approach and a longshore variation in wave height, although $\partial H/\partial y$ was not measured and in utilizing the data to test Equations 12.13 and 12.14 it was assumed that $\bar{v}_1$ resulted from the oblique wave approach alone. This might account for much of the scatter of the data seen in Figure 2.4.

The agreement between the field data and Equation 2.14 is actually somewhat better than is indicated in Figure 2.4. This is because Figure 2.4 combines data from five separate field studies and, in addition to being at different locations, there might also be systematic differences as to how the data are collected and analysed. The best example of this is the data of Harrison and coworkers (1968) where the maximum longshore current just shoreward from the breakers was measured, rather than the lesser current at mid-surf. As a result, as can be seen in Figure 2.4, their data plot somewhat higher than the measurements obtained in the other studies. Replotting the Harrison and coworkers (1968) data by itself in Figure 2.5, it can be seen to form a reasonable trend but with a slope coefficient of 3·5 rather than the 2·7 value of Equation 2.14 based on all the data taken together. Similarly, the field data of Putnam and coworkers (1949) and Komar and Inman (1970) have been replotted alone in Figures 2.6 and 2.7. These two data sets agree with the 2·7 slope coefficient. The data of Inman and Quinn

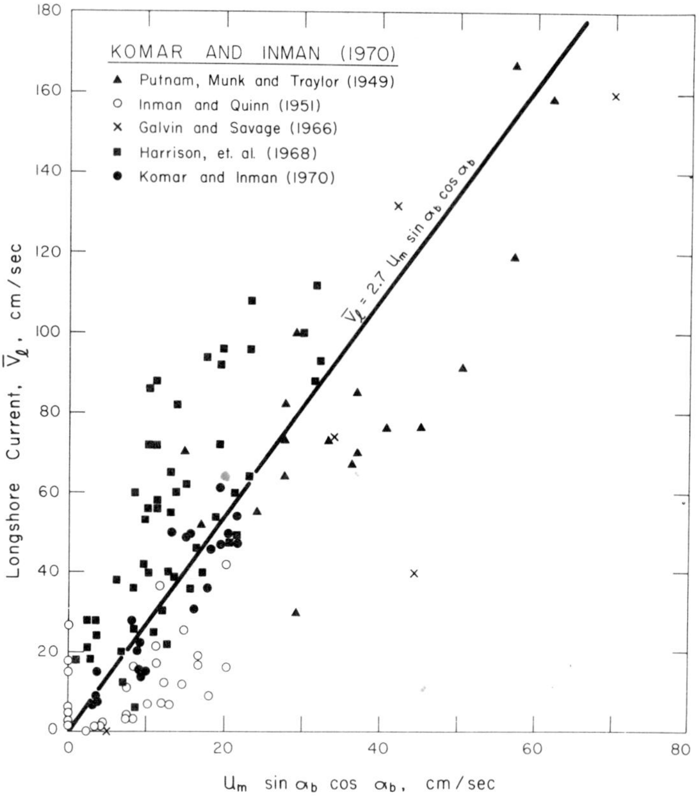

Figure 2.4. A replot of the field data shown in Figure 2.3, testing the longshore current relationship, Equation 2.14, suggested by Komar and Inman (1970). A comparison with Figure 2.3 indicates that $\tan \beta \cos \alpha_b / c_f$ is approximately constant.

(1951) consists of very low velocities so that a comparison by itself is not meaningful. It can be seen that individually the various sets of field data tend to confirm the relationship of Equation 2.14 or its theoretically equivalent equation (2.13) so long as an increase in $\tan \beta$ is offset by an increase in c_f (2.15). In a recent review of all the equations that have been proposed for the longshore current generated by an oblique wave approach, the writer has found (Komar, in preparation) that the agreement with Equation 2.14 shown in Figure 2.4 surpassed, by far, any other theory.

Figures 2.8 to 2.12 compare Equation 2.14 or slight variations in form with the available laboratory data as given in Galvin and Nelson (1967). The laboratory data of Putnam and coworkers (1949), Figure 2.8, are of interest because they include breaker angles ranging from 10° to as large as 58°. It is apparent that there is general agreement with

$$\bar{v}_l = 2{\cdot}7 u_m \sin \alpha_b \cos \alpha_b \tag{2.16}$$

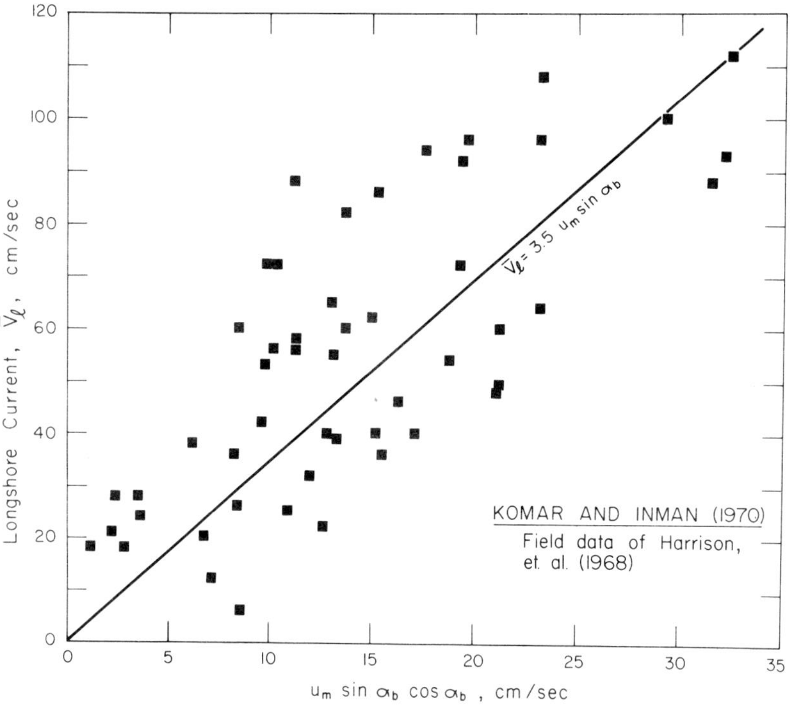

Figure 2.5. Test of the Komar and Inman (1970) longshore current relationship utilizing only the field data of Harrison and coworkers (1968). The slope coefficient is 3·5 rather than 2·7, probably because the maximum velocity was measured rather than the mid-surf velocity.

until the breaker angle exceeds 45° at which point the generated current is much greater than predicted. The data also suggest the inclusion of the cos α_b factor so that Equation 2.14 becomes modified to Equation 2.16. In the field data, the breaker angles are in all cases small so that cos $\alpha_b \simeq 1$ and its inclusion or exclusion is not significant. Figure 2.9 shows the data of Saville (1950) on longshore currents, collected during his studies of littoral sand transport. This comparison is of special interest because the study is the only laboratory investigation in which a sandy beach was utilized rather than an evenly sloping solid beach. With the exception of two points of anomalously low generated current velocities, the data agree approximately with Equation 2.16 but with a coefficient of 2·5 rather than 2·7. Figures 2.10 and 2.11 include the laboratory data of Brebner and Kamphuis (1963). Figure 2.10 compares the data with Equation 2.16 and it is seen that there is a good trend, but a straight-line fit does not pass through the origin. Including a second cos α_b factor in Figure 2.11 improves this since it rotates the data, the larger $\bar{v}_l$ data at higher α_b values being rotated more. The straight line through the origin gives

$$\bar{v}_l = 3{\cdot}8 u_m \sin \alpha_b (\cos \alpha_b)^2 \tag{2.17}$$

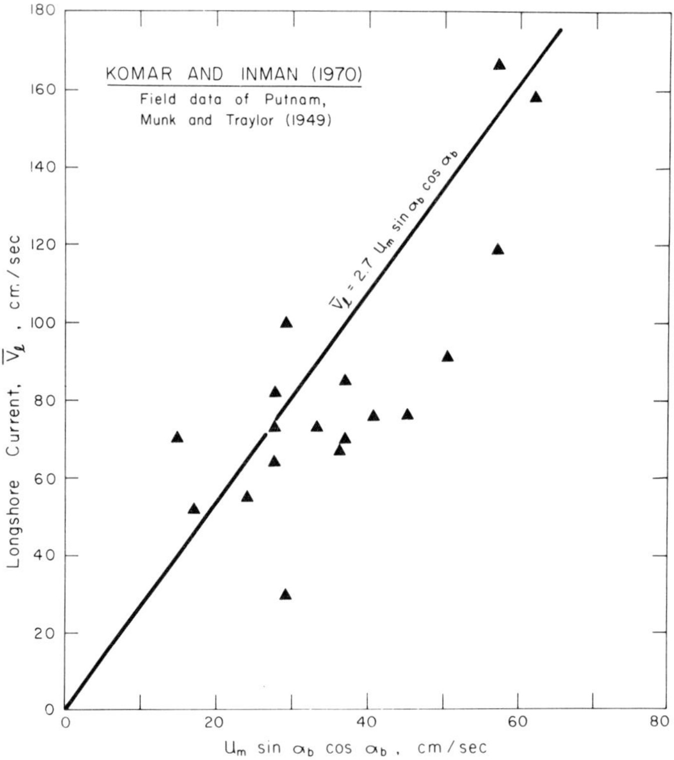

Figure 2.6. Examination of the Komar and Inman (1970) longshore current relationship utilizing only the field data of Putnam and coworkers (1949).

but it can be seen that there is still a slight indication that a best-fit straight line would not pass through the origin. Including yet another cos α_b factor would further improve the trend, but it is not felt to be warranted. From a theoretical standpoint (Equation 2.12), only a single cos α term is justifiable. Wave refraction really should be taken into consideration when α becomes large, as in many of these laboratory cases. The final data test is with the laboratory data of Galvin and Eagleson (1965), Figure 2.12, and it is apparent that there is little relationship. This is somewhat surprising, in that Galvin (1967) in his review of the data concluded that these data are the best available on longshore currents. The spread is not the result of variations in the beach slope since $\tan\beta = 0{\cdot}109$ throughout. The values shown next to the data points are breaker angles and it can be seen that there is some grouping according to angle, the group to the left farthest from the curve, is for $\alpha_b < 5°$, the middle set is for $6° < \alpha_b < 14°$, and the set closest to the line is for generally larger angles. The form of Equation 2.17 is used with $(\cos\alpha_b)^2$; the inclusion of only a single cos α_b factor gives still greater spread. The reasons for the lack of agreement with the data of Galvin and Eagleson (1965) and the spread according to the breaker angle are not known.

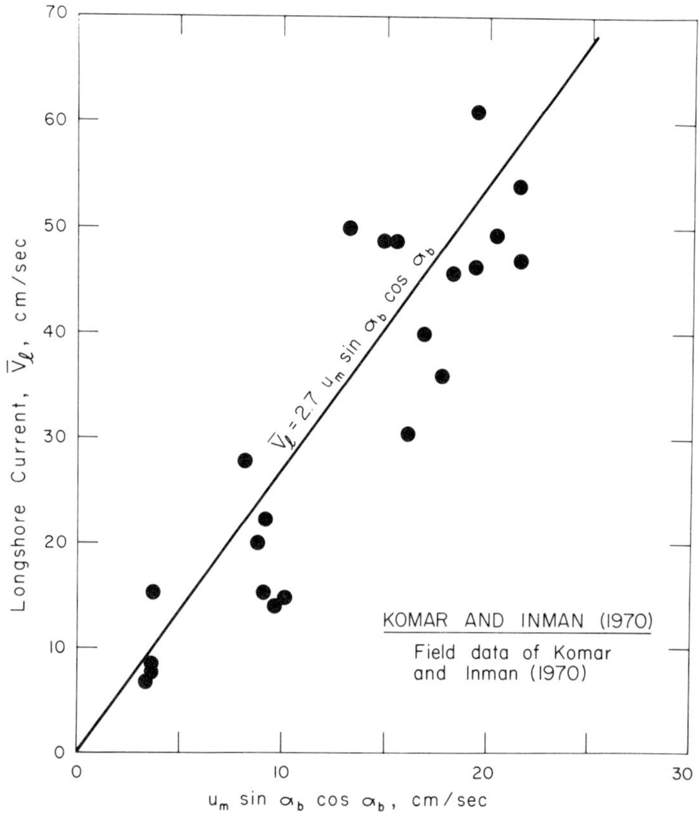

Figure 2.7. Agreement between the longshore current relationship suggested in Komar and Inman (1970) with our own field data.

It is apparent from this review that more and better measurements of longshore current generation are required. These should be field rather than laboratory measurements since there appears to be some difficulty with the laboratory data. In some cases this may be the effect of the finite length of the laboratory beach not permitting the current to accelerate to its equilibrium value. Another difficulty may be that the wave basin with its fixed beach face slope does not adequately represent the natural conditions in that the beach slope cannot reach a balance with the drag coefficient c_f according to Equation 2.15 or with the horizontal eddy mixing across the surf zone. To test Equation 2.12, the field measurements should include data on the longshore variation in the wave height as well as the angle of wave approach and other wave parameters.

In the light of the results from the data comparison, Equation 2.12 can be modified to

$$\bar{v}_l = 2{\cdot}7 u_m \sin\alpha_b \cos\alpha_b - \frac{\pi\sqrt{2}}{c_f \gamma_b{}^3}\left(1 + \frac{3\gamma_b{}^2}{8} - \frac{\gamma_b{}^2}{4}\cos^2\alpha\right) u_m \frac{\partial H_b}{\partial y} \tag{2.18}$$

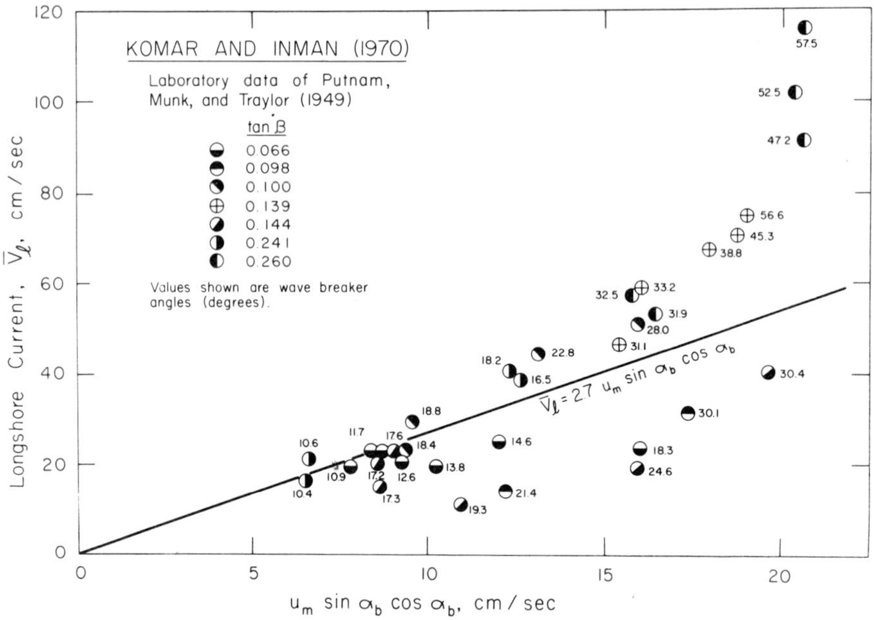

Figure 2.8. Test of the Komar and Inman (1970) longshore current relationship with the laboratory data of Putnam and coworkers (1949). Agreement is good until the breaker angles exceed 45°.

As before, the second term on the right side of the equation has been modified such that the longshore current mid-way through the surf zone is related to the breaking wave conditions. The parameters γ_b (2.4) and c_f remain to be evaluated in the relationship. Iverson (1951) in wave channel experiments found that γ_b varies between approximately 0·7 and 1·20, the value depending on the deep water wave steepness and on the beach slope. His curves may be consulted for the proper selection of γ_b. From the available data on longshore currents, Longuet-Higgins (1970a) gave

$$
\begin{aligned}
c_f &= 0{\cdot}036B \qquad \text{(field)} \\
c_f &= 0{\cdot}045B \qquad \text{(laboratory)}
\end{aligned}
\tag{2.19}
$$

where B depends on the horizontal mixing and is between 0·167 and 0·50. B increases with increasing horizontal mixing length across the surf and with decreasing surf zone width. These latter factors both decrease with increasing beach slope so that B and therefore c_f might remain approximately constant. It can be seen that c_f might be expected to be larger in the laboratory than in the field. This fact will be used later. Longuet-Higgins (1970a, p. 6784) reviewed the available estimates of c_f and arrived at $c_f \simeq 0{\cdot}01$ as a reasonable value. Sonu (1972a) gives $c_f = 0{\cdot}014$ as the best value for his numerical solutions. Later in this paper Equation 2.16 will be compared with the solution for the velocity

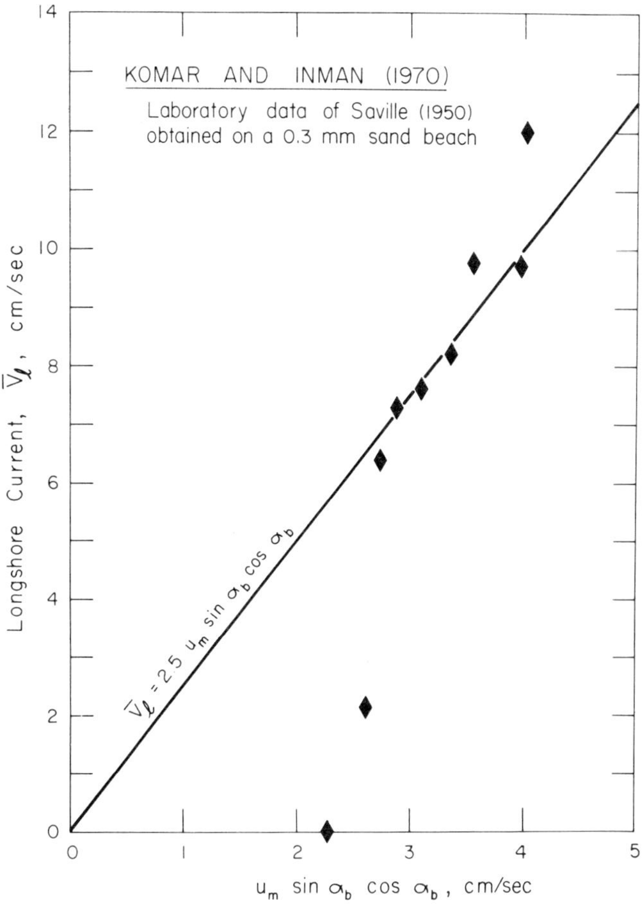

Figure 2.9. Examination utilizing the data of Saville (1950) from a laboratory sand beach.

distribution across the entire surf zone; such a comparison indicates that $c_f \simeq$ 0·017 to 0·018 is the most reasonable value for the drag coefficient under field conditions (and approximately 0·0225 for laboratory longshore currents).

Longshore current balance and the cuspate shoreline

According to Equation 2.18, the longshore current $\bar{v}_l$ mid-way through the surf depends on the combination of angle of wave approach, α_b, and the longshore variation in breaker height, $\partial H_b/\partial y$, as well as on the value of u_m evaluated at the breaker zone with

$$u_m = (2E_b/\rho h_b)^{1/2}$$

where E_b is the wave energy of the breakers and h_b is the depth at breaking. When

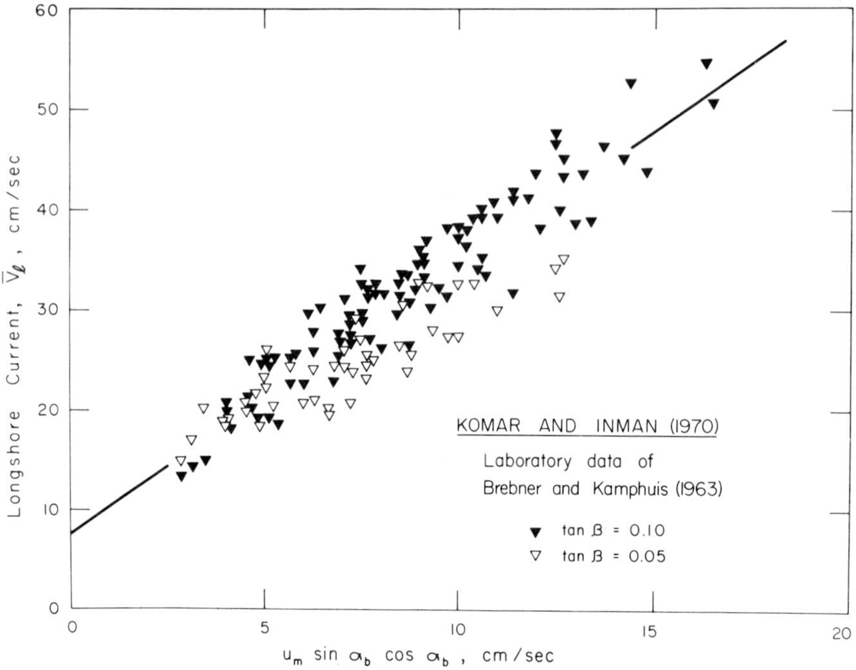

Figure 2.10. Test with laboratory data of Brebner and Kamphuis (1963) of $u_m \sin \alpha_b \cos \alpha_b$ relationship. Compare results with Figure 2.11.

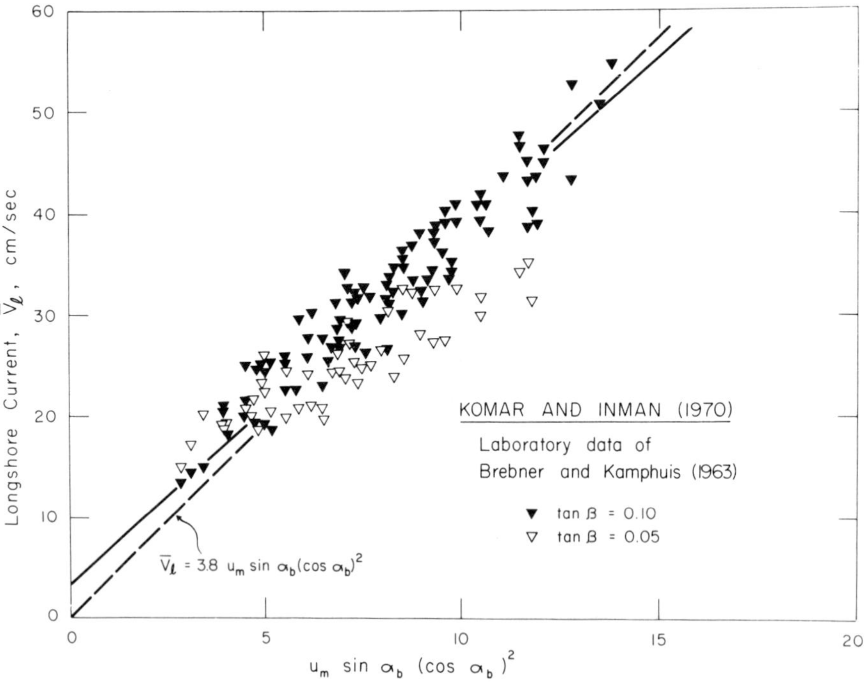

Figure 2.11. Test with laboratory data of Brebner and Kamphuis (1963) of $u_m \sin \alpha_b (\cos \alpha_b)^2$ showing that the inclusion of a second cosine factor permits a straight line relationship through the origin. Compare with Figure 2.10.

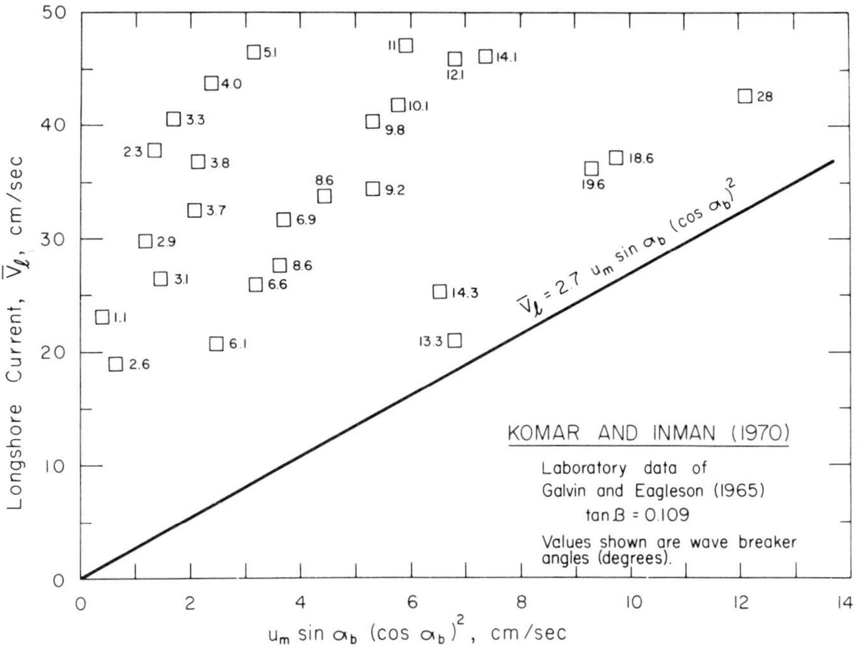

Figure 2.12. Examination of the Komar and Inman (1970) longshore current relationship with the laboratory data of Galvin and Eagleson (1965). The departure between this data and the equation appears to be governed by the size of the breaker angle, it being greatest for the smallest angles.

$\partial H_b/\partial y$ is negative it supports the current produced by the oblique wave approach; when $\partial H_b/\partial y$ is positive it opposes. A negative value for $\bar{v}_l$ in Equation 2.18 indicates that the current flows alongshore in a direction opposite to the oblique wave longshore component. For $\bar{v}_l = 0$, the $\partial H_b/\partial y$ term must exactly oppose and balance the thrust resulting from the oblique wave approach. This balance is obtained by placing $\bar{v}_l = 0$ in Equation 2.18 yielding,

$$\frac{1}{c_f}\frac{\partial H_b}{\partial y} = \frac{2{\cdot}7\gamma_b^3}{\pi\sqrt{2}}\frac{\sin\alpha_b\cos\alpha_b}{(1 + 3\gamma_b^2/8 - \gamma_b^2/4\cos^2\alpha_b)} \tag{2.20}$$

which is slightly different in form to that presented by Komar (1971, 1972). This balance between $\partial H_b/\partial y$ and α_b is graphed in Figure 2.13 for $\gamma_b = 0{\cdot}8$ and $1{\cdot}0$. With this graph the wave height gradient required to oppose and balance a given angle of wave breaking can be evaluated. Since the second term on the right of Equation 2.18 has not been tested against data, the balance of Equation 2.20 and Figure 2.13 might require some modification once this has been done.

An example of special interest in the balance between $\partial H_b/\partial y$ and α_b along the shoreline such that no longshore current is produced is the cuspate shoreline shown in Figure 2.14. Such a balance was first recognized by Komar (1971), working with Price and Willis, in experiments conducted in the large 16 m by 58 m wave basin at the Hydraulics Research Station, Wallingford, England; A granular coal beach was used, smoothed to an initially straight shoreline. Soon

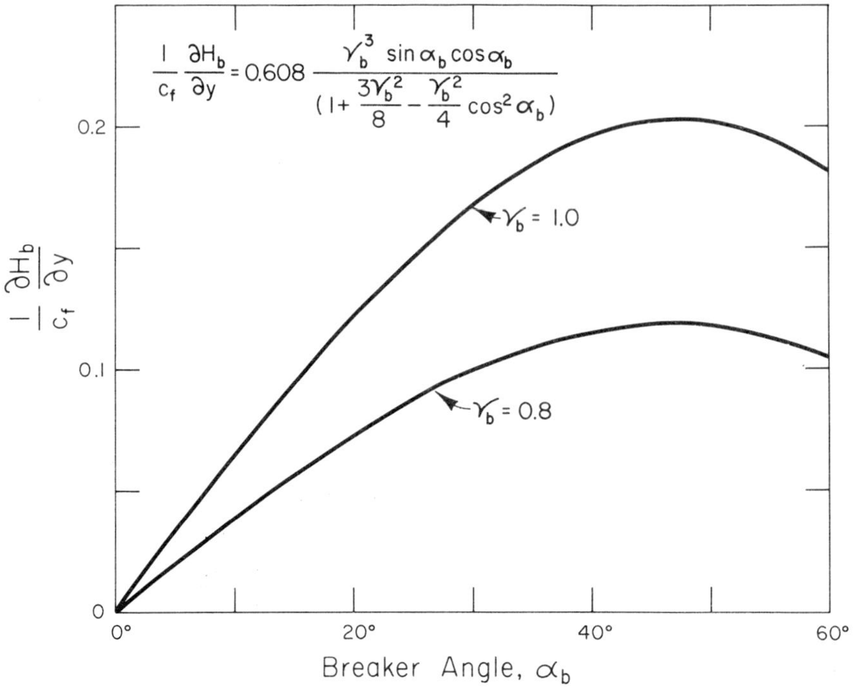

Figure 2.13. Curves obtained from Equation 2.20 for the equilibrium balance of the oblique wave approach of the waves and the longshore variation of breaker height such that $\bar{v}_l = 0$. The range of γ_b values includes those at which waves become unstable and break.

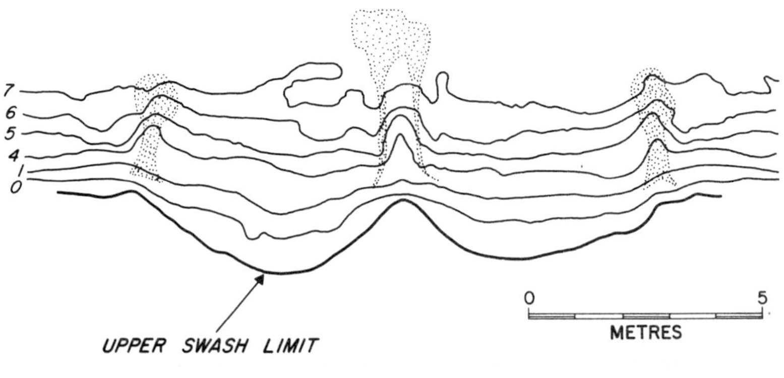

Figure 2.14. The configuration of the granular coal beach brought about by cell circulation after two hours of wave action. Contours are in inches below still-water level (0). The heavy contour represents the scarp notched into the beach by the upper limit of wave swash. The stippled areas denote the positions and respective sizes of the three rip currents that produced the cusps (after Komar, 1971).

after waves were generated in the basin three rip currents developed, a strong central rip flanked by weaker rips equally spaced to either side. Permanent cusps soon developed in the lee of the rip currents, Figure 2.14, a larger cusp forming in the lee of the stronger central rip current. The shoreline assumed this configuration within the first half an hour of wave action, by which time the rip currents suddenly disappeared and all longshore currents ceased to exist. Since there were no longshore currents there was no further sand transport alongshore and the cusps remained stable over the remainder of the two hours of wave action. A stable state of equilibrium had been achieved in spite of waves striking the flanks of the cusps at appreciable angles, a condition which of course normally generates longshore currents and produces a longshore transport of sand. The waves at the points of the cusps were observed to be smaller than those present in the embayments to either side throughout the entire experiment. Initially this variation in wave height produced the cell circulation and rip currents but the variation remained even after the rip currents disappeared. The wave height variations were apparently produced by edge waves, the edge waves remaining even after equilibrium had been achieved and the rip currents were no longer observable. Komar (1971, p. 2649) hypothesized that in the equilibrium state the two mechanisms or forces related to α_b and $\partial H_b/\partial y$ exactly opposed and balanced one another according to Equation 2.20 as discussed above. Once the cusps had developed sufficiently, both currents ceased to exist. One significance of this equilibrium condition is that it is possible for cusps to have been produced by rip currents, though the circulation is no longer present or is present in a much weakened state at the time of cusp observation.

Equation 2.20 defines the balance inherent in the observed equilibrium cuspate shoreline condition, the longshore variation in wave height acting opposite to the angle of wave approach to the shoreline. A series of theoretical equilibrium cuspate shoreline forms have been calculated assuming a sinusoidal longshore variation in the wave height and thus in the gradient of the wave height given by

$$\frac{\partial H_b}{\partial y} = \frac{\pi \Delta H_b}{2\lambda} \sin(\pi y/\lambda) \tag{2.21}$$

where ΔH_b is the difference in the wave breaker height over the longshore distance λ; 2λ would be the longshore distance between rip current positions and therefore cusp positions. In the examples shown in Figure 2.15 it has been taken that $\lambda = 2{\cdot}5$ m and $\Delta H_b = 0{\cdot}5$, $0{\cdot}75$ and $1{\cdot}0$ cm. The λ value corresponds to the cusp spacing in Figure 2.14 obtained in the laboratory. The drag coefficient in Equation 2.20 was set at $c_f = 0{\cdot}0225$ which is a factor $1{\cdot}25$ times the value $0{\cdot}018$ which, as is shown below, is the best value for the field data. The $1{\cdot}25$ factor is based on Equation 2.19, the difference between field and laboratory data noted by Longuet-Higgins (1970a). This choice for a drag coefficient seems to be further justified by the correspondence between the theoretical cusp of Figure 2.15 and the laboratory cusp of Figure 2.14. The writer also tried to use values of $c_f = 0{\cdot}010$ and $0{\cdot}018$ to produce the theoretical cusp, but the cusp projection was

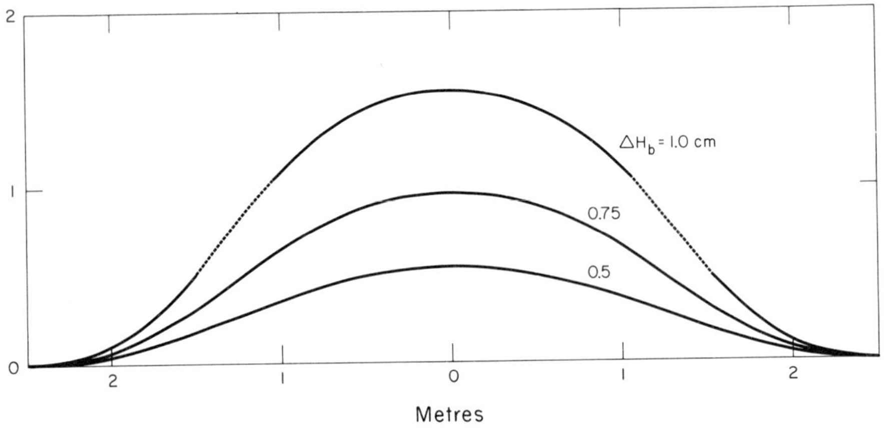

Figure 2.15. A series of theoretical equilibrium cuspate shoreline forms developed by assuming the sinusoidal form of the longshore gradient in wave height given by Equation 2.21 and taking $\lambda = 2{\cdot}5$ m, $\gamma_b = 1{\cdot}0$ and $c_f = 0{\cdot}0225$.

then much too small for acceptable values of ΔH_b over the distance $\lambda = 2{\cdot}5$ m. It can be seen in Figure 2.15 that the greater the value of ΔH_b, and therefore the greater the value of $\partial H_b/\partial y$ from Equation 2.21, the further the equilibrium cusp projects seaward; for $\Delta H_b = 0{\cdot}75$ cm the cusp projects nearly one metre seaward from the embayments. For $\Delta H_b = 1{\cdot}0$ cm the values of $\partial H_b/\partial y$ locally exceed the value at which balance can be achieved even by a breaker angle of the order of 45° so that no balance could be achieved throughout the dashed portion of the cuspate shoreline in Figure 2.15. Therefore, the shoreline shown for $\Delta H_b = 1{\cdot}0$ cm is not entirely in equilibrium. There would be a weak current flowing along the dashed portion of the shoreline toward the apex of the cusp but turning seaward to form a 'rip' off from the flank of the cusp. In other instances the breaker angle could exceed that necessary for equilibrium and the current would flow, however slowly, from the cusp apex toward the embayment where it would turn seaward as a rip current. Such reversals in current direction following cusp formation have been observed in the laboratory experiments (D. H. Willis, personal communication, 1971).

The central cusp in Figure 2.14, observed in the wave basin experiments, is somewhat more pointed than the theoretical cuspate shoreline in Figure 2.15. This is probably attributable to the fact that the original purpose of the experiment was to simulate the effects of an offshore shoal on the shoreline shape. The initial wave crests were therefore not entirely straight but rather each crest reached the originally straight shoreline, first at the margins of the stretch of the beach, and somewhat later in the centre. There was, therefore, an overall tendency to drive longshore currents and sand towards the centre of the beach. This is the reason why the central rip current and finally the central cusp were larger than those to the sides. The observed and theoretical cusps would also differ if the real wave height variations were not sinusoidal as has been assumed in developing the cusps of Figure 2.15. Longshore variations in tan β, γ_b, and breaker type could also affect the observed equilibrium conditions.

The wave basin experiments described above and also in Komar (1971) indicate the possibility of an equilibrium cuspate configuration being produced by rip currents although the currents are no longer present at the time of cusp observation. Depending upon the lack of a total balance between the driving forces, the longshore current in a weakened condition could flow either towards the apex of the cusp or towards the embayments. This may explain why rip currents are sometimes observed at the apex of the cusps and at other times within the embayments (Komar, 1971).

Such an origin may account for some series of cusps along a shoreline, beach cusps of medium scale as well as giant cusps. The relationship between rips and cusps with a cusp in the lee of each rip, suggested by the wave tank experiments, is described by Shepard (1963, p. 195) at Scripps Beach, La Jolla, California. At that location the rip currents are positioned by the effects of offshore submarine canyons on wave refraction and remain approximately stable in position most of the time. The high tide summer shoreline configuration and relationship to the rip currents is very similar to that found in the wave basin experiments. In Komar (1971, p. 2647) I described observations of small-scale beach cusps at Crail, Scotland. A series of six rip currents with an average spacing of 5·8 m was readily observable because of the presence of minute bits of kelp and sea grass which collected in the rips. In the lee of each rip current a definite cusp had formed. The cusps were composed of coarse shell debris and grit size rock fragments, noticeably coarser than the medium-grained sand of the remainder of the beach. The positions of the rip currents had apparently remained stationary for some time as the coarse cusp material was strung out down the beach face by the retreating tide. It is likely that most geomorphologists would consider the cusps at Crail as typical 'beach cusps'. The writer does not know of any field examples of cusp formation where rip currents were observed to form cusps and then disappeared, such as was the case for the laboratory experiments. Because of the irregularity of field conditions, most probably a complete balance would never be achieved, although the remaining longshore currents would be left in a much reduced state, as was the case for the cusps found at Crail, Scotland.

The above mechanism most assuredly does not account for all cases of cuspate shorelines. Many series of cusps are associated with rhythmic topography (Sonu, 1973) and with offshore crescentic bars (Bowen and Inman, 1971; Sonu, 1972b). In these two cases the principal topographic features are in the surf or even beyond the breaker zone, the cusps being only a minor expression of this offshore topography. Other sets of beach cusps are found between much more widely spaced rip currents; their origin has been suggested to be caused by edge waves of low mode (Bowen and Inman, 1969). The importance of edge waves in cusp formation has also been indicated by Komar (1973). It appears from all this that cuspate shorelines, from small beach cusps to widely spaced giant cusps or rhythmic topography, may be generated by several separate mechanisms. This may be why our search for 'the one mechanism' has been filled with disagreement and discouragement.

The complete solution

Although there may be a balance at mid-surf in the forces of longshore current generation, a balance given by Equation 2.20, there could still be non-equilibrium in the other portions of the surf zone (Bakker, 1971). This can be examined only by a complete solution of the velocity distribution across the surf, a solution which includes horizontal eddy mixing. Such horizontal eddies transfer and redistribute the driving thrust of the longshore current and therefore are important in determining the velocity distribution across the surf. Bowen (1969b) first included this effect in his analysis of longshore currents from an oblique wave approach. Longuet-Higgins (1970b) and Thornton (1970) used a somewhat different and preferable form for the horizontal eddy viscosity, $\rho N x (gh)^{1/2}$, where N is a numerical constant with the limits $0 < N < 0{\cdot}016$ (Longuet-Higgins, 1970b, p. 6791). Including this term for the horizontal eddy exchange, Equation 2.11 is modified to

$$px^{3/2} - qx - r\frac{\partial}{\partial x}\left[x^{5/2}\frac{\partial \bar{v}}{\partial x}\right] - sx^{1/2}\bar{v} = 0 \qquad (2.22)$$

where

$$p = \frac{5}{16}\rho g \gamma^2 (\zeta \tan\beta)^2 [g\zeta \tan\beta]^{1/2}\left[\frac{\sin\alpha}{C}\right]\cos\alpha$$

$$q = \rho g \zeta \tan\beta\left(1 + \frac{3\gamma^2}{8} - \frac{\gamma^2}{4}\cos^2\alpha\right)\frac{\partial \bar{\eta}}{\partial y}$$

$$r = N\rho(g\zeta\tan\beta)^{1/2}\zeta\tan\beta$$

$$s = \frac{\gamma}{\pi}\rho c_f (g\zeta\tan\beta)^{1/2}$$

where ζ is the constant term

$$\zeta = \frac{1}{1 + 3\gamma^2/8}$$

For a given situation where the beach slope and wave conditions are specified, the parameters p, q, r and s become constants. $\partial\bar{\eta}/\partial y$ is the longshore gradient in the wave set-up within the surf, the longshore slope of the water surface. This must be evaluated from the longshore variation in the breaker height.

The solution of Equation 2.22 yields

$$\bar{v} = B_1 x^{P_1} + A_1 x + A_2 x^{1/2} \qquad \text{for } X_b > x > 0 \qquad (2.23a)$$

$$\bar{v} = B_2 x^{P_2} \qquad \text{for } \infty > x > X_b \qquad (2.23b)$$

where x is taken as its absolute value, still zero on the shoreline, and where

$$P_1 = -\frac{3}{4} + \left[\frac{9}{16} + \frac{s}{r}\right]^{1/2}$$

$$P_2 = -\frac{3}{4} - \left[\frac{9}{16} + \frac{s}{r}\right]^{1/2}$$

$$A_1 = \frac{-1}{(1 - P_1)(1 - P_2)}\left(\frac{p}{r}\right)$$

$$A_2 = \frac{1}{(0{\cdot}5 - P_1)(0{\cdot}5 - P_2)}\left(\frac{q}{r}\right)$$

$$B_1 = \frac{A_1(1 - P_2)X_b^{1-P_1}}{P_2 - P_1} + \frac{A_2(0{\cdot}5 - P_2)X_b^{0{\cdot}5-P_1}}{P_2 - P_1}$$

$$B_2 = \frac{A_1(1 - P_1)X_b^{1-P_2}}{P_2 - P_1} + \frac{A_2(0{\cdot}5 - P_1)X_b^{0{\cdot}5-P_2}}{P_2 - P_1}$$

so long as $r/s \neq 2/5$. Equation 2.23a gives the variation in velocity $\bar{v}$ across the surf zone of width X_b while Equation 2.23b gives the velocity distribution outside the breaker zone. The velocity distribution is continuous at the breaker zone and goes to zero at the shoreline, although this was not a boundary condition in the solution of Equation 2.22.

Figure 2.16 includes typical examples from Equation 2.23 for a range of $\partial\bar{\eta}/\partial y$ values. In all cases $H_b = 10$ m and $\alpha_b = 20°$, the beach slope being 0·100. The surf zone width X_b is obtained from $h_b = H_b/\gamma_b$ and $X_b = h_b/\tan\beta$, having used $\gamma_b = 0{\cdot}8$ in this example. Furthermore, $N = 0{\cdot}005$ and $c_f = 0{\cdot}017$, which will be justified below. With $\partial\bar{\eta}/\partial y = 0$, the longshore current is due entirely to the oblique wave approach, the solution being equivalent to that obtained by Longuet-Higgins (1970b, Equation 21). With $\partial\bar{\eta}/\partial y = -0{\cdot}0005$ (Figure 2.16), the water slopes downward in the positive y direction, the direction of the longshore component of the oblique wave approach, so that the two forces combine to produce stronger longshore currents. For positive values of $\partial\bar{\eta}/\partial y$ the water slope opposes the oblique wave approach and the current is diminished. For $\partial\bar{\eta}/\partial y = 0{\cdot}0025$ (Figure 2.16), the forces are close to balancing and the resulting longshore current is very small. The strongest residual current is close to the shoreline and is flowing in the opposite direction to the current beyond the mid-surf zone. This is because $\partial\bar{\eta}/\partial y$ is constant across the surf zone and is smaller than the thrust due to the oblique waves in the outer surf zone but is greater within the inner surf. This value $\partial\bar{\eta}/\partial y = 0{\cdot}0025$ was rather arbitrarily selected; another value would probably produce a better balance and smaller $\bar{v}$ values. Considering the assumptions involved, such as a constant beach slope, a constant γ value independent of x, N independent of x, and so on, the balance between the forces yielding diminished longshore currents is surprisingly close. Although the balance is not complete over the entire surf zone, the results indicate that velocities can be diminished to near zero. This conforms with our laboratory observations.

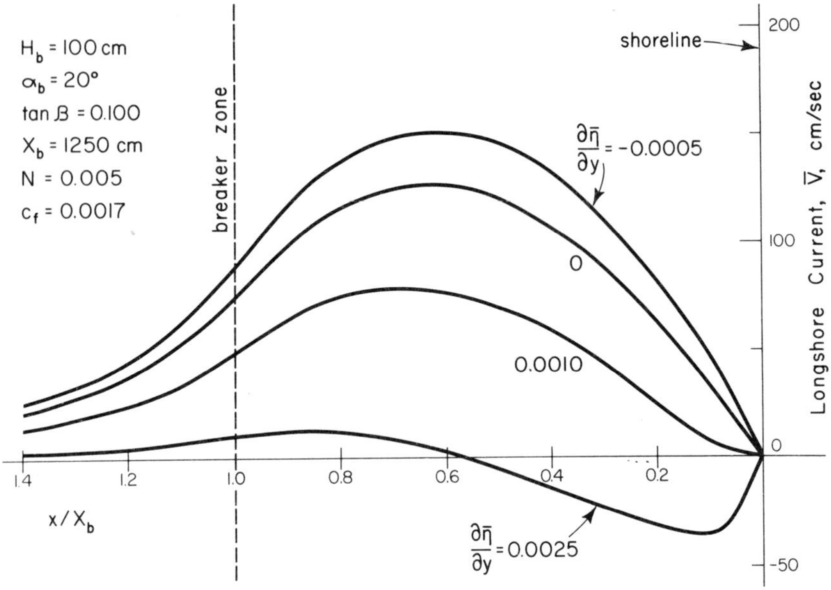

Figure 2.16. Examples of complete solutions of the distribution of longshore current velocities through the surf zone obtained with Equation 2.23 for a series of values for the longshore variation in the wave set-up ($\partial\bar{\eta}/\partial y$). With $\partial\bar{\eta}/\partial y =$ 0·0025 the set-up slope in the longshore direction nearly opposes and balances the thrust due to the oblique wave approach and the velocities are greatly weakened.

It remains now to compare Equation 2.23, the complete solution, with Equation 2.16 which is semi-empirical, the 2·7 coefficient having been obtained from the data comparison. This is done by balancing the value obtained with Equation 2.16 against that calculated with the complete solution at the mid-surf position ($x = X_b/2$). A range of α_b and H_b values are utilized. The eddy coefficient N and drag coefficient c_f are varied within reasonable ranges until the two solutions agree. This procedure is basically calibrating the complete solution against Equation 2.16 so that it, too, agrees with the available data, at least at the mid-surf position. The results are shown in Table 1.1. It turns out that changing N does not have too much effect on the mid-surf velocity, its main importance being noted beyond the breaker zone. An increase in N tends to reduce the current at mid-surf, however slightly, so that this requires a decrease in c_f so that $\bar{v}_{0.5}$ continues to agree with the semi-empirical $\bar{v}_l$ values. The drag coefficient c_f is principally responsible for bringing $\bar{v}_{0\cdot 5}$ from the complete solution into agreement with Equation 2.16. One example of this is shown in Figure 2.17. In this case $N = 0{\cdot}003$, and $c_f = 0{\cdot}0179$ provides agreement between the two solutions. For the beach slope 0·100 it is seen in Table 1.1 that N could vary over an order of magnitude with the drag coefficient changing only from about 0·014 to 0·018. Also shown in Table 1.1 is that a change in the beach slope requires a change in the drag coefficient to give agreement, the greater the slope the larger

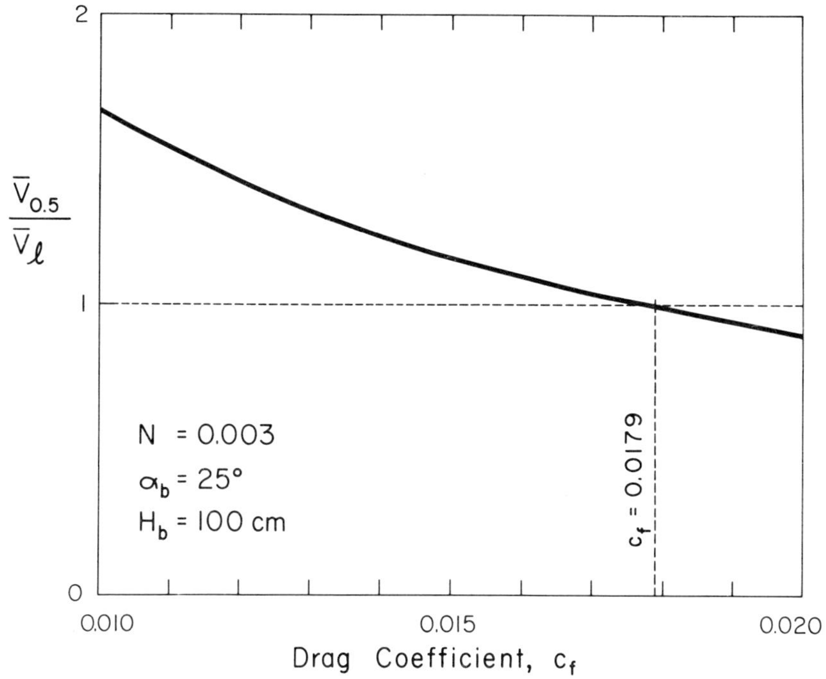

Figure 2.17. A comparison between the solution of Equation 2.16 for the velocity $\bar{v}_l$ at mid-surf and the complete solution of Equation 2.23 evaluated at the mid-surf position giving $\bar{v}_{0.5}$. When $\bar{v}_{0.5}/\bar{v}_l = 1$, the two solutions agree and, in this example, indicate that $c_f = 0{\cdot}0179$ for $N = 0{\cdot}003$.

the value of c_f. In fact, doubling the beach slope approximately requires a doubling of c_f such that the ratio $\tan\beta/c_f$ remains approximately constant. This conforms with the findings and conclusions of Komar and Inman (1970) which led to Equation 2.15. It indicates that eddy diffusion effects are not as important, and thus disagrees with Longuet-Higgins' (1972, p. 227) comments as to the reasons for the proportionality. Since in laboratory experiments the beach slopes are generally high, the results of Table 1.1 also explain why drag coefficients under such conditions are greater than under field conditions (Longuet-Higgins, 1970b).

The last column in Table 1 gives the value of the dimensionless parameter

$$P = \frac{\pi N \tan\beta}{\gamma_b c_f} \tag{2.24}$$

defined by Longuet-Higgins (1970b) as governing the overall longshore current distribution through the surf zone. The larger the value of P the greater the coupling between the driving forces within the surf zone and the offshore, and the greater the resulting longshore directed currents outside the breaker zone. Comparing his theoretical velocity distributions with the laboratory data of Galvin and Eagleson (1965), Longuet-Higgins (1970b) concluded that P is between 0·1 and 0·4. The writer's impression from the field is that P must be of the order of 0·1, or somewhat smaller, as larger values give excessive currents outside the breaker zone. The values of P in Table 1.1 can be expected to give reasonable longshore current distributions across the surf.

Table 1.1

$\tan\beta$	N	c_f	$\tan\beta/c_f$	$P = \pi N \tan\beta/\gamma_b c_f$
0·050	0·005	0·0083	6·02	0·118
0·100	0·001	0·0183	5·46	0·022
	0·003	0·0179	5·59	0·066
	0·005	0·0169	5·92	0·116
	0·010	0·0137	7·30	0·287
0·150	0·005	0·0250	6·00	0·112

Conclusions

The following conclusions are arrived at or are supported by this paper.

(1) The longshore current velocity $\bar{v}_l$ at the mid-surf position is related to the breaking wave orbital velocity $u_m = (2E_b/\rho h_b)^{1/2}$ and angle of approach α_b to the shoreline by

$$\bar{v}_l = 2{\cdot}7 u_m \sin\alpha_b \cos\alpha_b$$

This equation was first proposed by Komar and Inman (1970), prompted by the agreement between two independent estimates of the littoral sand transport, and has been derived theoretically by Longuet-Higgins (1970a) through application of radiation stress concepts. The 2·7 coefficient is based on the field and laboratory longshore current data. A complete comparison between theory and data has been undertaken in this paper and it has found that most of the data are in agreement with this equation and coefficient although there are some noteworthy exceptions. More field studies especially are needed.

(2) If there is a longshore gradient in the wave breaker height, $\partial H_b/\partial y$, as well as an oblique wave approach, then the resulting longshore current stems from the combined effects of the two driving forces. In this case Equation 2.18 predicts the longshore current $\bar{v}_l$ at mid-surf. At the present there are no data to test the contribution of the $\partial H_b/\partial y$ term.

(3) Under certain circumstances the thrust resulting from the oblique wave approach may oppose and balance the contribution of the longshore variation in wave height such that $\bar{v}_l = 0$ in Equation 2.18. One such case is a cuspate shoreline as described by Komar (1971) from wave basin tests, wherein rip currents generated cusps but then ceased to exist once the cusps reached an equilibrium development and the forces balanced according to Equation 2.20. It is therefore possible that some cuspate shorelines result from rip currents although the rips are no longer present at the time of observation. Greatly weakened residual longshore currents could persist if the balance is not exact, possibly even flowing in the opposite direction to the previous flow.

(4) Equation 2.23 gives the complete solution for the distribution of the longshore current velocity across the surf zone, resulting both from an oblique wave approach and from a longshore variation in wave height (set-up). It was found that a near-balance can be achieved across the entire surf zone giving $\bar{v} \simeq 0$, not only at the mid-surf position.

(5) A comparison between the complete solution of Equation 2.23 and the semi-empirical relationship for the mid-surf longshore current indicates that an increase in the beach slope requires a proportional increase in the drag coefficient c_f such that $\tan\beta/c_f$ remains approximately constant. The value of the horizontal eddy exchange is not as important to the mid-surf velocity. Values of $\tan\beta$ and the required c_f are given in Table 1.1.

Acknowledgements

I would like to thank M. K. Gaughan for helping me with the solution to the complete problem, including the horizontal eddy exchange, and A. J. Bowen for reading the manuscript, making many suggestions and corrections. This research was supported by the Oceanography Section, National Science Foundation, NSF Grant GA-36817.

Notation

C = friction coefficient
c_f = drag coefficient
E = wave energy density
E_b = wave breaker energy
F = total driving force of current
F_{cell} = total driving force of current for cell
g = gravitational force
H = wave height
H_b = breaker height
h = still-water depth
h_b = wave depth at breaking
k = 2π/(wavelength)
N = numerical constant, $0 < N < 0{\cdot}016$
N = eddy effect
n = wave group velocity/wave phase velocity
p = dimensionless parameter governing the overall longshore current distribution through the surf zone
u_m = maximum value of breaking wave orbital velocity, assumed to be sinusoidal
$\bar{v}$ = velocity across surf zone of width X_b
$\bar{v}_l$ = longshore current velocity
x = coordinate normal to shoreline, positive onshore (direction of wave advance)

y = coordinate parallel to shoreline (and wave crests)
α = angle of wave crest to the shoreline
α_b = wave breaker angle
$\tan\beta$ = beach slope

$$\gamma = \frac{H}{\bar{\eta} + h}$$

ΔH_b = difference in wave breaker height over longshore distance, λ
$\bar{\eta}$ = wave set-up
λ = longshore distance where 2λ = distance between successive rip current or cusp positions
ρ = water density

$$\zeta = \text{constant}\,\frac{1}{1 + 3\gamma^2/8}$$

τ_y = longshore current

References

Bakker, W. T., 1971, Littoral drift in the surf zone. *Rÿkswaterstaat Dir. Wat. en Wat. afd. Kust.* Studierapport 70, 16.

Bowen, A. J., 1969a, Rip currents, 1, Theoretical investigations, *J. geophys. Res.*, **74**, 5467–5478.

Bowen, A. J., 1969b, The generation of longshore currents on a plane beach, *J. mar. Res.*, **37**, 206–215.

Bowen, A. J., D. L. Inman and V. P. Simmons, 1968, Wave 'set-down' and 'set-up', *J. geophys. Res.*, **73**, 2569–2577.

Bowen, A. J., and D. L. Inman, 1969, Rip currents, 2, Laboratory and field observations, *J. geophys. Res.*, **74**, 5479–5490.

Bowen, A. J., and D. L. Inman, 1971, Edge waves and crescentic bars, *J. geophys. Res.*, **76**, 8662–8671.

Brebner, A., and J. W. Kamphuis, 1963, Model tests on the relationship between deep-water wave characteristics and longshore currents, Queen's University, Canada, *Research Report No. 31*, 25 pp.

Galvin, C. J., 1967, Longshore current velocity: A review of theory and data, *Reviews of Geophysics*, **5**, 287–304.

Galvin, C. J., and P. S. Eagleson, 1965, Experimental study of longshore currents on a plane beach, Coastal Engin. Res. Center (US Army Corps of Engineers), *Tech. Memo. 10.*

Galvin, C. J., and R. A. Nelson, 1967, Compilation of longshore current data, Coastal Engin. Res. Center (US Army Corps of Engineers), *MP 2-67.*

Harrison, W., E. W. Rayfield, J. D. Boon III, G. Reynolds, J. B. Grant and D. Tyler, 1968, A time series from the beach environment, Land and Sea Interaction Lab., ESSA Research Lab., *Tech. Memo. No. 1*, 28 pp.

Inman, D. L., and W. H. Quinn, 1951, Currents in the surf zone, *Proc. 2nd Conf. cst. Engng., Houston, Texas*, pp. 24–36.

Iverson, H. W., 1951, Studies of wave transformation in shoaling water, including breaking, *Gravity Waves, National Bureau of Standards Circular 521.*

Komar, Paul D., 1971, Nearshore cell circulation and the formation of giant cusps, *Bull. geol. Soc. Am.*, **82**, 2643–2650.

Komar, P. D., 1972, Nearshore currents and the equilibrium cuspate shoreline, Oregon State University Department of Oceanography, *Tech. Report 239*, 20 pp.

Komar, P. D., 1973, Observations of beach cusps at Mono Lake, California, *Bull. Soc. Am.*, **84**, 3593–3600.

Komar, P. D., (in preparation), *Beach Processes and Sedimentation*, Prentice-Hall, Englewood Cliffs, N.J.

Komar, P. D., and D. L. Inman, 1970, Longshore sand transport on beaches, *J. geophys. Res.*, **75**, 5914–5927.

Longuet-Higgins, M. S., 1970a, Longshore currents generated by obliquely incident sea waves, 1, *J. geophys. Res.*, **75**, 6778–6789.

Longuet-Higgins, M. S., 1970b, Longshore currents generated by obliquely incident sea waves, 2, *J. geophys. Res.*, **75**, 6790–6801.

Longuet-Higgins, M. S., 1972, Recent progress in the study of longshore currents, in R. E. Meyer (Ed.), *Waves on Beaches*, Academic Press, New York, pp. 203–248.

Longuet-Higgins, M. S., and R. W. Stewart, 1960, Change in the form of short gravity waves on long waves and tidal currents, *J. Fluid Mech.*, **8**, 565–583.

Longuet-Higgins, M. S., and R. W. Stewart, 1963, A note on wave set-up, *J. mar. Res.*, **21**, 4–10.

Longuet-Higgins, M. S., and R. W. Stewart, 1964, Radiation stress in water waves, a physical discussion with applications, *Deep Sea Res.*, **11**, 529–563.

Putnam, J. A., W. H. Munk and M. A. Traylor, 1949, The prediction of longshore currents, *Trans. Am. geophys. Un.*, **30**, 337–345.

Saville, T. Jr., 1950, Model study of sand transport along an infinitely long straight beach, *Trans. Am. geophys. Un.*, **31**, 555–565.

Shepard, F. P., 1963, *Submarine Geology* (2nd ed.), Harper and Row, New York, 557 pp.

Shepard, F. P., and D. L. Inman, 1950a, Nearshore circulation related to bottom topography and wave refraction, *Trans. Am. geophys. Un.*, **31**, 555–565.

Shepard, F. P., and D. L. Inman, 1950b, Nearshore circulation, *Proc. 1st Conf. cst. Engng., Long Beach, California*, pp. 50–59.

Sonu, Choule J., 1972a, Field observation of nearshore circulation and meandering currents, *J. geophys. Res.*, **77**, 3232–3247.

Sonu, C. J., 1972b, Comment on paper by A. J. Bowen and D. L. Inman, 'Edge wave and crescentic bars', *J. geophys. Res.*, **77**, 6629–6631.

Sonu, C. J., 1973, Three-dimensional beach changes, *J. Geol.*, **81**, 42–64.

Thornton, E. B., 1970, Variations of longshore current across the surf zone, *Proc. 12th Conf. cst. Engng., Washington*, pp. 291–308.

CHAPTER THREE

Wave Asymmetry in the Nearshore Zone and Breaker Area

P. H. KEMP

Abstract

In coastal processes the asymmetry of the forces acting on the shoreline or on the nearshore sediments is a dominant factor both in the short-term and long-term changes which these forces produce. Wave shape asymmetry starts outside the breaker zone. The wave shape can be described in a number of ways which allow the profile distortion to be evaluated. The ratio of the slope of the front face of the wave to the rear face, or the ratio of the length of the crest area at mean water level to the corresponding length of the trough, are examples of classification systems which can be used. The wave shape asymmetry is associated with asymmetry of the velocity field within the wave and this influences the movement of the sediment on the bed.

Once the wave has broken, the water surges up the beach and subsequently returns as backwash. Although this motion is related to the period of the waves, the times of uprush and backwash are not the same. The general character of the flow for the two parts of this cycle also differs. The ratio of time-of-uprush to wave period, or 'phase difference', is related to both the incident wave characteristics and to the beach slope. The beach slope, especially near the beach crest, is dependent on the grain size, except under conditions of severe wave attack. The grain size and the associated permeability of the beach undoubtedly have a great influence on the beach configuration, absorption of wave energy and phase difference.

One has, therefore, the situation in which two asymmetrical flow fields associated with wave action are developed close to the shore, the two zones being separated by the discontinuity in wave characteristics which occurs at the breakers. However, the conditions outside the breaker zone appear to be influenced by the backwash of the broken wave. Model tests show that this variation from the results found under normal backwash conditions decreases seaward and tends to disappear at a water depth to wavelength ratio of about 0·14. In a similar way, the ratio of the forward to backward velocities of the water close to the bed, outside the breaker zone, is higher if the backwash is eliminated.

Studies suggest that the elimination or reduction of the backwash has the effect of increasing the shoreward movement of bed material. The result of reducing the backwash on a steep beach is to produce velocity asymmetries closer to those found on a flatter beach. Thus the velocity patterns for flat beaches resemble those which occur in the absence of backwash. This further explains why beaches flatten in order to achieve stability. The reduction in backwash is equivalent to an increase in beach permeability.

Outside the breaker zone

Introduction

In the study of fluid mechanics, water waves represent an example of unsteady motion which can be described by the Navier-Stokes equations. The most direct solution to these equations is obtained by making simplifying assumptions which include a presumption that the flow is inviscid and irrotational, that the wave slope is small and that the water depth is uniform. These assumptions lead to the first order, or Airy, theory which is remarkably good but does not predict asymmetry of velocity or wave shape. Stokes' second-order solution for finite wave steepness indicates that the mean level of the wave surface lies above the still-water level, the crests are steeper and the troughs are flatter than predicted by the simpler first order, or linear, theory, and that there is a non-periodic or net drift of fluid, or mass transport, in the direction of wave advance. In the upper layers of the fluid this transport can be adequately described by the prediction that the water particle orbits are not closed circles or ellipses, even when the theory assumes that the potential theory applies. However, close to the bed, the similar progression of the fluid in the direction of wave advance has been shown to be due to the effect of viscosity (Longuet-Higgins, 1953). This net transport of water does not necessarily imply a corresponding net transport of sediment, unless the sediment motion exactly corresponds with that of the surrounding fluid. This could be said of low concentrations of fine sediment in suspension. For higher concentrations the fluid motion is modified by the suspension and a complete physical description of this interaction has yet to be made. For larger particles, which do not react completely to the fluid motion, the net transport becomes increasingly dependent on the absolute values of the orbital velocities in the region of the bed. In other words the shear stress which results from the velocity must exceed the critical value for first movement of a given grain (Grass, 1970). This is similar to the situation which exists for uni-directional flow whether in a stream or due to the tide. Should ripples be formed, an entirely new and more complex situation arises and, although recent progress has been made in this field, research into the mechanics of separated flows is still continuing (Williams and Kemp, 1971, 1972).

It is not the purpose of this introduction to define the conditions under which various wave theories are applicable in practice, since the paper mainly relates to the results of experiments and observations made in the laboratory and in the field. The asymmetry of orbital motion referred to in the previous paragraph is, however, well illustrated by the second-order expression for the horizontal component of orbital velocity

$$u = \frac{\pi H}{T} \cdot \frac{\cosh (z + d)/L}{\sinh 2\pi d/L} . \cos 2\pi\left(\frac{x}{L} - \frac{t}{T}\right)$$

$$+ \frac{3}{4}\frac{\pi^2 H^2}{LT} \cdot \frac{\cosh 4\pi(z + d)/L}{\sinh^4 2\pi d/L} . \cos 4\pi\left(\frac{x}{L} - \frac{t}{T}\right)$$

This is of the form

$$u = A \cos 2\pi X + B \cos 4\pi X$$

and as noted by Inman and Nasu (1956) the first term is positive under the wave crest and negative under the trough, whereas the second term is positive under both crest and trough, and negative at $L/4$ and $3L/4$ from the crest. The effect is to increase the magnitude but shorten the duration of the orbital velocity under the crest, and to decrease the magnitude but lengthen the duration of the offshore velocity under the trough. Although the trend is correct, the theory does not seem to correspond very well with measured values in shallow water.

In shallow water the cnoidal wave theory seems to offer a better description of large amplitude waves (Mehauté and coworkers, 1968). Svendsen and Brink-Kjaer (1972) also consider that this theory gives a more accurate estimate of the change in wave height with shoaling. Biesel (1951) has produced a theory which predicts the asymmetry of the slope of the front and rear face of the wave crest. Measurements by Adeyemo (1970) seem to substantiate the theory and to show that the deformation of a shoaling wave, whether measured in terms of the asymmetry of height, crest slope or crest-to-trough length, can be used as a measure of the asymmetry of water particle velocities. Iwagaki and Sakai (1972) have produced a theory for finite amplitude long waves on constant beach slopes which predicts both wave height and wave shape asymmetry for values of d/L less than 0·05.

The limitations imposed by the assumptions made in all the relevant wave theories, whether in relation to the hydrodynamic equations or the boundary conditions, necessitate continuing observations both in the laboratory and in nature. This is especially so where interest is centred on areas close to the breaker zone, where the influence of wave deformation, beach slope and backwash is considerable. The following comments are related exclusively to this zone and mainly to the influence of the backwash on the wave velocity asymmetry just outside the breakers and to the inferences which can be drawn in relation to the onshore–offshore movement of coarse sediments. The experimental apparatus and procedure designed by the author are as described by Adeyemo (1970).

Velocity asymmetry close to the breakers

A good deal has been written on the classification of breaker types into spilling plunging or surging types (e.g. Wiegel, 1964). It is evident that, in a qualitative sense, the change in wave profile depends largely on the rate of change of water depth and the ability of the wave to conform to this change. This effect is accentuated when the beach is sufficiently steep for the front and rear faces of the crest to be in significantly different depths of water. On such steep beaches the backwash from a broken wave has a higher velocity and a correspondingly greater effect on the succeeding wave. The following observations resulted from a study of the behaviour of a wave with a deep water wavelength of 1 m and a deep water steepness of 0·037, on slopes varying from 1 in 4 to 1 in 18. The beaches were plane and impermeable but the uprush or swash of the wave could, by adjusting the top of the beach slope, be allowed to overtop the beach and thus reduce the backwash. Two of the slopes studied were used for a comparison of orbital velocities close to the bed. The two slopes, 1 in 9 and 1 in 18, were chosen on the basis of previous experiments which indicated that they were respectively representative of characteristically steep beaches with plunging breakers and flat beaches with spilling breakers. The velocities were measured using a hydrogen bubble technique at a height of 5 mm above the bed, which was outside the wave-induced boundary layer.

The measured velocities are significant in a number of ways. The actual velocity is relevant to the initial movement of grains of a given size. The ratio v_m of the magnitude of the maximum onshore velocity to that of the offshore velocity indicates the differential effect on grains of different sizes. This can be termed the 'velocity magnitude asymmetry'. The ratio of the duration of onshore velocity to offshore velocity is correspondingly called the 'velocity time asymmetry'. Thus, a proportionally high onshore velocity lasting for a small proportion of the wave cycle would have a value of v_m greater than unity and a value of v_t less than unity.

As would be expected, the actual velocities increased for both beach slopes as the breaker zone was approached from the offshore area. The actual maximum velocities were greater on the 1 in 9 slope than on the 1 in 18 slope, as shown in Figure 3.1. On the other hand, the values of v_m were greater on the flatter slope, as shown in Figure 3.2. Figure 3.3 shows that the onshore movement lasts for a proportionately shorter time on the steeper slope. The three figures also show curves indicating the effect on the velocities of an artificial reduction in the backwash. Under these conditions the velocity magnitude asymmetry v_m was higher

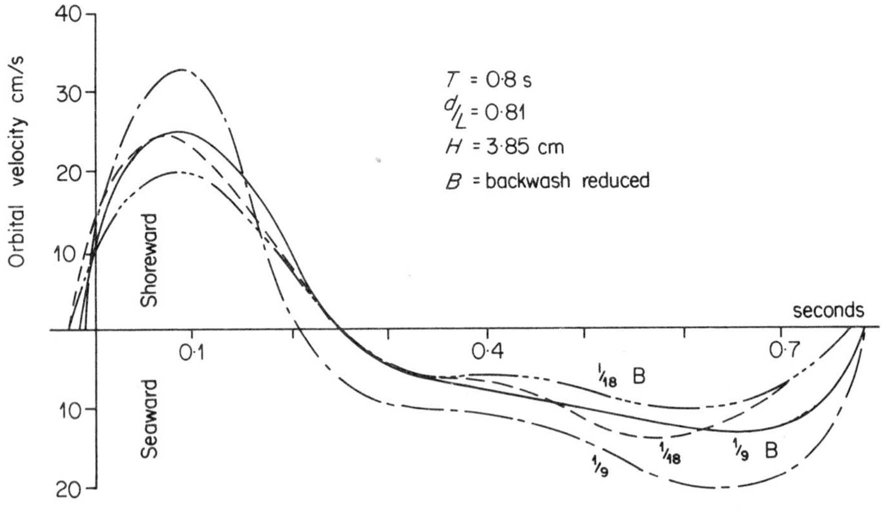

Figure 3.1. Orbital velocities close to the bed on beach slopes of 1 in 9 and 1 in 18, showing the effect of reducing the backwash.

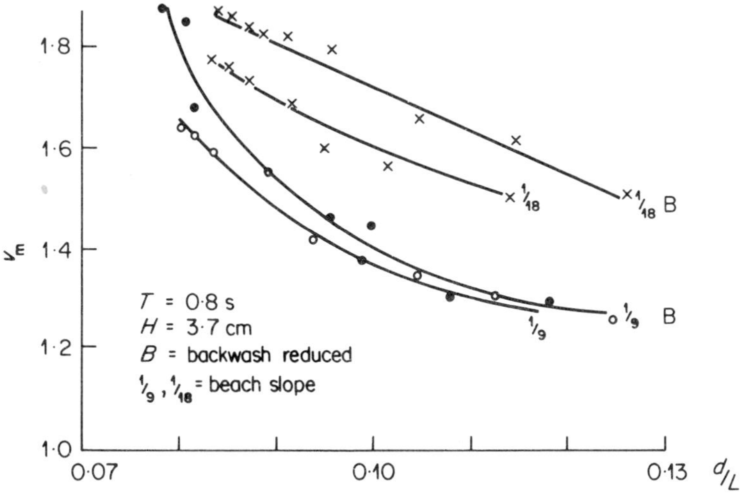

Figure 3.2. Velocity magnitude asymmetry of orbital motion on beaches of slope 1 in 9 and 1 in 18, showing the effect of reducing the backwash.

than under normal backwash conditions. The shoreward motion was also found to be of slightly longer duration.

The magnitude and the duration of the shoreward and seaward orbital velocities can be qualitatively interpreted in relation to the movement of sediment. The net effect of reducing the backwash would be to increase the shoreward movement of bed material. The result of reducing the backwash on a steep beach is to produce velocity asymmetries closer to those found on a flatter beach. Thus, the velocity patterns for flat beaches seem to resemble those which occur in

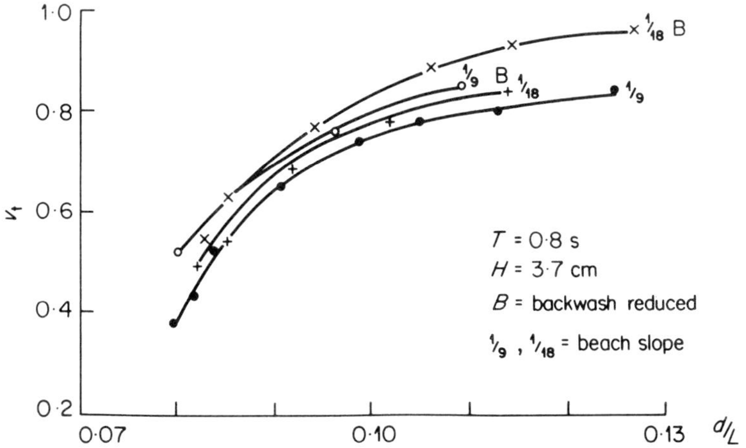

Figure 3.3. Velocity time asymmetry of orbital motion on beaches of slope 1 in 9 and 1 in 18, showing the effect of reducing the backwash.

the absence of backwash. This may help to explain why beaches flatten in order to achieve stability, for not only is wave energy absorbed over greater distances, but there is an increased tendency for material to be moved shorewards. This is particularly evident if the beach in an hydraulic model is artificially flattened below its equilibrium slope or if it is flattened by storm waves. The rate at which tracer material moves shorewards in the subsequent period of accretion under the original conditions is far greater than on the original steeper slope.

The reduction in backwash is evidently equivalent to an increase in beach permeability and it is interesting to conclude that the permeability of the beach face can have an effect on the wave velocity asymmetry outside, but close to, the breakers.

Figure 3.4 gives an indication of how a time–velocity curve could be used to assess the potential grain transport. The hatched area indicates that material moved only by velocities in this zone would not be moved in the seaward direction when the velocity reversed. If it were possible to make a quantitative calculation of this type, the gravitational effect in the seaward direction resulting from the beach slope would have to be taken into account. The two lines, a–a and b–b, define zones in which the onshore and offshore velocities are equal. In this case, it could be argued that particles moved by this velocity would migrate seaward under the action of the longer period of offshore motion, in the manner suggested by Grant (1943).

The experiments described above were carried out on a plane impermeable beach in order to overcome the difficulty of controlling the reduction in backwash on a beach composed of loose grains, and also because it would have been difficult to maintain a variety of beach slopes for a wave of given characteristics. Previous model experiments by Scott (1954) using a quartz sand of 310 microns median diameter produced a rippled bed. He used neutrally buoyant droplets to examine the water motion. The results showed considerable asymmetry of

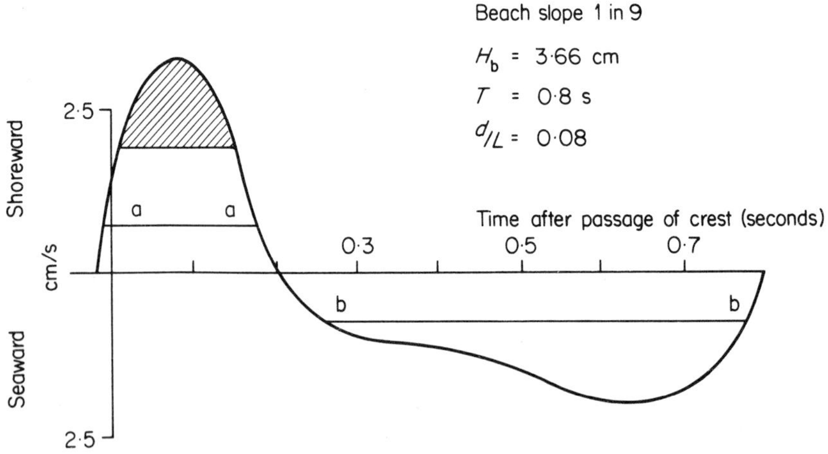

Figure 3.4. Potential grain transport.

orbital motion which he attributed to the distortion of the wave as it moved into shallower water. Scott's velocity measurements were made by following individual droplets in space and were thus of the Lagrangian type, whereas, in the experiments previously described, velocities were measured at stationary points close to the bed over a complete wave cycle. Since, in addition, the height above the bed of the droplets used by Scott is not stated, it is not possible to correlate the measurements, but the qualitative agreement is good. It would be expected that the velocity magnitude asymmetry found using Scott's analytical method would be greater than the single point method, since during the onshore movement droplets would be progressing into shallower water and thus into a higher velocity zone, whereas the opposite would hold during the offshore part of the wave cycle.

Finally, it should be stated that whereas wave-induced asymmetrical motion close to the bed in the offshore zone is a dominant factor in relation to the movement of coarse sediments, the movement of particles which are readily put into suspension will be influenced by many other factors. These include the effects of tidal currents, the distribution of mass transport velocity with height above the bed, wind-induced currents and local circulations associated with bed topography or rip currents. The structure of the fluid motion which results from the interaction of wave-induced and current-induced turbulent boundary layers has yet to be defined even for hydraulically smooth boundaries. In the presence of ripples the resultant motion is even more complex. A great deal can, therefore still be learnt by the experienced observer with a good knowledge of the relative importance of the many forces acting in a given situation.

So far the paper has been concerned with the internal water velocities associated with waves which are approaching the shore but which have not yet broken. They may be envisaged as climbing a sloping beach which is progressively reducing their rate of advance and at the same time changing their surface geometry. On relatively steep beaches the waves will eventually break and, from then onwards, the velocity patterns associated with their progress up the beach are markedly different from those in the offshore zone. For this reason the velocities in the uprush and backwash zone are considered separately in the following section.

Flow shoreward of the breakers

Introduction

The flow pattern and internal velocities in the breaker zone are important since the quantity of beach material in motion is primarily a function of the local velocities.

It has been shown by Kemp (1960) that the ratio of the time of uprush of a wave t to the wave period T can be used to characterize the flow conditions on a beach. This ratio t/T Kemp defined as the 'phase difference', and on this basis classified the wave–beach relationship into three categories with characteristically different flow regimes. The classification was based on the observation that for low values of phase difference the broken wave was able to surge up the beach to the limit of uprush and return as backwash to the breaker point before the succeeding wave had broken. Under these conditions the flow shoreward of the breakers was distinctly oscillatory in character and the beaches were steep and plane. As the height of the incident wave increased the uprush distance l_b increased and the time of uprush increased, so that a point was reached at which the backwash could not be completed before the next wave broke. This condition involved interference between the backwash of one wave and the uprush of the next, and the oscillatory nature of the flow gave way to a transition flow regime with some increase in the interchange of water between the zones landward and seaward of the breakers. As the phase difference was further increased to values greater than unity, the transition phase gave way to 'flow' conditions in which successive lines of breakers continually spilled water into the inshore zone, producing a corresponding seaward return flow. These three categories were designated 'surge', 'transition' and 'surf' conditions, the surge condition being associated with waves surging or breaking on steep beaches, the transition condition with plunging breakers, and the surf condition with spilling breakers on flattish beaches.

By making certain simplifying assumptions based on observations of the profile of the wave during uprush and backwash, the surging wave is open to quantitative description. The essentially oscillatory nature of the uprush and backwash of waves under these conditions bears a close resemblance to the phenomena of standing waves systems on rigid impermeable slopes. This problem has been studied by Miche (1944) and it was thus possible to compare the results of the investigation with an equation based on Miche's solution for low waves on steep slopes. The theory assumed the water movement on the beach to be sinusoidal and also that the uprush and backwash were identical in nature but of opposite sign. The phase difference according to Miche would thus always be 0·5. This is not true in the case of beaches and, in the study of water velocities on the beach outlined below, this was taken into account.

It is essential to separate uprush and backwash, since the initial boundary conditions are so different. The backwash starts from zero velocity everywhere on the beach and a 'wedge' of water then moves down the beach, the water surface remaining essentially plane, from the instantaneous position of the waterline on the beach to the breakpoint, throughout the backwash duration. In the case of the uprush the kinematics of the breaking wave are involved. For very low phase differences (say $t/T < 0{\cdot}3$) the uprush is approximately the reverse of the backwash but to a different time scale. However, as the phase difference approaches the limit for the surging condition, the uprush starts and may remain in the form of a bore. Thus, there are two possible methods of solution.

The general characteristics of the flow on a beach under surging and plunging conditions are shown in Figures 3.5, 3.6, 3.7 and 3.8 which illustrate both model and field conditions. For surging conditions the retardation of the uprush and subsequent acceleration of the backwash are equal and continuous over the

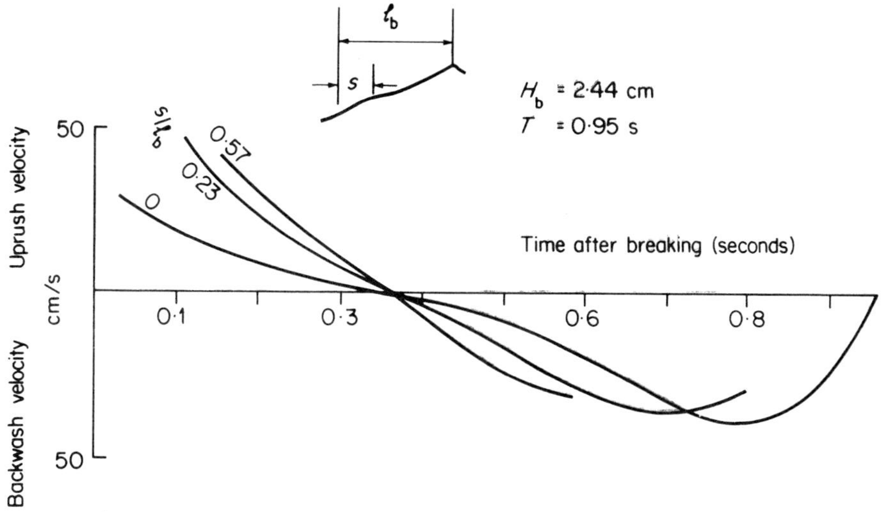

Figure 3.5. Water velocities in the model under surging conditions.

point of zero velocity, and are progressively greater at points further from the breakpoint. Most significantly, the velocity is zero at all points simultaneously and the backwash is complete before the following wave breaks (Kemp and Plinston, 1968).

For plunging conditions the uprush velocity is not zero at all points simultaneously. The water nearer the breakers begins to move seaward before the water further up-beach comes to rest. It is also clear that the backwash is not complete before the following wave breaks.

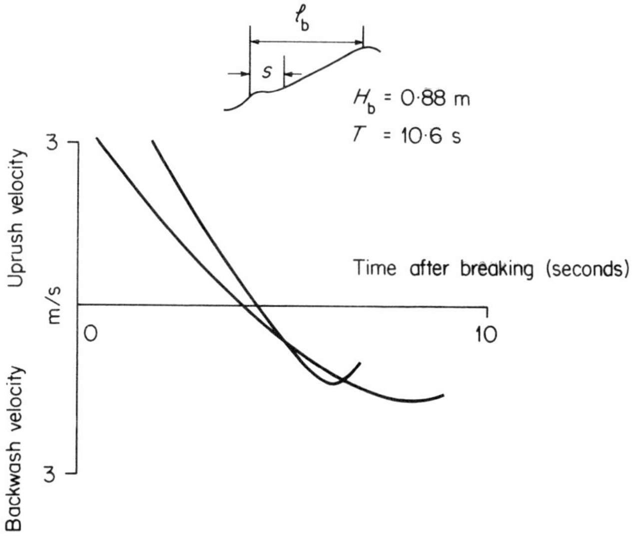

Figure 3.6. Water velocities in the prototype under surging conditions.

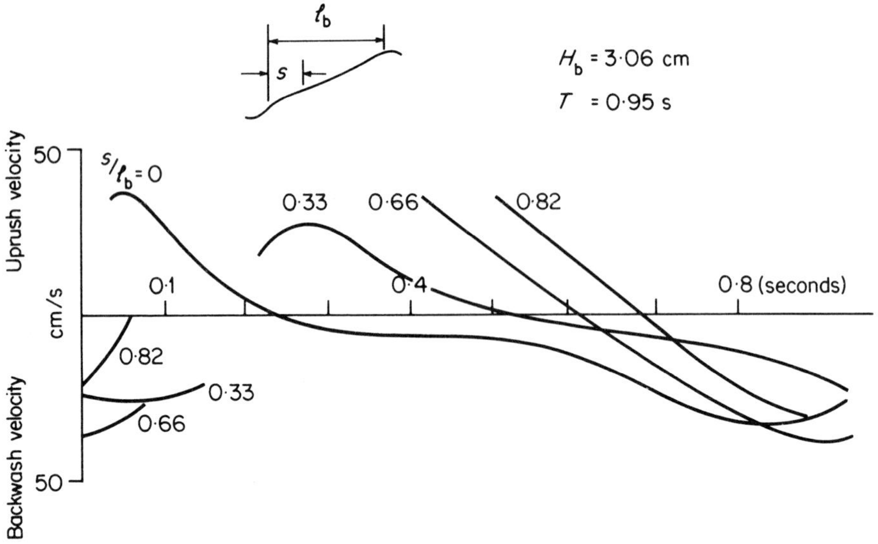

Figure 3.7. Water velocities in the model under plunging conditions.

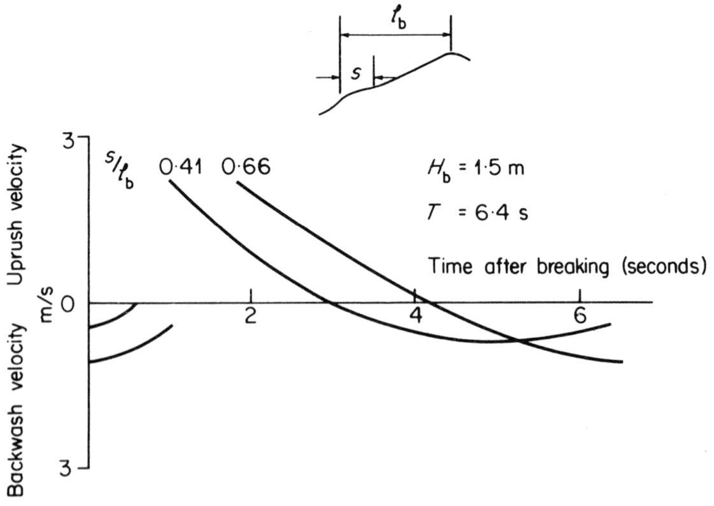

Figure 3.8. Water velocities in the prototype under plunging conditions.

Flow under surging conditions

As already mentioned, Miche's theory described a symmetrical cycle of uprush and backwash and, on the basis of his analysis, the component of velocity along the beach slope at a time θ measured from the instantaneous still-water level is given by

$$v = \frac{1}{\sin\alpha}\cdot\frac{\pi H}{2}\cdot\left(\frac{g}{\alpha L}\right)^{1/2}\cos\left[\left(\frac{2\pi g}{L}\right)^{1/2}\theta\right]$$

The movement of the water's edge on the beach is also sinusoidal and the distance s' from the breaker position to the edge can be expressed as

$$s' = \frac{l_b}{2}\left[1 - \cos\left(\frac{2\pi\theta}{T}\right)\right]$$

In fact, the movement of the water on the beach for the low, long waves postulated by Miche is asymmetrical. However, a sinusoidal motion can be maintained by treating the uprush and backwash as two independent cycles of period t for the uprush, and $T - t$ for the backwash. For the uprush the time θ is measured from the breaking point, and for the backwash from the limit of uprush.

Thus for the uprush

$$\frac{s'}{l_b} = \frac{1}{2}\left(1 - \cos\frac{\pi\theta}{t}\right) \tag{3.1}$$

and for the backwash

$$\frac{s'}{l_b} = \frac{1}{2}\left(1 + \frac{\cos\pi\theta}{T - t}\right) \tag{3.2}$$

By comparing these curves with the data from model experiments, Kemp and Plinston (1974) found that the sine curve appears to be satisfactory for the backwash but that a better description of the uprush can be developed by introducing the characteristics of a bore into the uprush equation. The results of this modification are outlined later in this section.

Backwash velocities. The derivation of the instantaneous velocities from the changing geometry of the water level on the beach is made possible by making assumptions based on observations of surging waves. They are:

(1) The beach is plane, of slope $\tan\alpha$.
(2) The velocity of water on the beach is everywhere zero at the temporal limit of uprush.
(3) There is no loss or gain of water into or out of the beach.
(4) The water surface remains plane from the limit of uprush to the breakpoint throughout the backwash period, although the slope can vary.
(5) The velocity is uniform from bed to water surface, for a given position S from the breakers.

(6) The position of the 'edge' of backwash, s' from the breakpoint, can be described by a function of l_b, θ and $T - t$.
(7) The backwash is complete in the time $T - t$, i.e. surging conditions only are considered.
The symbols used are defined in the Notation and in Figure 3.9.

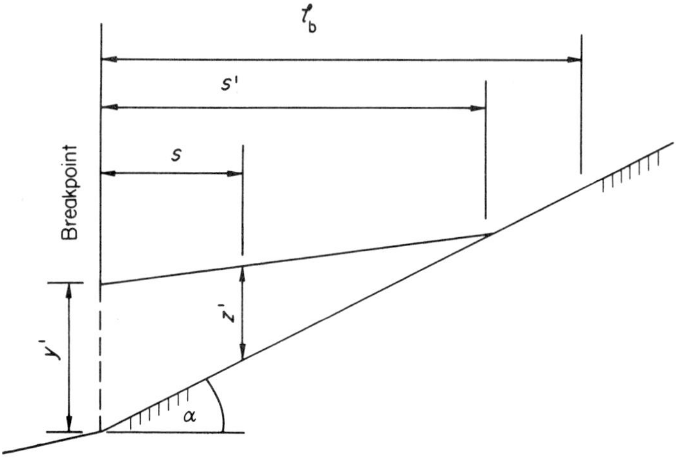

Figure 3.9. Definition diagram of dimensions relating to the wedge of water on the beach.

The expressions for the internal velocities associated with both uprush and backwash are based on the normal equation of continuity, on the geometry of the system and on the observed variations in water level at the breakers during the wave cycle. For a full development of the equations, reference should be made to Kemp and Plinston (1974).

The continuity equation is given by

$$\frac{\partial(z'v')}{\partial s} = \frac{\partial z'}{\partial \theta}$$

and if it is assumed that the beach and water surface are plane, then $z' = y'(1 - S/s')$, and since the distance from the breakers to the edge position is given by

$$s' = \frac{l_b}{2}\left(1 + \cos\frac{\pi\theta}{T - t}\right) \tag{3.3}$$

these equations can be used to derive an expression for the instantaneous velocity v' parallel to the beach, in the form

$$v' = \frac{1}{\cos\alpha}\cdot\frac{\pi l_b}{4(T - t)}\left[\frac{3}{2}\cdot\frac{\sin\pi\theta}{T - t} + \frac{S}{l_b}.\tan\frac{\pi\theta}{2(T - t)}\right] \tag{3.4}$$

This is independent of the initial water depth at the breakpoint, since the rate of change of depth is proportional to the depth.

When the backwash water edge passes any chosen point on the beach at which the velocities are being studied, the beach becomes dry and the velocity is zero. In these circumstances, it can be shown that the internal velocity just before the beach becomes dry is given by

$$v' = \frac{1}{\cos\alpha}\cdot\frac{\pi l_b}{2(T-t)}.\sin\frac{\pi\theta}{T-t} \tag{3.5}$$

and the time of occurrence is obtained from

$$\frac{S}{l_b} = \frac{s'}{l_b} = \cos^2\frac{\pi\theta}{2(T-t)} \tag{3.6}$$

Figure 3.10 shows velocity curves plotted non-dimensionally against time for various positions S/l_b on the beach.

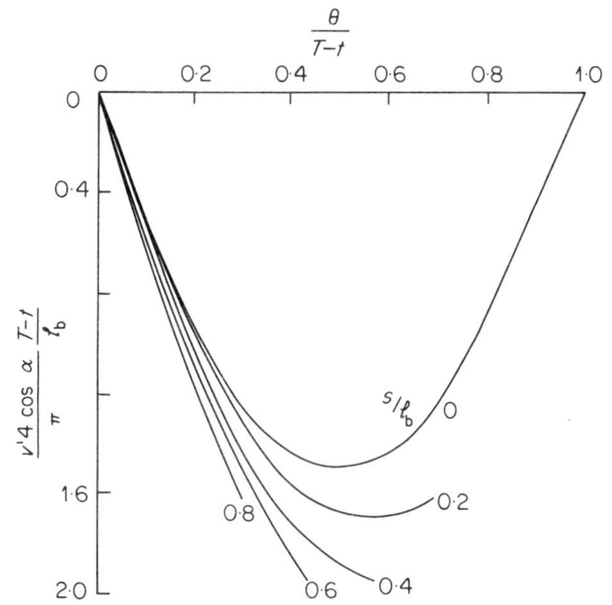

Figure 3.10. Backwash velocity curves.

Equation 3.4 can also be used to examine how the maximum backwash velocity at a point on the beach varies with its distance from the breakpoint. Abbreviating Equation 3.4 for convenience to

$$V = \frac{3}{2}\sin 2R\theta + \frac{S}{l_b}.\tan R\theta \tag{3.7}$$

where

$$R = \frac{\pi}{2}\cdot\frac{1}{T-t}$$

and

$$V = v' \cos \alpha . \frac{4(T - t)}{\pi l_b}$$

For a particular value of S/l_b in Equation 3.7, $\partial V/\partial \theta = 0$ will give the value of $R\theta$ for maximum velocity

$$\frac{\partial V}{\partial \theta} = 3R \cos 2R\theta + \frac{RS}{l_b} \sec^2 R\theta = 0$$

This produces line ABC on Figure 3.11.

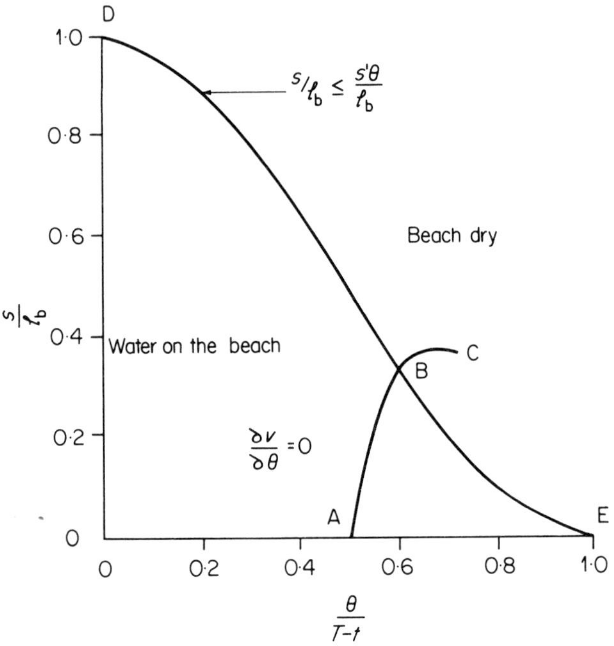

Figure 3.11. Condition for which the backwash velocity is a maximum.

The condition that there is water on the beach is given by Equation 3.6 and this is plotted as line DBE in Figure 3.11. The intersection of these two curves indicates that it is only for values of S/l_b less than that given by point B that the backwash velocity reaches a maximum before the beach becomes dry. The line ABD gives the time at which the maximum backwash velocity occurs at any point on the beach.

Uprush velocities. Uprush velocities can be in the form of

(a) a simple surge similar to the backwash in reverse;
(b) a bore;

(c) a combination of (a) and (b), the bore form giving way to a simple wedge-type surge.

If the uprush is in the form of a wedge, the result can be shown to be

$$v' = -\frac{1}{\cos\alpha}\cdot\frac{\pi l_b}{4T}\left[\frac{3}{2}\sin\frac{\pi\theta}{t} + \frac{S}{l_b}\tan\frac{\pi\theta}{2t}\right] \tag{3.8}$$

This is the expression for backwash velocity with $T - t$ replaced by t.

For the case of a bore-type uprush, it is assumed that the height of the bore front decreases linearly with distance up the beach (Kemp, 1960). If the velocity of the bore is given by $v' = k(gz')^{1/2}$, then it is found that

$$k(gy_0)^{1/2} = 2\frac{l_b}{t} \quad \text{and} \quad \frac{s'}{l_b} = \frac{2\theta}{t} - \left(\frac{\theta}{t}\right)^2$$

When this curve was plotted together with the sine curve adapted from Miche's theory, it was found that the experimental points from model observations fell between the two curves, indicating that the actual uprush was a combination of the two conditions.

Kemp and Plinston (1974) show that the internal velocity associated with the bore-type uprush is given by

$$v' = \frac{2l_b}{(\cos\alpha)t}\cdot\frac{(1 - \theta/t)^3}{(1 - S/l_b)} \tag{3.9}$$

Figure 3.12 shows the two possible uprush patterns for the position $S/l_b = 0$, i.e. at the breakpoint. Figure 3.13 shows typical results from model tests compared with the theoretical curves for a bore-type and wedge-type uprush. Figure 3.14 illustrates predicted and observed backwash velocity patterns. Field

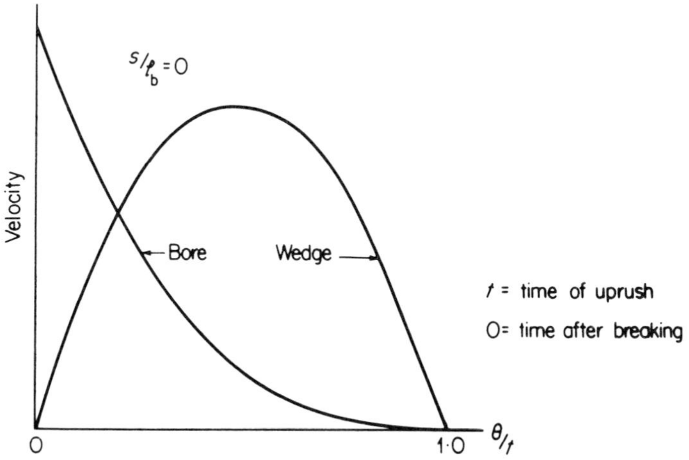

Figure 3.12. Comparison of bore-type and wedge-type uprush velocity patterns.

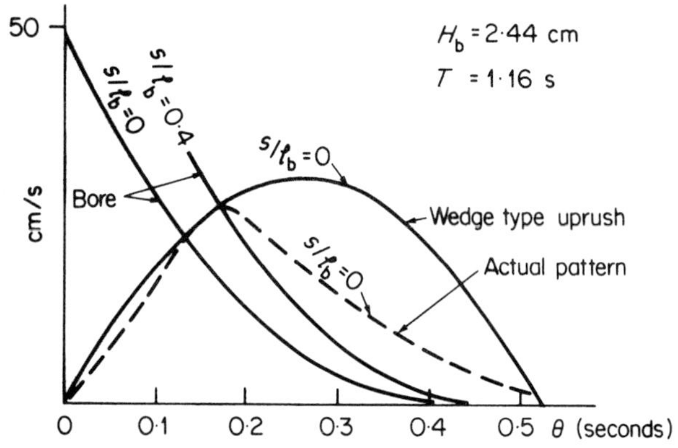

Figure 3.13. Comparison of actual uprush velocities in the model with the bore-type and wedge-type velocity patterns.

observations made with small propeller current meters fixed to stakes set into the beach showed that the flow broadly followed the same pattern.

The limitations of the predicted velocities both for uprush and backwash are those imposed by the assumptions. The backwash seems to be adequately described by the sine function expression. So far as the uprush is concerned, long, low waves seem to be described by the wedge form. The complete bore form is achieved towards the limit of surge conditions. Between the two there is a wide range when both forms are combined. Observations on the coast showed that the uprush was of the bore form during the observational periods.

Backwash measurements on the coast gave good qualitative agreement with theory, but generally the maximum predicted velocities were higher than those observed. For practical purposes, the significant wave period and mean uprush time t were used in the theory and individual wave velocity measurements could

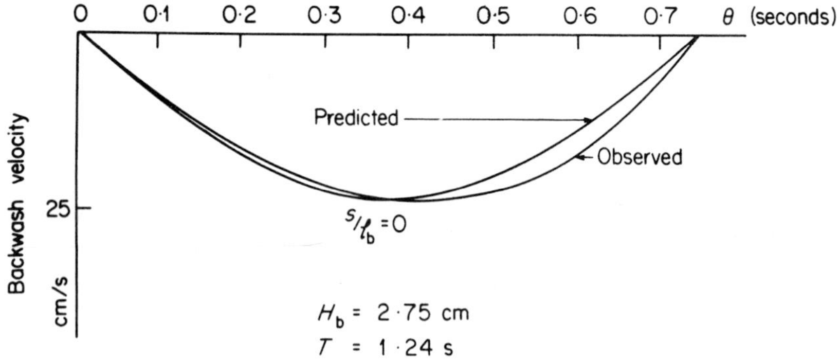

Figure 3.14. Comparison of observed and predicted backwash velocity patterns in the model.

be expected to show considerable variations from the mean predicted values. The variations from wave to wave were of the same order as the differences between measured and predicted values.

Movement of beach material

Observations in a model demonstrated the marked effect of the asymmetrical nature of the flow upon the movement of beach material.

Records obtained by high-speed cine film showed that the movement of material takes a different form on the uprush than on the backwash, though in both cases, for surging conditions, the movement is predominantly as bed load. On the uprush the rapid acceleration, coupled with the high local velocity when the wave front passes the breakpoint, sets a mass of material in motion. This material is carried forward on the front of the uprush almost to the crest of the beach. Once this 'wave' of material has passed, the movement for the remainder of the uprush is relatively slight. However, on the backwash which starts from rest, both the high initial acceleration and the high initial velocity are absent and the movement of material is essentially a gradual shift down the beach, aided by any water running out of the beach partially fluidizing the bed. Therefore material at the breakpoint at the beginning of an uprush–backwash cycle may be carried in one motion to a point near the crest and will then take several cycles to return to the breakpoint.

The above description applies to the surging condition. In the plunging case turbulence is extensive with the result that much of the wave energy is dissipated and a lot of the bed material is put temporarily into suspension in the vicinity of the plunge point. Some of this material is carried up to the crest together with a certain amount of bed material. The movement of material from the breakpoint or step area, as in the surging case, is interrupted by the apparent discontinuity of motion as the wave crest plunges almost vertically onto the beach. It is not clear what proportion of the material reaching the crest originated from the step or from the plunge hole. On the backwash the movement is similar to the surging case, though now the material from the upper part of the beach is deposited in the plunge hole and the material from the lower part of the beach replenishes the step. It would seem that while more material is in motion in the plunging case, more material is moving directly up and down the beach in the simpler, surging, case.

On the formation of the equilibrium beach, exploratory tests at the beginning of the experimental programme showed that, unless the preformed beach slope was excessively steep, the same waves always formed the same profile irrespective of the slope of the initial preformed beach. Material is always moved shorewards as a result of the asymmetry of orbital motion offshore of the breakers, the higher shoreward velocity under the wave crest exceeding the seaward velocity under the trough. When the beach crest has become fully developed the zone offshore of the breakers tends to a horizontal shelf at a depth corresponding to the maximum depth of material movement under the prevailing waves. The formation of a step generally occurs immediately wave action commences. It is connected with the discontinuity of motion at the breakpoint between the oscillatory motion offshore and the predominantly horizontal motion shoreward of the breakpoint (Kemp, 1960). For very long, low waves sometimes used in the model, the step feature was absent though there was a clear discontinuity in the profile at the breakpoint. Here the asymmetry of motion offshore was so well developed that, in fact, there was no discontinuity of flow on either side of the breakpoint. This, perhaps, explains the absence of a step.

Notation

A = constant
B = constant
d = water depth from M.W.L.
g = force per unit mass due to gravity
H = wave height (trough to crest)
k = constant defined in text
l_b = breaker length
L = wavelength
S = distance from the breakpoint
s' = variable distance from breakpoint
T = wave period
u, v = velocity, defined in text
v_m = velocity magnitude asymmetry
v_t = velocity time asymmetry
t = time of uprush/time
x = horizontal coordinate
y = water depth at breakpoint
y_0 = water depth at breakpoint at temporal limit of uprush
z = vertical coordinate
α = beach slope at M.W.L.
θ = time
$'$ = denotes variable quantity

References

Adeyemo, M. D., 1970, Velocity fields in the wave breaker zone, *Proc. 12th Conf. cst. Engng., Washington, D.C.*, Vol. 1, pp. 435–460.

Biesel, F., 1951, Study of wave propagation in water of gradually varying depth, *Gravity Waves Circular No. 521*, Nat. Bur. Standards, Washington D.C.

Grass, A. J., 1970, Initial instability of fine bed sand, *J. Hydraul. Div. Am. Soc. civ. Engrs.*, **96**, 619–632.

Grant, U. S., 1943, Waves as a sand transporting agent, *Am. J. Sci.*, **241**, 117–123.

Inman, D. L., and N. Nasu, 1956, Orbital velocity associated with wave action near the breaker zone, *Tech. Memo. Beach Eros. Bd. U.S. 79*.

Iwagaki, Y., and T. Sakai, 1972, Shoaling of finite amplitude long waves on a beach of constant slope, *Proc. 13th Conf. cst. Engng., Vancouver, Canada*, Vol. 1, pp. 365–383.

Kemp, P. H., 1960, The relation between wave action and beach profile characteristics, *Proc. 7th Conf. cst. Engng., The Hague*, pp. 262–276.

Kemp, P. H., and D. T. Plinston, 1968, Beaches produced by waves of low phase difference, *J. Hydraul. Div. Am. Soc. civ. Engrs.*, **94**, 1183–1195.

Kemp, P. H., and D. T. Plinston, 1974, Internal velocities in the uprush and backwash zone, *Proc. 14th Conf. cst. Engng., Copenhagen* (in press).

Mehauté, B. Le, D. Divoky, and A. Lin, 1968, Shallow water waves: a comparison of theories and experiments, *Proc. 11th Conf. cst. Engng., London*, Vol. 1, pp. 86–107.

Longuet-Higgins, M. S., 1953, Mass transport in water waves, *Phil. Trans. R. Soc. A*, **245**, 535–581.

Miche, M., 1944, Mouvements ondulatoires de la mer, *Annls. Ponts Chauss.*, **114**, 131–164.

Scott, T., 1954, Sand movement by waves, *Tech. Memo. Beach Eros. Bd. U.S. 48.*

Svendsen, I. A., and O. Brink-Kjaer, 1972, Shoaling of cnoidal waves, *Proc. 13th Conf. cst. Engng., Vancouver, Canada*, Vol. 1, pp. 365–383.

Wiegel, R. L., 1964, *Oceanographical Engineering*, Prentice-Hall, Englewood Cliffs, N.J.

Williams, P. B., and P. H. Kemp, 1971, Initiation of ripples on flat sediment beds, *J. Hydraul. Div. Am. Soc. civ. Engrs.*, **97**, 1078–1079.

Williams, P. B., and P. H. Kemp, 1972, Initiation of ripples by artificial disturbances, *J. Hydraul. Div. Am. Soc. civ. Engrs.*, **98**, 1057–1070.

Discussion

J. Orford, Department of Geography, University of Reading. The wave velocity profile showing asymmetry tends to resemble single waves deformed by shoaling. What is the effect on the wave velocity asymmetry profile when the wave phase value t/T is higher than 1·0 (i.e. surf conditions prevail) and high impedance by backwash of incoming swash occurs? Also one cannot take a single wave as being representative. Can you comment on the nature of periodicity/frequency that may appear in the changing velocity profiles during impeding flow?

The studies so far carried out have been restricted to surge conditions and, to some extent, transition conditions with phase differences of up to 0·7. Higher values usually imply flatter beaches and the subsequent history of a wave after breaking may take the form illustrated by Scott (1954), where the intermittent shoreward motion due to the passage of isolated wave fronts has superimposed on it a fairly uniform return flow. The resultant velocity profile is thus generally shoreward at the surface and seaward at the bed. This qualitative flow pattern is readily seen in an hydraulic model if dye is introduced in the run-up zone. Solid flow tracer particles also demonstrate that the seaward flow is strongest at the bed even in the plunge hole at the breaker position. From personal observations in models the magnitude of the seaward flowing current under surf conditions is greater than would be predicted from an approach based on the mass transport associated with, say, solitary waves. The seaward flow increases towards the breakers and appears to increase at the successive points of breaking in the surf zone.

A paper by Cook and Gorsline (1972) again highlights the variability of current velocities in nature due to the more complicated wave situation and the influence of wind and topographic factors. In shallow water the wind-induced currents are proportionally more significant than in the offshore zone and may dominate the flow pattern. Spilling breakers can also reverse the bottom drift.

Waves not infrequently break down into shorter periods in the nearshore area and the picture is further complicated by the differential shoaling effect on natural beaches where a whole range of wave periods are present. The paper by Huntley and Bowen, Chapter 4 in this volume, is relevant in this respect.

Reference. Cook, D. O., and D. S. Gorsline, 1972, Field observations of sand transport by shoaling waves, *Mar. Geol.*, **13**, 31–55.

J. Orford. On a mixed sand/shingle beach where the seaward slope is flatter than the shingle bank, the tidal translation in a limited distance allows for changing asymmetry of waveform as a function of depth. This changing asymmetry will probably be reflected in a changing breaker type. If one extends the velocity profile into the swash zone, such that asymmetry of backwash becomes the extension of the velocity profile past breaking, then how does the changing breaker type reflect energy expenditure as seen in the velocity profiles at different points within the swash zone?

The changing depth of water due to the tide can alter the type of breaker. This is a factor which comes into the design of engineering structures and, in some cases, the wave impact due to the near vertical face of the breaking wave has to be reduced by modifying the water depth close to the structure. This is normally achieved by studying the problem in an hydraulic model. However, on beaches this specific effect is of less importance. On the other hand the influence on run-up and backwash will be generally as described in the answer to the previous question. As mentioned in the text, continuity of asymmetrical motion between the offshore zone and the surf zone, is most marked in the case of very low surging waves, when the conditions closely resemble reflection from a sloping surface. Some tentative conclusions on energy losses were made by Kemp and Plinston (1968). An attempt was also made to predict the phase difference by a determination of the asymmetry of orbital motion just offshore of the breakers using the Solitary Wave Theory and Stokes Second Order Theory. Although there was some qualitative agreement it was concluded that phase difference could not be determined satisfactorily in this way.

Dr. G. Evans, Department of Geology, Imperial College, London. Would you please explain why the grain size of the beach zone inside the breakers, but excluding the breaker zone itself, usually shows a grain size pattern which increases up to the beach crest, which is the limit of the swash. Does not the velocity gradient decrease in this direction?

The behaviour of shingle beaches depends on the relative size of the waves and of the beach material. Conditions can therefore vary between the situation in which the waves are too small to move any of the material, in which case the beach is acting as a permeable breakwater, through to the condition in which all the material is put in motion by the uprush of the wave. On natural shingle beaches the variability in height of successive waves produces a variability in run-up. Towards the limit of each uprush the material carried towards the crest is deposited as the water velocity decreases.

Part of the uprush is then absorbed by the permeable beach crest, so that the backwash depth in this area is reduced. In addition the backwash starts with zero velocity. The larger fraction of the material tends to be stranded near the limit of uprush. Larger waves which slightly overrun the crest accentuate this effect. Only when the beach is subjected to very large waves, combined perhaps with high tides, does the shingle crest become saturated, as the rate of percolation becomes small in relation to the quantity of water in the uprush. At this point the differential effect of beach material size becomes unimportant, and shingle beaches of differing sized material behave in a similar way (Kemp, 1968).

Reference. Kemp, P. H., 1968, A field study of wave action on natural beaches, *Proc. Int. Ass. Hydraul. Res., London*, pp. 131–137.

CHAPTER FOUR

Comparison of the Hydrodynamics of Steep and Shallow Beaches

D. A. HUNTLEY and A. J. BOWEN

Abstract

A first step towards understanding the complexities of nearshore sediment dynamics must be to study the nearshore motion of the water itself, and this is the aim of the present research programme. Beaches of widely different slopes have been studied and, in particular, a steep (slope 0·13) shingle beach and a shallow (slope 0·014) sand beach have produced results of considerable interest.

This paper describes times series for the two horizontal components of water velocity, measured at different distances from the shoreline on the two beaches, and discusses the probability distributions of the velocities, the mean current patterns close to the shore and the frequency dependence of the wave energy.

The hydrodynamics of the steep and shallow beaches differ widely. In particular, periodic longshore variations of the nearshore velocity field were observed on the steep beach but not on the shallow beach. On the steep beach, edge waves with a period twice that of the incident waves and with a longshore wavelength of 32 m were observed and it is suggested that swash interaction in the narrow surf zone was responsible for generating these waves. The mean flow field on other occasions also revealed the existence of nearshore circulation cells on the steep beach. On the shallow beach, on the other hand, no short period edge waves were observed and the measured mean flow was a steady longshore current generated by obliquely incident waves.

On the shallow beach, low frequency wave motion became increasingly important in the velocity field as the shoreline was approached, causing a long period variation in run-up at the shoreline itself. Observations suggest that this may be due to interaction between breakers in the wide surf zone.

Within the surf zone on both beaches, the contribution to the velocity field made by motion at frequencies greater than the incident wave frequency is found to have the functional form $E(\sigma) = E_0 \exp(-p\sigma)$, where $E(\sigma)$ is the velocity field energy at frequency σ, and p is an exponential decay constant, which is found to vary as (the local depth)$^{5/2}$. The physical significance of this observation is not clear. The dependence of this effect on beach slope also needs further investigation.

4.1 Introduction

One of the fundamental questions facing coastal geomorphologists is: Why are some beaches steep and others shallow? Obviously the answer lies to a considerable extent in the size of the material making up the beach, with large diameter sands producing steep beaches and fine sands shallow beaches; but there is also a marked dependence of beach slope on exposure to wave action. Weigel (1964) plots results from beaches in North America which show, for example, that, with a median sand diameter of 0·5 mm, a well protected beach might have a slope of 0·2 while an exposed beach could have a slope of less than 0·07, one third of the sheltered beach slope. Furthermore, changes of wave climate incident on a beach can change the beach slope on a time scale of the order of hours or days and often produce cyclic changes of beach profile over a long period (see, for example, Sonu, 1968). The mechanics controlling this dependence of beach slope on wave action are at present very poorly understood, but clearly there is a complex interaction whereby the velocity field close to the shore influences beach slope, through sediment transport, and beach slope, in turn, influences the nearshore velocity field.

Sediment transport under waves in shallow water has been fairly extensively studied in recent years. Empirically, long, low waves approaching a beach are found to be generally depositional in nature, building up a beach and smoothing over any undulations, while steep storm waves tend to erode the beach, taking material from the foreshore and often depositing it in the form of a bar close to the breakpoint of the waves. Laboratory experiments suggest that, for a given beach slope, there exists a critical steepness above which the incident waves are able to remove material from the beach (Iwagaki and Noda, 1963) and below which they deposit material onto the beach. Other laboratory experiments (King, 1972) show that onshore or offshore drift varies across the surf zone and may be onshore just inside the breakpoint even when the average motion inside the surf zone is offshore. Theoretical understanding of these empirical results is, however, relatively meagre and the relevance of 2-D laboratory experiments to actual field conditions is, in any case, doubtful.

The shoreward drift of material outside the surf zone can be understood in terms of 'mass transport' of the waves progressing towards the shore; as the shoaling waves become progressively steeper, the orbital motions of the water particles under the waves, which in deep water are closed circles, become distorted and fail to close, resulting in a net movement of water in the direction of the wave motion (Lamb, 1932; Eagleson and Dean, 1961).

The possibility of seaward motion of sediment is more of a problem. Longuet-Higgins (1953) showed theoretically that, in order to give no net onshore flow of water, the shoreward drift under incident waves at the bottom of the water column is balanced by a seaward drift at higher levels. Although this is known to occur in narrow laboratory flumes (Russell and Osorio, 1958), the return flow at intermediate depths has not been observed in the ocean. In fact on many natural beaches horizontal circulation patterns are commonly set up, with the seaward flow concentrated in a few narrow 'rip' currents (see Section 4.4 of this paper) and it is often these which control the flow of sediment away from the beach (Bowen and Inman, 1969; Komar, 1971). Frequently, however, erosion of beach material occurs relatively uniformly along the coast and is unlikely to be the result of such regular horizontal circulation patterns. Various explanations of the destructive nature of steep waves have been suggested, including the effect of increased set-up (Weigel, 1964) and swash asymmetry (Bagnold, 1940; King, 1972) but none has yet been universally accepted.

Other theoretical studies have shown that some periodic nearshore sedimentary features, in particular submerged longshore sand bars (Carter and coworkers, 1973; Lau and Travis, 1973) and crescentic bars (Bowen and Inman, 1971) can be attributed to the action of the drift velocity in the boundary layer of the wave motion close to the seabed.

Most of these studies have tended to assume that sediment movement is directly related to the nearshore drift velocities. However, Wells (1967) has compared empirical formulae for bed load transport and suggests that the fourth power of the velocity is the important parameter, a suggestion supported by the theoretical work of Bagnold (1963). To retain directionality Wells writes the fourth power as a product of the cube and root mean square of the velocity. By taking an assumed form for the nearshore velocity field, he can explain the existence of the frequently observed 'neutral line' of zero sediment transport some distance offshore of the breakpoint, with shoreward transport inshore of the line and seaward transport seawards of the line, in terms of a change of sign of the cubic term.

The influence of beach slope on nearshore velocity fields is even less well understood than the effect of waves on the beach sediment. Nevertheless the significance of the interaction is clearly seen in the different breaker types which occur on beaches of different slope (Galvin, 1972). Figure 4.1(a) shows the variety of breaker types that can occur. Spilling breakers form at the breakpoint with foaming crests and their waveform changes only slowly as they propagate

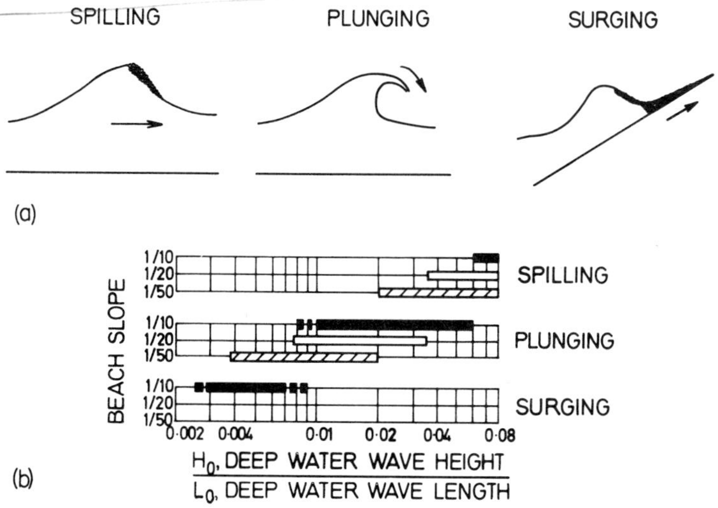

Figure 4.1. (a) Schematic diagrams of breaker types. (b) A suggested scheme relating breaker type to beach slope and deep water wave steepness.

through the surf zone. Plunging breakers on the other hand change their form radically near the breaker line as the top of the wave curls over and plunges into the water ahead. Surging breakers form on steep beaches when the base of the deforming wave rushes up the slope as a thin layer of water and the breaker collapses. Several estimates of the interdependence of the three nearshore parameters, beach slope, wave steepness and breaker type, have been proposed. For example, Figure 4.1(b) shows the scheme suggested by Patrick and Weigel (1955). It seems likely that there are significant differences in the velocity fields of these breaker types on beaches of different slopes, but very few direct measurements of the velocity fields under different breakers have previously been made.

The authors have conducted experiments to measure the velocity fields on beaches of widely different slopes, under both plunging and spilling breakers. In particular, results from two beaches in Devon, England, are discussed. Slapton Sands in the south is a steep (0·13) shingle beach facing eastwards along the English Channel and Saunton Sands in the north is a shallow (0·014) sandy beach facing westwards towards the Celtic Sea and Atlantic. The results from these beaches show the influence of beach slope on nearshore hydrodynamics and thus throw useful light on the whole problem of interaction between water motion and beach profile.

4.2 Shoaling and breaking waves

Steep beach

During the experiments on the steep beach at Slapton, the surf zone was formed by plunging breakers. Figure 4.2 shows typical time series for the onshore velocities at different distances from the shoreline. The instrument used to make these measurements was a two-component electromagnetic flowmeter (Tucker and coworkers, 1970) mounted 15 cm above the seabed and oriented to measure the horizontal components of flow. The time series shown in Figure 4.2 were obtained from a single fixed instrument, the change of shoreline position during a tidal cycle being used to obtain records at different distances from the shoreline. Although the records are therefore not synoptic, they were all taken within a 4-hour period during which wave and wind conditions remained essentially constant and it can therefore be assumed that the records are representative of the true variation of velocity field normal to shoreline.

The breakpoint during the measurements was estimated to be 5–6 m from the shoreline, with breaker height approximately 0·3 m. Series A and B are therefore seaward of the breakpoint, in depths of 1·4 m and 1·2 m respectively. Series A show low amplitude waves of a rather broad spectrum of frequencies. There is some evidence of peaking of the onshore velocity (positive) and broadening of the offshore velocity (negative) as expected for steepening waves. Nevertheless the waveform is generally symmetrical about the peak, with the increase towards maximum onshore current being of approximately the same duration as the

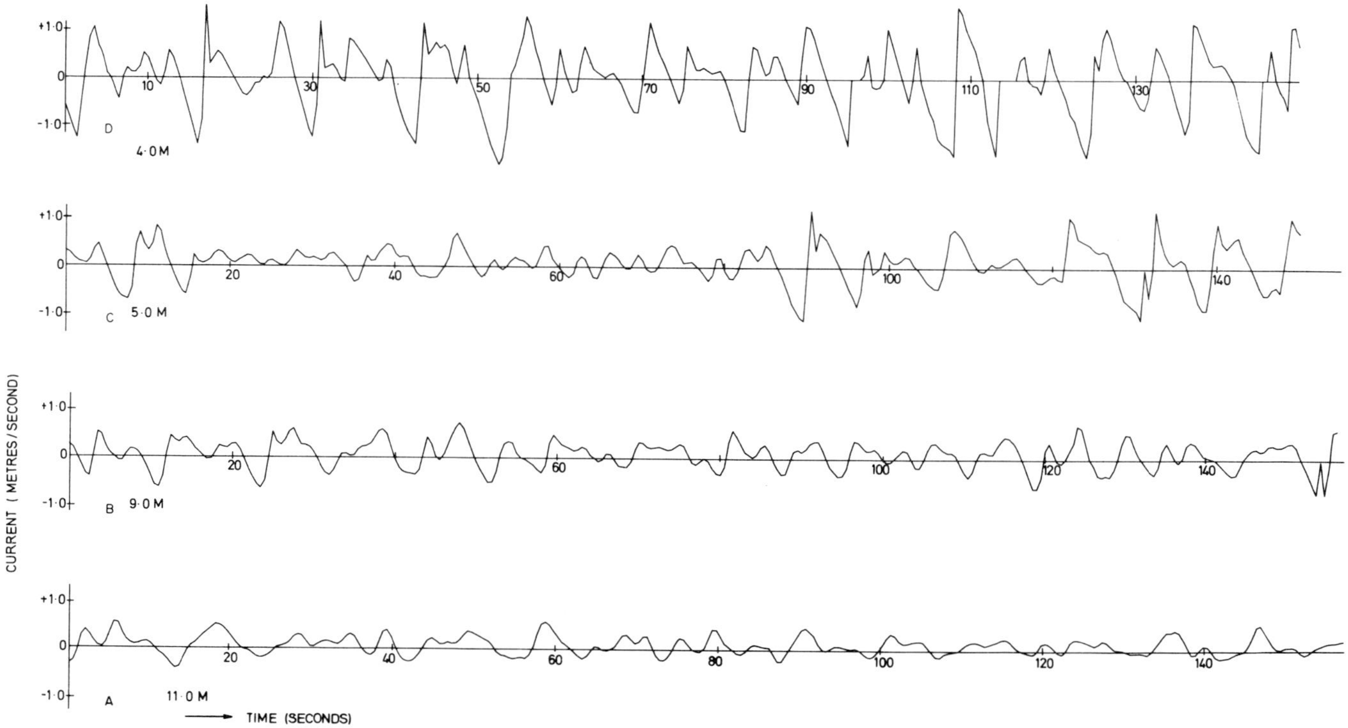

Figure 4.2. Time series of onshore currents on the steep beach at different distances from the shoreline; series **A**, **B**, **C**, **D** are respectively 11·0 m, 9·0 m, 5·0 m, and 4·0 m offshore. The breakpoint was estimated to be 5–6 m offshore.

decrease on the other side of the peak. By the time the waves reach position B their amplitudes have increased considerably. For the larger waves the acceleration towards maximum onshore velocity has become rapid compared to a relatively gentle fall towards maximum offshore velocity. This behaviour, characteristic of steep fronted waves, might be expected as a stage in the formation of plunging breakers. Record C is taken very close to the breakpoint itself. Here we must distinguish between the breakpoint, where a breaker finally overhangs and becomes irreversibly unstable, and the plunge point, where the falling jet from the top of the plunging breaker hits the water or seabed and the whole wave becomes highly turbulent. Clearly when the incident waves form a broad spectrum of wave heights and frequencies the breakpoint and plunge point change position continually. Different parts of record C therefore show characteristics of the velocity field both near the breakpoint and near the plunge point. Close to the breakpoint smaller onshore velocities just before breaking are shown in some small waves in record C. Inside the plunge point the movement consists of a sustained and strong backwash changing abruptly to a shorter inrush as the collapsed breaker surges forward; this behaviour is seen in the high waves of record C. Record D, at an average depth of 0·5 m, is well inside the plunge point and shows clearly the violent alternation between uprush and backwash close to the shoreline.

It is instructive to compare these results with those obtained by Nagata (1964) for a steep beach in Japan. The broad features of the wave field and the plunging breakers are similar on both beaches, but Nagata reports backwash velocities consistently 5–6 times higher than uprush velocities approximately at the plunge point, while our results show the velocities to be essentially similar in magnitude throughout. The reason for this difference may be the different wave heights for the two experiments; on Slapton beach the breakers were low (0·3 m), while the breakers observed by Nagata were considerably larger (2·0–2·5 m). The dominance of backwash for the larger breakers would then be consistent with the observation, previously discussed, that, beyond a certain critical steepness, steep waves erode beach material. On Slapton beach the low waves were unlikely to be eroding and therefore do not show the backwash dominance. These observations tend to support the hypothesis that swash asymmetry is a primary cause of foreshore erosion in steep waves (Bagnold, 1940; King, 1972).

During measurement of the velocities at 5 m from the shore a film was taken of plunging breakers in the vicinity of a graduated staff placed near the flowmeter. This has been used to obtain the simultaneous measurements of onshore current and elevation shown in Figure 4.3 and has also revealed many details of the surf zone motion which could not have been deduced from the flowmeter records alone.

One feature observed in this way is the occasional presence of a number of secondary peaks or wavelets following a steep breaker face. This is shown in Figure 4.3 at, for example, the positions marked I and II; at these positions oscillations in the onshore velocity component are not clearly resolved immediately behind the steep breaker face because of the 0·5 second measuring

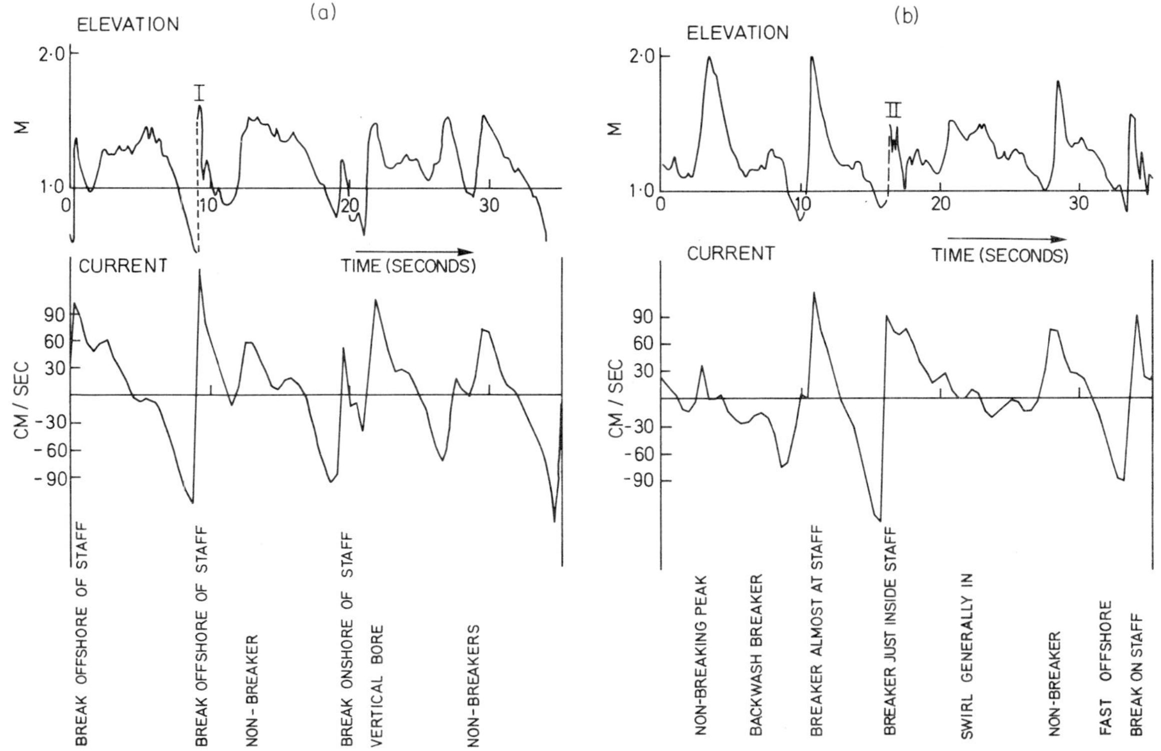

Figure 4.3. (a) and (b) Simultaneous time series of onshore current and surface elevation (arbitrary zero for elevation scale). The breaker positions and other comments are from the film taken concurrently with these measurements.

period of the digital record. Figure 4.4 is a frame from the film showing a breaker with wavelets approaching the staff. Wavelets of this type have been observed in the surf zone by Nagata and notably by Gallagher (1972). They are well known in rivers and hydraulic channels where the steep face and following wavelets form what is known as an undular bore. These undular bores are formed when the energy flux through the bore face is too large to be dissipated by the turbulence at the bore face itself and must therefore be radiated behind the bore in the form of wave motion. Laboratory experiments (Favre, 1935; Binnie and Orkney, 1955; Sandover and Zienkiewicz, 1957) show that the efficiency of the front in dissipating energy depends upon h_2/h_1, the ratio of water depths behind (h_2) and in front of (h_1) the bore. For $1{\cdot}0 < h_2/h_1 < 1{\cdot}28$ the bore moves forward with wave motion following and without breaking; for $1{\cdot}28 < h_2/h_1 < 1{\cdot}75$ wavelets are still present but the bore face is breaking as it moves; as h_2/h_1 tends towards the critical value 1·75 the bore becomes fully breaking, no further energy is left for radiating and the wavelets are no longer present.

Gallagher found that non-breaking undular bores were formed from breakers in the Hawaiian Islands because of either a rapid decrease in bottom slope or an increase in bottom-roughness as the waves passed over a reef. Our results suggest that undular bores can also occur on steep beaches with relatively simple topography and that this may be due to backwash interaction. Backwash in a

Figure 4.4. A breaker approaching the shore with two wavelets following. Notice the strong backwash at the range pole, causing a turbulent wave seaward of the pole.

surging breaker zone can effectively increase the water depth in front of an incoming wave after it has begun to steepen or break thus causing the wave front to weaken, as h_2/h_1 decreases, and to radiate energy in the form of wavelets; both breaking and non-breaking undular bores may be formed by this process. Another possible effect of backwash is to increase the effective speed of the incoming wave relative to the rapidly moving backwash in front of it and hence increase the rate of energy transfer across the wave front. Initially the wave front may be too weak to dissipate this energy and the excess will be radiated back as wavelets. Since backwash is likely to be generally important on steep beaches, these observations suggest that the formation of wavelets in the waves incident on a steep beach may be relatively common.

Another feature observed in the records of Figure 4.3 is the predominance of high breakers at about a 10-second period; this is approximately twice the average period of the incident wind waves observed offshore (Figure 4.2(A)). These records and the concurrent film show that a common sequence of events in the surf zone is as follows. A steep breaker plunges into very shallow water in the surf zone and surges strongly up the beach. Before the backwash has begun a second wave arrives and travels forward into the deep shoreward moving water as an unbroken surge. It finally collapses at the shoreline rather like a surging breaker. A strong backwash then occurs which interacts with the third wave to form a steep breaker and leaves only a shallow covering of water in the surf zone. The cycle is then repeated, resulting in a sequence of strong breakers at the first subharmonic of the incident wave period. An approximate calculation, for a beach of this slope and observed surf zone width, shows that the natural period of run-up is close to the observed subharmonic of the incident waves and our observations therefore suggest that this natural period dominates the surf zone motion on this steep beach. Since the natural run-up period depends upon surf zone width, and hence breaker height, subharmonic resonance should exist only when both the amplitude and period of the incident waves are right (Huntley and Bowen, 1975). These observations may have significance for edge wave generation (Section 4.5 of this paper). Emery and Gale (1951) describe similar interference effects between successive swashes on a number of natural beaches on the Pacific coast of North America and they find that the observed swash period increases reasonably consistently with decreasing beach slope. A swash period of 10 seconds for Slapton beach fits well with their observations. These swash interaction observations on natural beaches complement laboratory experiments, for example those of Adeyemo (1970) on the effect of backwash on the asymmetry of individual waves, and lead to the conclusion that the behaviour of the swash is of considerable importance in the surf zone, at least on steep beaches.

Another interesting effect observed in plunging breakers is the rebound of the plunging breaker face. As the plunging jet of water hits the shallow water and beach it rebounds upwards and forwards and plunges a second time further in front of the breaker. Figure 4.5 is a frame from the film which shows the plunging water splashing upwards in front of the plunging face. Two or more plunging

Figure 4.5. A breaker plunging on the range pole and sending a second cascade of water shorewards (to the left-hand side of the photograph). This rebound effect occurred frequently for the plunging breakers on the steep beach.

faces formed in this way from a single plunging breaker have also been observed on laboratory beaches by other workers. It seems likely that such a violent process will have important implications for sediment motion in plunging breakers, lifting material which can be moved forward in the uprush.

Shallow beach

As expected from Figure 4.1, on the shallow beach at Saunton, with an average slope of 0·014, spilling breakers were formed in the surf zone. In sharp contrast with the steep beach the waves at Saunton begin to break several hundred metres from the shore and move forward through a wide surf zone. At first their form resembles breaking solitary waves and slow transformation of the waveform takes place as it progresses towards the shore so that the final form is closer to a low, steep fronted, breaking bore. Also in contrast to the steep beach, the shallow beach has many breakers in the surf zone at any given time.

Records of onshore velocity at different distances from the shoreline are shown in Figure 4.6. These records were obtained in a similar manner to those in Figure 4.2, but are in this case all taken within the very broad surf zone and are therefore directly comparable only with record D (Figure 4.2) of the steep beach.

For records A and B (Figure 4.6) peaking of the onshore velocity and broadening of the offshore velocity can be clearly seen. Generally the front and rear

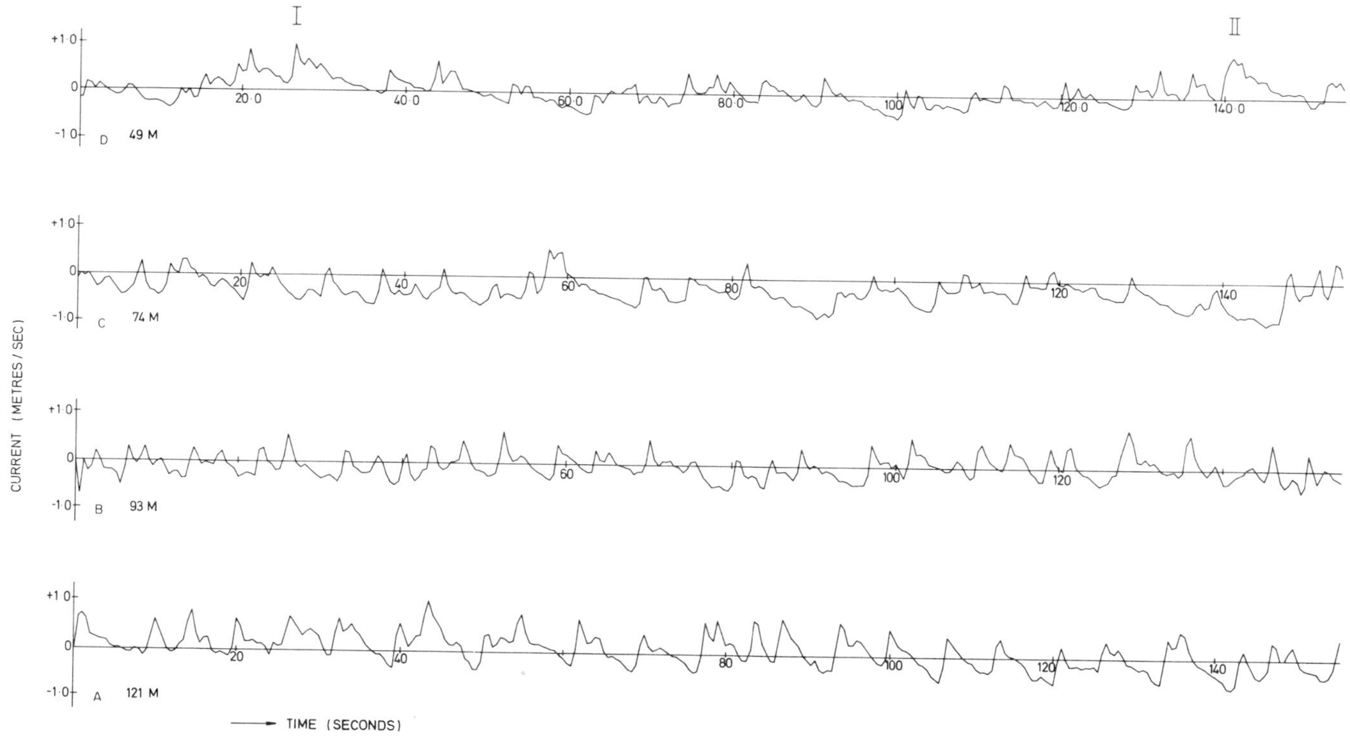

Figure 4.6. Time series of onshore currents on the shallow beach at different distances from the shoreline; series **A**, **B**, **C**, **D** are respectively 121 m, 93 m, 74 m, and 49 m offshore. All measurements are inside the surf zone.

faces of the velocity curves are symmetrical, with the increase in velocity towards an onshore peak taking essentially the same time as the decrease towards an offshore maximum, a feature which is characteristic of a spilling solitary wave (Inman and Nasu, 1956). As a wave progresses towards the shore its front steepens as the breaker tends towards a bore and this results in the saw-toothed velocity curves observed frequently in records C and D. In this respect the transformation of wave shape inside the surf zone on the shallow beach is similar to that occurring outside the surf zone on the steep beach. The slow decrease in amplitude as the waves approach the shore is also clear in Figure 4.6. There is also some evidence in record D of the existence of undular bores in the surf zone for example at I and II.

Also present in these records is the suggestion of a shift towards dominance of lower frequency wave motion as the shoreline is approached. On this shallow beach swash interaction does not seem to provide a satisfactory explanation for the effect since backwash is so small. A much more plausible possibility lies in the interaction between the many irregularly spaced breakers in the broad surf zone. While it is possible that certain waveforms can maintain their identity in the surf zone even when interacting (Zabusky and Galvin, 1970), the evidence from Saunton suggests that merging of breakers effectively reduces the average wave period as the shoreline is approached. Measurements of peak run-up times at Saunton using a stop-watch confirm this trend towards lower frequencies; there was a strong indication of long period oscillations in run-up, with high run-up occurring at intervals of around 30 seconds. An attempt was made to relate the times of maximum run-up to the concurrent measurements of the velocity field in the surf zone but in no case was satisfactory correlation found between times of high velocity in the surf zone and observed times of high run-up. This is apparently inconsistent with the existence of a coherent low frequency wave motion across the surf zone and thus supports the suggestion that breaker interaction is the dominant cause.

The observed 30-second swash period on this shallow beach is consistent with the empirical results of Emery and Gale (1951); similar low-pass filtering effects of shallow beaches have been observed more recently by Sonu and coworkers (1973). Webber and Bullock (1969) also describe laboratory experiments on run-up where they found that it was impossible to relate run-up measurements directly to individual peaks travelling from the surf zone. For a beach slope of 0·1 they found about 30% fewer run-up crests than waves, and this percentage appeared to increase with decreasing beach slope.

4.3 Statistical distribution of currents

It is useful to quantify some of the features of the observed velocity time series in terms of the probability distribution of velocities. Figure 4.7 shows examples, from both the shallow and steep beaches, of histograms of probability for longshore and onshore currents. The shape of such probability distributions is

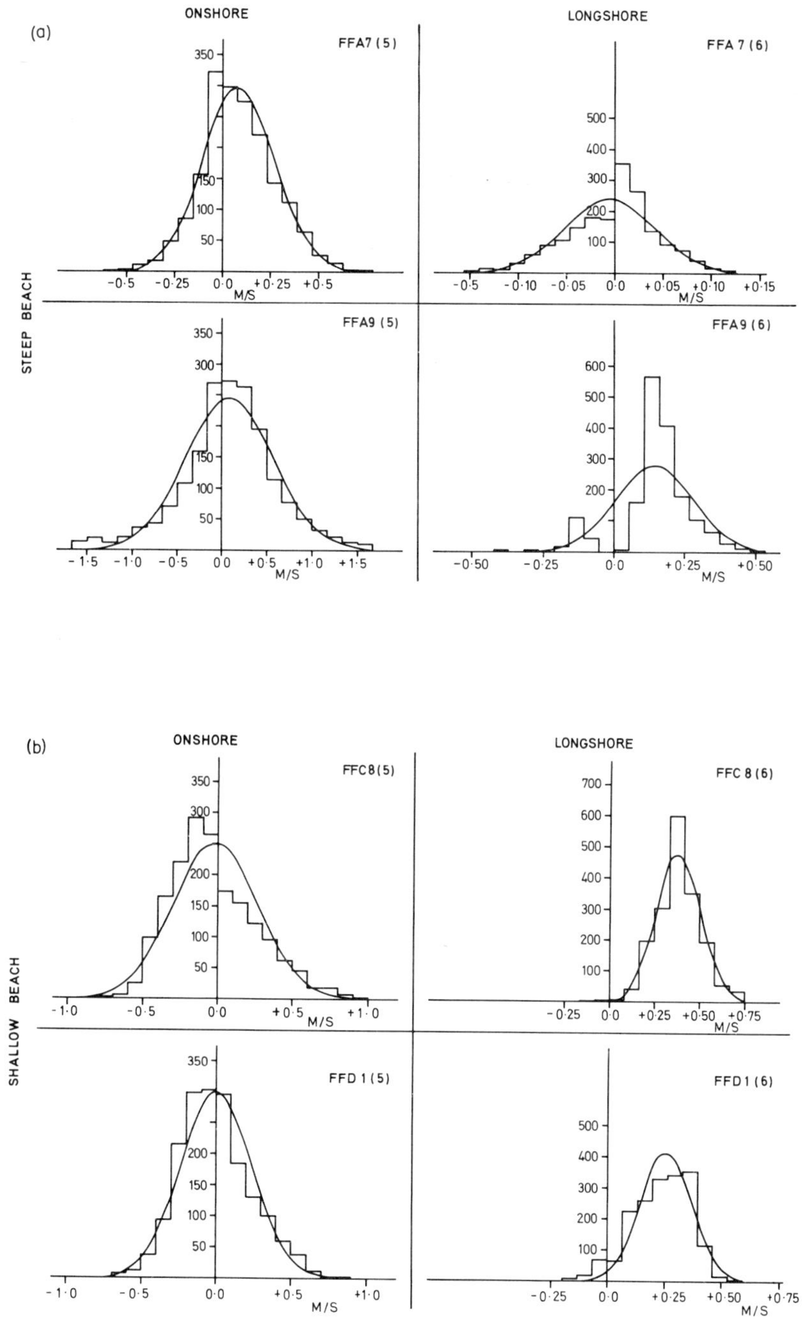

Figure 4.7. Histograms of the probability distributions of onshore and longshore currents (a) for the steep beach and (b) for the shallow beach. The smooth curves are the Gaussian distributions for the observed means and variances, shown in Table 4.1.

normally described in terms of the moments of the distribution about the mean, where, for example, the rth moment of the onshore velocity is defined as

$$\mu_r = \int_{-\infty}^{\infty} P(u)(u - \bar{u})^r \mathrm{d}u = \lim_{\tau \to \infty} \frac{1}{2\tau} \int_{-\tau}^{\tau} (u - \bar{u})^r \mathrm{d}t$$

where $P(u)$ is the probability of current value u, $\bar{u}$ is the arithmetic mean of u, 2τ is the length of the time series of u and r is an integer.

Table 4.1 summarizes some of the characteristics of the probability distributions of sample records from both beaches; the square root of the second moment $\sqrt{\mu_2}$ is shown, being essentially the root mean square of the wave orbital velocities, and the third and fourth moments are shown in non-dimensional form $\mu_3/(\mu_2)^{3/2}$ and $\mu_4/(\mu_2)^2$, known respectively as the skewness and kurtosis of the distributions.

The r.m.s. onshore current amplitudes shown in Table 4.1 have been calculated in two different ways. In the fifth column are values obtained using the peak amplitudes and average crest and crossing periods in the manner prescribed for deep water waves by Draper (1963); the sixth column shows the values obtained from direct averaging of the total time series. It can be seen that these two techniques give results which agree to within 15% even within the surf zone, showing that the crest and zero crossing analysis gives a reasonable estimate of the r.m.s. velocities even where the theory justifying the technique has become inappropriate.

Differences between the steep and shallow beaches are again clear from these values of orbital velocity. On the steep beach, waves approaching the breakpoint steepen rapidly and, at the breakpoint itself, horizontal velocities of the wave motion are large; the measured r.m.s. onshore velocity of about 0·5 m/s at the breakpoint, equivalent to an average peak amplitude of 0·7 m/s, occurred for waves of only about 0·15 m amplitude just before breaking, suggesting that much large orbital velocities can occur under storm waves. The much smaller, but still appreciable, longshore currents are due to the oblique incidence of the incoming breakers of about 4·5 seconds period. The measured onshore to longshore current ratio suggests that the waves approached at about 17° from the normal, and this is consistent with the estimated breaker angle.

On the shallow beach, the velocities of the wave motion tend to be smaller than on the steep beach, but are sustained over a much longer distance. As expected, the onshore component decays slowly as the waves approach the shoreline through the wide surf zone, but, surprisingly, the longshore currents remain approximately constant. It is unlikely in this case that the longshore component of the wave motion is due to oblique incidence of long-crested waves approaching the breakpoint. Such waves would have to approach the shoreline at a large, and increasing, angle to the normal in order to produce the observed longshore to onshore current ratios, whereas refraction of incident waves in the shallow water predicts a small and decreasing ratio. The presence of a longshore drift current, increasing from zero at the shore to about 35 cm/s at 120 m from shore, suggests that incident waves did approach the breakpoint

Table 4.1. Characteristics of wave motion measured on steep and shallow beaches

			Onshore Component						Longshore Component				
				r.m.s. amplitude of oscillations $(m/s)^2$									
Run identification	Distance from shore m	Depth m	mean current m/s	Draper* 1	direct average	skewness	kurtosis	T_b $(m/s)^4$ $\times 10^3$	mean current m/s	r.m.s. amplitude of oscillations $(m/s)^2$	skewness	kurtosis	T_b $(m/s)^4$ $\times 10^3$
(a) Shallow beach: Saunton													
FFC 7	121	1·33	−0·017	0·28	0·29	+0·60	3·08	+3·01	+0·33	0·12	−0·05	3·72	+18·0
FFC 8	115	1·27	−0·019		0·28	+0·61	3·08	+2·60	+0·37	0·12	+0·14	3·35	+28·0
FFC 5	93	1·02	+0·020	0·24	0·26	+0·52	3·07	+3·68	+0·27	0·12	−0·55	6·90	+8·4
FFD 1	74	0·82	−0·002	0·26	0·24	+0·50	3·26	+1·43	+0·26	0·11	−0·61	3·87	+7·4
FFD 2	49	0·54	+0·019		0·23	+0·27	3·20	+1·49	+0·15	0·11	−0·17	4·00	+1·6
(b) Steep beach: Slapton													
FFA 7	11	1·65	+0·08	0·19	0·18	+0·06	3·25	1·77	−0·005	0·05	−0·18	3·10	−0·003
FFA 8	9	1·28	+0·08	0·27	0·28	−0·15	2·80	4·58	+0·05	0·08	−0·54	2·48	+0·07
FFA 9	5	0·52	+0·08	0·55†	0·48	−0·27	4·10	13·40	+0·14	0·14	−0·88	3·76	+1·7

* Draper (1963).
† Estimated from maximum amplitude.

obliquely, but the angle of approach estimated from this drift current is several times smaller than is needed to produce the observed r.m.s. current ratios. The rather erratic and somewhat large values of the third and fourth moments of the distribution of the longshore velocities tend to support the suggestion that oblique incidence of the dominant waves cannot be responsible. A more likely explanation is that the broad spectrum of incident waves resulted in short-crested waves which set up variations in breaker height along the breaker line. These longshore variations, irregular both in time and distance along the breakers, are expected to produce irregular longshore currents (Bowen, 1969) and these may have persisted throughout the surf zone on the shallow beach.

The observed orbital velocities are, of course, related to the elevation amplitude of the wave motion. If we assume that the horizontal current vector u_w measured along the direction of wave propagation, varies sinusoidally with a single frequency σ,

$$u_w = u_0 \sin(kx_w - \sigma t)$$

where k is the wave number of the wave, x_w is the direction of wave propagation and u_0 is the amplitude of the horizontal component of orbital velocity, then, linear wave theory gives u_0 in terms of a, the wave elevation amplitude, as

$$u_0 = a\sigma \frac{\cosh k(z + h)}{\sinh kh} \qquad (4.1)$$

where z is the vertical coordinate, measured with positive distances upwards from an origin at the water level, and h is the local water depth.

Equation 4.1 reduces to

$$u_0 = a\sigma e^{kz} \qquad \text{in deep water} \qquad (4.2)$$

and

$$u_0 = \frac{a\sigma}{kh} = \frac{ac}{h} = a\sqrt{\frac{g}{h}} \qquad \text{in shallow water} \qquad (4.3)$$

where c is the phase velocity of the wave motion.

Inside the surf zone, although non-linearity of the wave motion is clearly very important, the linear theory has been found still to provide a good approximation to some features of the motion (see, for example, Komar and Gaughan, 1972). Furthermore measurements have shown that the ratio of breaker height to local water depth $2a/h = \gamma$ remains approximately constant for breakers in a surf zone, the value of γ lying between 0·6 for shallow beaches and 1·2 for steep beaches. In the surf zone we can therefore write Equations 4.2 and 4.3 in the form

$$u_0 = \frac{\gamma}{2} h\sigma e^{kz} \qquad \text{in deep water} \qquad (4.4)$$

$$u_0 = \frac{\gamma}{2} c = \frac{\gamma}{2}\sqrt{gh} \qquad \text{in shallow water} \qquad (4.5)$$

In the linear approximation, the total energy of the wave field—in unit area of the water surface—is given by

$$E_T = \tfrac{1}{2}\rho g a^2 = \tfrac{1}{2}\rho g u_0{}^2\left(\frac{kh}{\sigma}\right)^2 \quad \text{in shallow water} \qquad (4.6)$$

$$E_T = \tfrac{1}{2}\rho g a^2 = \tfrac{1}{2}\rho g\left(\frac{u_0}{\sigma}\right)^2 e^{-2kz} \quad \text{in deep water} \qquad (4.7)$$

where ρ is the density of water.

In some of the following sections we shall be concerned with the mean square of the measured wave velocity in a given direction considered as a function of frequency. For conciseness, this function will be called the 'spectral energy' of the wave velocity and designated $E(\sigma)$. The spectral energy for the current component in the direction of wave propagation and at a given frequency σ is then related to the total wave energy $E_T(\sigma)$ at that frequency by a relationship which, in the linear approximation, is given by Equations 4.6 and 4.7.

No direct measurements of the wave elevation were made on either of the beaches, so that it is not possible to predict the velocity field outside the surf zone from these equations. Inside the surf zone, however, Equation 4.5 can be used to predict the velocity field when only the local water depth is known, once an estimated value of γ is chosen. We have therefore compared the measured onshore velocities in the surf zone at Saunton with those predicted by Equation 4.5.

Clearly the r.m.s. onshore velocities at Saunton given in Table 4.1(a) do not fall off towards the shoreline as quickly as $\sqrt{h}$. However, results in much closer agreement with Equation 4.5 are obtained if we consider only the sum of orbital velocity components at frequencies equal to or greater than the predominant incident wave frequency; these frequencies will contain the energy of the incident wave motion even when the individual waves are distorted in the surf zone, but will not include any interaction effects leading to lower frequencies. Figure 4.8 shows the mean square of the onshore velocities at frequencies greater than 0·17 Hz, a cut-off frequency chosen to cover most of the incident wave frequencies; the measured values show a reasonable fit to the linear decay to zero at the shoreline predicted by Equation 4.5. Nevertheless the values of r.m.s. velocity predicted by Equation 4.5 are more than three times larger than the observed velocities and a number of different factors may contribute to this discrepancy. The peaked breakers, with shallow troughs, are expected to produce considerably smaller r.m.s. velocities than a sine wave of the amplitude of the breaker; in fact a similar calculation assuming a train of solitary waves in the surf zone predicts velocities less than twice the observed velocities. The irregularity of the wave train will also result in a lower r.m.s. velocity if only a proportion of the incident waves are large enough to break; for a completely random wave field the r.m.s. orbital velocity will be 1·6 or 2·05 times lower if only the highest one-third or one-tenth waves respectively are breaking.

These results suggest that in the surf zone on a shallow beach the incident wave energy decays to zero at the shoreline but is in part used to generate lower

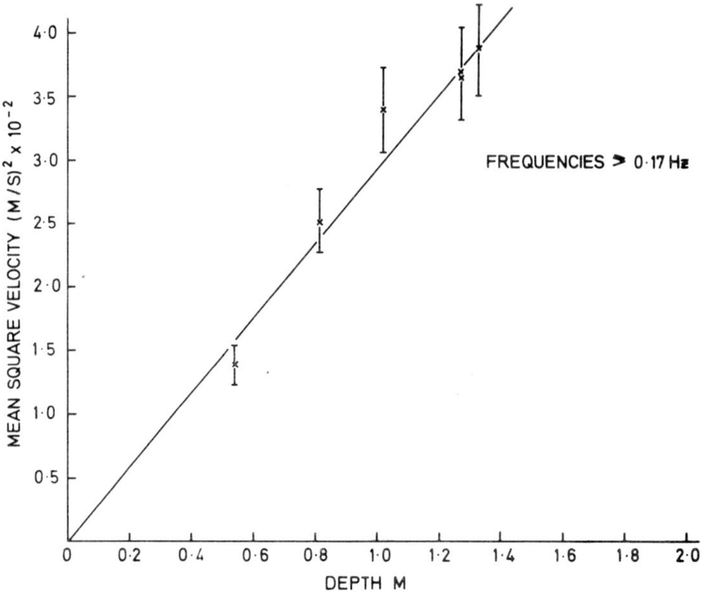

Figure 4.8. The mean square velocity of the breakers on the shallow beach plotted against mean water depth, for velocity components with frequency greater than 0·17 Hz.

frequency oscillations in such a way that the total r.m.s. velocity remains almost constant. This therefore supports the suggestion that there is an increase in low frequency motion as the shoreline is approached.

If it is assumed that the wave motion in the sea is made up of a linear combination of waves of different, uncorrelated frequencies and directions of propagation, then, by the central limit theorem, the probability distribution of wave velocities should be Gaussian, with the shape completely determined by the mean velocity and the r.m.s. amplitude of velocity fluctuations about the mean (Kinsman, 1965). In deep water the wave motion is generally found to fit a Gaussian distribution very well, but close to the shore, non-linear interactions between different frequency components become large, and the distributions depart from the Gaussian shape. In Figure 4.7 the solid curves show the Gaussian distribution for the measured mean and r.m.s. velocities given in Table 4.1. It can be seen that although the distributions generally bear some resemblance to a Gaussian form, discrepancies are often large. Qualitatively it might be expected that, for finite amplitude waves travelling towards the coast with flattened troughs and peaked crests, the deviation from a Gaussian distribution will take the form of a skewing of the maximum towards a negative current (offshore) with a higher than Gaussian distribution of large onshore velocities and a lower than Gaussian distribution of large offshore velocities. The resulting distribution will have a positive skewness. In Figure 4.7 the onshore velocities at

Saunton show just this behaviour. On the steep beach, on the other hand, the large backwash currents and sharp changes to onshore motion close to the break-point may produce peaking in the offshore direction which would be responsible for the observed negative skewness of both onshore and longshore velocities. Other measurements of skewness for nearshore wave motion have been reported by Kinsman (1960) and Hasselman and coworkers (1963) but were obtained in much deeper water than the present results; their values, in the range 0·09 to 0·35, were generally smaller than our results and were always positive.

The observed saw-toothed asymmetry of the wave velocity curves, leading to bore formation, contributes nothing to the skewness, since the resulting probability distribution is the same for both positive and negative velocities. It will nevertheless contribute to the fourth moment, or kurtosis, of the distribution. Since saw-toothed motion has a higher probability of near-zero velocities relative to near-peak velocities than a sinusoidal motion, we expect qualitatively that saw-toothed asymmetry will result in a distribution with higher than Gaussian probability near zero and lower than Gaussian probability at extreme velocities. Such distributions are described as 'leptokurtic' and have a value of kurtosis greater than 3. The distributions shown in Figure 4.7 and the values of kurtosis shown in Table 4.1 show that, except for one set of records on the steep beach, the distributions are all leptokurtic as expected.

Higher order moments of these distributions may be important but have not been calculated since their statistical significance would be low for the 15-minute time series.

4.4 Nearshore drift currents

As waves propagate into shallow water and steepen and break, net drift currents are generated which are usually confined to a narrow strip along the shoreline several times the surf zone width. Such wave-induced drift currents were observed on both the steep and shallow beaches, where they formed a significant part of the total velocity field.

On the steep beach at Slapton the low incident waves resulted in generally small drift velocities of the order of 10 cm/s, but the narrow surf zone and rapid steepening of the waves gave rapid changes of direction and magnitude of the longshore velocity along a line normal to the shoreline.

An initial indication of the nearshore drift was obtained on one occasion by studying the drift of dye patches at various distances from the shoreline. Injection of dye just shoreward of the breaker line revealed a slow northerly movement inside the surf zone, measured as about 10 cm/s, and, close to the injection point, a relatively narrow, localized band of weak offshore flow penetrating seaward through the breaker line. Foam lines seaward of the breaker line supported the impression that this offshore flow was a poorly developed rip current (Shepard and Inman, 1950; Bowen, 1969; Bowen and Inman, 1969). Once seaward of the breaker line the dye moved southerly at a rate, close to the breaker line, of the order of 2 cm/s. These values of longshore currents compare reasonably with the drift currents of 6 cm/s southerly inside and 2·5 cm/s northerly outside the surf zone, measured by the flowmeter, although precise comparison is difficult because dye patches spread rapidly normal to the shore so that only an average longshore drift across the width of the patch could be measured. The weak rip current did not flow over the flowmeter, which recorded only very small onshore currents of the order of ± 3 cm/s, close to the limit of resolution of the onshore measurements in the presence of large orbital velocities.

The dye patch and concurrent flowmeter experiments were insufficient on this occasion to describe the nearshore drift current pattern. However, a more complete set of flowmeter records was obtained on the previous tide and the mean longshore and onshore currents are plotted against distance from the shoreline in Figure 4.9; results seaward of 13 m from the shoreline were measured with a second flowmeter. The results from both flowmeters show a consistent trend of both longshore and onshore drift velocities with distance from the shore and are in reasonable agreement with the data for the subsequent tide, although the zero crossing of longshore current is somewhat further offshore. The measured onshore velocity is much larger than can be explained by tidal motion. However, the observation of a weak rip current suggests that the measured mean currents are part of a horizontal nearshore circulation pattern. Figure 4.10(a) shows schematically a possible nearshore circulation system; the onshore

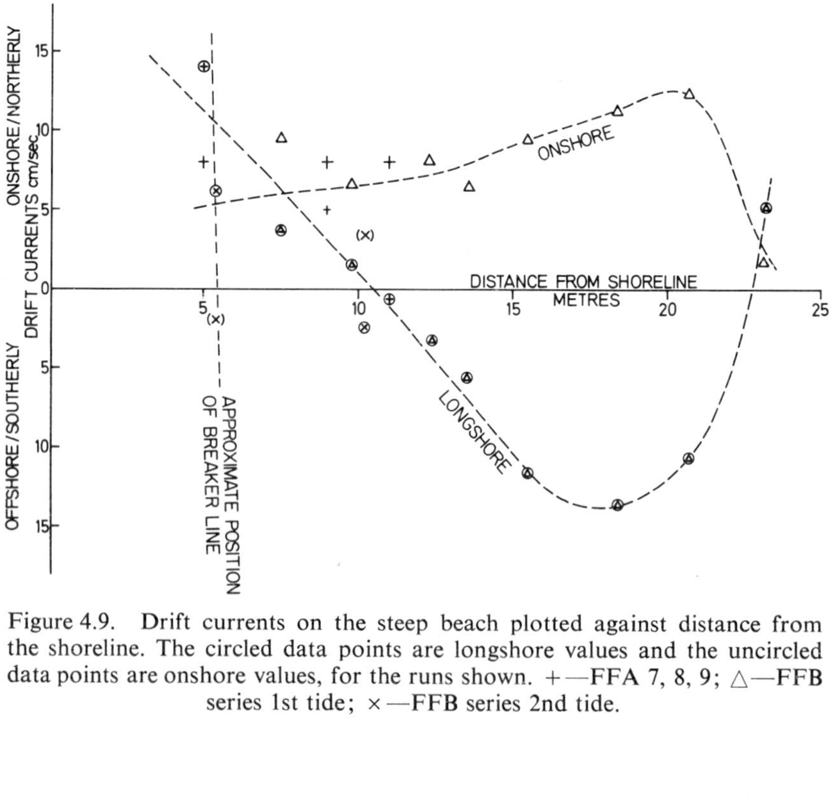

Figure 4.9. Drift currents on the steep beach plotted against distance from the shoreline. The circled data points are longshore values and the uncircled data points are onshore values, for the runs shown. +—FFA 7, 8, 9; △—FFB series 1st tide; ×—FFB series 2nd tide.

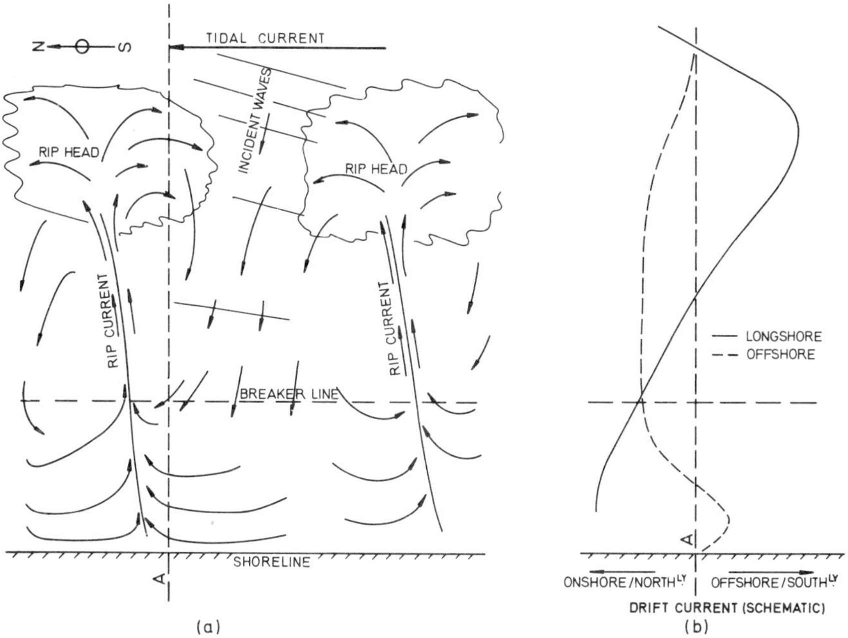

Figure 4.10. (a) Schematic diagram of the proposed nearshore circulation system on the steep beach. The mean onshore drift under the incident waves is returned seaward in a series of regularly spaced narrow rip currents fed by longshore currents inside the surf zone. Some distance seaward of the breaker line each rip current broadens into a rip head and the seaward flow ceases. (b) The expected dependence of longshore and onshore drift currents on distance from the shore, along a line A normal to the shoreline, just south of a rip current.

drift under the incident waves turns parallel to the shore inside the surf zone and feeds a series of regularly spaced seaward-flowing rip currents. If flowmeter measurements are made close to, but just south of, a rip current, for example on the line A in Figure 4.10(a), then the variation of onshore and longshore currents with distance from the shoreline should look qualitatively like Figure 4.10(b). Comparison of Figure 4.10(b) with the measured drift currents (Figure 4.9) shows that the observations are qualitatively consistent with the suggested circulation pattern. The large southerly flow is caused by the expanding head of the rip current and a reversal to a northerly flow further offshore connects with the expected strong northerly tidal flow in the centre of the bay.

Nearshore circulation systems of this kind have been extensively studied in field, laboratory and theoretical investigations (Shepard and coworkers, 1941; Shepard and Inman, 1951; McKenzie, 1958; Harris, 1964; Bowen, 1969; Bowen and Inman, 1969; Tam, 1973). The range of sizes and strengths of the nearshore circulation cells studied has been large, with rip currents reaching more than 2 m/s in the strongest systems, but the basic features are similar at all scales. In particular the position of the rip head, though varying with rip current strength, is commonly around three surf zone widths from the shoreline. In agreement with this observation, the results shown in Figure 4.9 suggest a rip head around 15–20 m from the shoreline, or three or four surf zone widths offshore.

Although no longshore periodicity of the drift velocities was observed with the limited flowmeter and dye drift experiments, strong indications of such periodicity were found in the existence on other occasions of shoreline beach cusps; Komar (1971, 1972) has shown that such cusps are generated by regularly spaced circulation cells and rip currents along the shore, Furthermore, Bowen (1969) and Bowen and Inman (1969) have suggested that the longshore spacing of rip currents is determined by the longshore variation of breaker height due to the presence of edge waves at the incident wave frequency (see Section 4.5). The observed beach cusp wavelength was consistent with such edge waves and rip currents on the steep beach (Huntley and Bowen, 1975).

On the shallow beach a different pattern of drift velocities was measured. The low frequency motion present on this beach made the estimates of steady drift currents, obtained from an average of the 15-minute long records, very unreliable under about ± 5 cm/s. The measured onshore drift currents were all less than 2 cm/s and were found to fluctuate widely between runs. They could not therefore be considered physically meaningful. The longshore currents on the other hand were considerably larger than the onshore drifts. In Figure 4.11 the measured values are plotted against distance from the shoreline and show a steady increase away from the shore; the two values at 115 m offshore were obtained on the ebb and flood tide respectively and therefore suggest a reasonably steady drift pattern over more than 4 hours. Also plotted on this graph are drift velocities obtained by measuring the longshore displacements of floats over periods of order 2 minutes at different distances from the shoreline; the results are seen to be in satisfactory agreement with the flowmeter data. The observed southerly flow in this shallow beach surf zone, rising to 35 cm/s and probably continuing to increase further offshore, was inconsistent with the very small (≈ 10 cm/s) tidal stream expected from the Admiralty chart of the area.

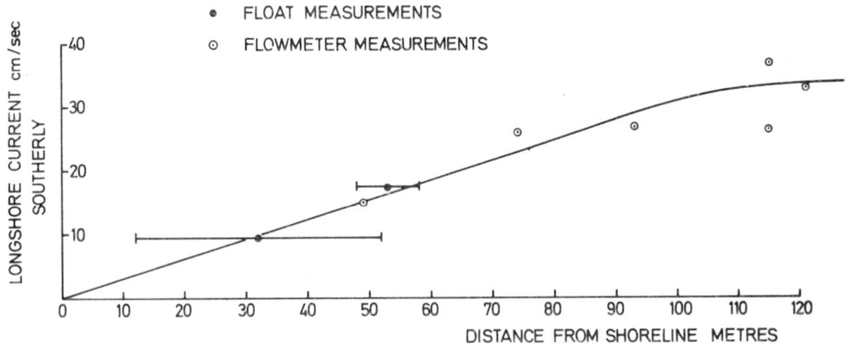

Figure 4.11. Longshore drift currents on the shallow beach. The exact position of the breaker line was not measured, but is estimated to have been about 250 m from the shore.

It was also inconsistent with a wind-induced drift since the flow was in the opposite direction to the wind. It is possible therefore that obliquely incident swell waves were responsible for the observed longshore drift. No observations were made of the angle of approach of the incident waves, but a rough calculation based on the theories of Bowen (1969) and Longuet-Higgins (1970a, 1970b) suggests that waves approaching the breaker line at about 10°–15° to the normal could generate longshore currents of the observed magnitude in the surf zone.

While the transport of suspended sediment on beaches is probably determined by these net drift velocities, the transport of sediment as bed load is likely to be more complex and, as has been shown, Wells (1967) has suggested a dependence of bed load motion on the fourth power of the total velocity u in the form

$$\text{Bed load motion} \sim \langle u^3 \rangle [\langle u^2 \rangle]^{1/2} = T_b \qquad (4.8)$$

where $\langle f(u) \rangle$ means, in this case, the time average of $f(u)$.

In the absence of a drift velocity, T_b is simply a combination of the skewness and variance of the distribution of the orbital velocities and the direction of bed load movement is determined by the sign of the skewness. However, where drift velocities form a significant part of the velocity field, as on both beaches in the present study, the sign of T_b can be the opposite of the skewness, as can be seen in Table 4.1. In fact the onshore values of T_b are all positive, suggesting an onshore movement of bed load sediment, and this is expected for accreting summer waves on both beaches. The longshore drift currents also dominate the values of T_b on both beaches, suggesting a longshore movement of bed load everywhere in the direction of the drift velocity.

4.5 Low frequency wave motion

Introduction

Although much of interest can be gained from a study of the time series of the longshore and onshore velocity field, it is instructive also to consider more carefully the spectral energy of the velocity field as a function of frequency. The data have therefore been analysed in two different ways to separate the velocity field into its constituent frequencies, Fourier analysis to produce energy spectra and band-pass filtering of the original time series to produce time series of the velocity variation in a chosen frequency range.

A typical spectrum of the wave motion outside the surf zone is shown in Figure 4.12. The spectral energy density, normalized to unit bandwidth, is plotted against the frequency in hertz for the frequency range 0 to 1·0 Hz. The upper frequency limit of the plot is determined by the sampling time of the original digitized time series. The resolution of the spectrum is determined by the total length of the time series and by the characteristics of a weighted averaging filter (Tukey) applied to the spectral values; the filter characteristics are chosen by balancing increasing resolution against decreasing statistical confidence limits for the spectral values.

The spectrum of Figure 4.12 is for the onshore component of the velocity field 9 m from the shoreline at Slapton and shows the general features common to wind wave spectra. There is a clear separation of the energy into different frequency bands, with a fairly sharp peak at around 0·1 Hz and a broad peak at the frequency of the observed incident waves (0·24 Hz), followed by a gently curving fall of energy with increasing frequency. The small peak at about 0·47 Hz may be the first of the harmonics of the incident wave frequency expected to be present in the distorted waves. In contrast Figure 4.13 shows frequency spectra for wave motion inside the surf zone, with Saunton and Slapton compared. Clearly, on the steep beach at Slapton the spectrum has changed considerably inside the surf zone, with the incident wave peak now, as expected,

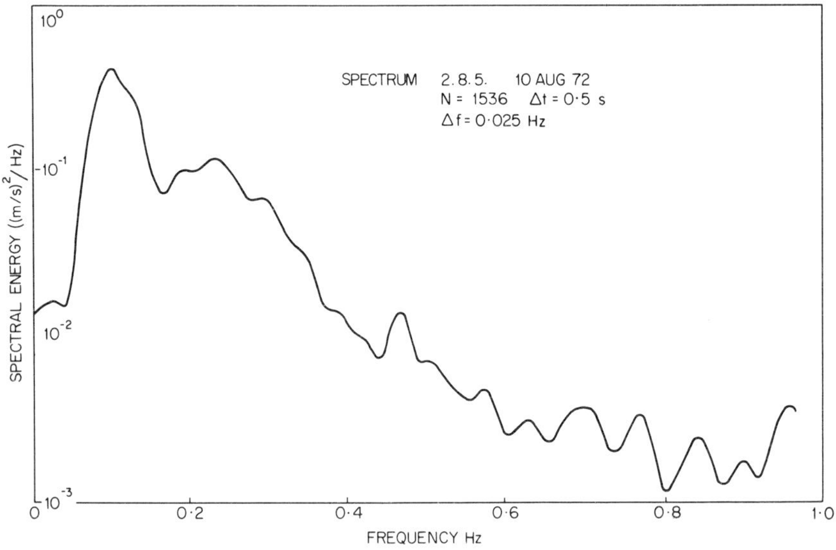

Figure 4.12. A typical onshore current spectrum outside the surf zone on the steep beach.

submerged beneath an increase in the high frequency energy; the peak at 0·1 Hz has increased. There is also a marked contrast between the surf zone spectrum on the steep beach at Slapton and that on the shallow beach at Saunton. The low frequency energy in the surf zone at Saunton, seen in the original velocity time series, is now clearly resolved and contrasts with the rapid drop in energy at low frequencies at Slapton. The Saunton spectrum is also relatively featureless, with no peak corresponding to the 0·1 Hz peak at Slapton. One feature common to both spectra however is the linear decay of energy, plotted on a logarithmic scale, with increasing frequency, at frequencies above about 0·35 Hz.

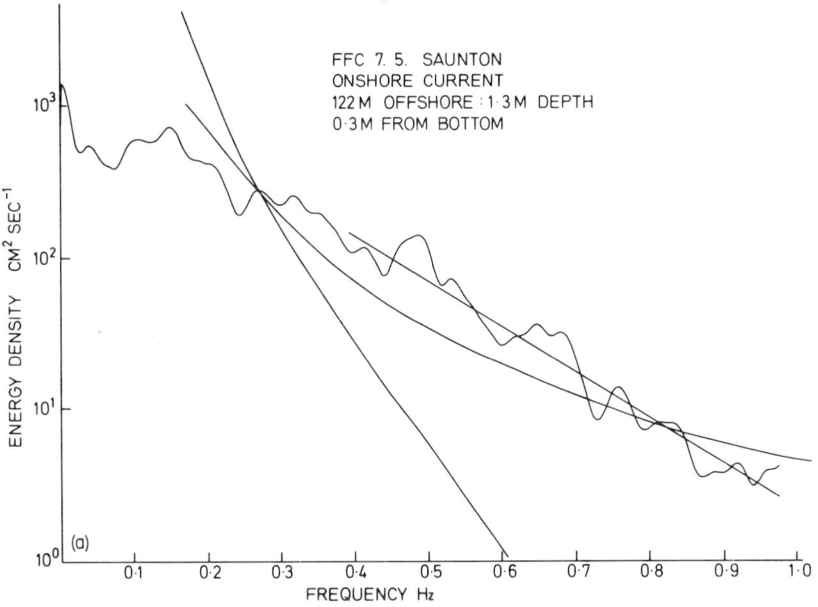

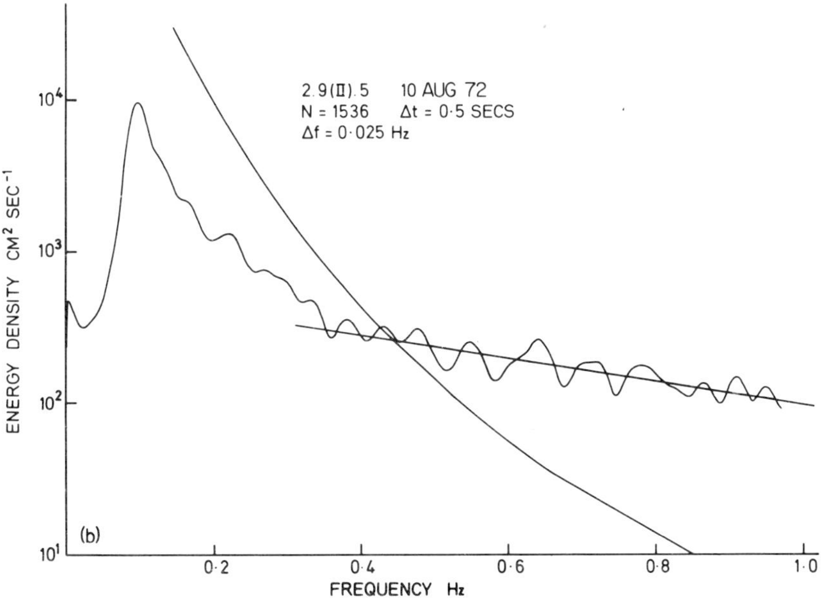

Figure 4.13. Typical onshore current spectra inside the surf zone (a) on the shallow beach and (b) on the steep beach. In (a) the steep line shows the expected dependence on frequency of the equilibrium wave spectrum (Phillips, 1966). The shallower curve shows a suggested $(\text{frequency})^{-3}$ dependence for turbulence. Also shown is a straight line through the data. In (b) the predicted equilibrium curve and a straight line fit to the data are shown.

Spectra from the steep beach

A prominent distinguishing feature of the spectra from the steep beach is the presence of a fairly sharp peak of energy at about 0·1 Hz. This can be seen clearly in spectra both outside (Figure 4.12) and inside (Figure 4.13(b)) the surf zone and is in fact also present in spectra of the longshore component of the velocity field. Surprisingly, this energy increases rapidly as the shoreline is approached, even within the surf zone, as can be seen from Figure 4.14, where the amplitudes of both longshore and onshore components are plotted against distance from the shoreline.

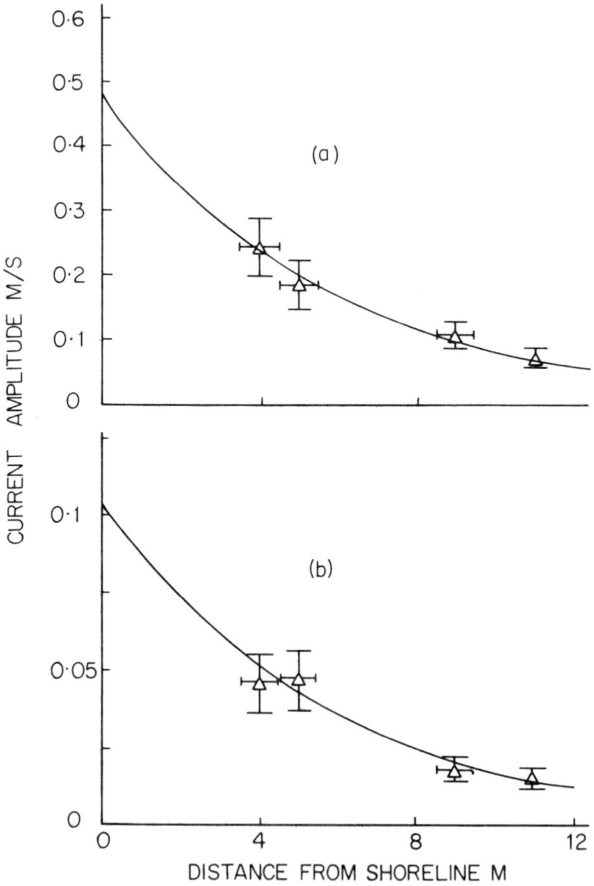

Figure 4.14. Graphs of (a) onshore and (b) longshore current amplitude at 0·1 Hz plotted against distance from the shoreline. The solid line shows the expected decay for an edge wave of the predicted wavelength of 32·0 m. The measured values are shown as triangles with error bars corresponding to the 80% confidence limits in the spectral analysis; horizontal errors are a consequence of averaging amplitudes during an incoming and receding tide.

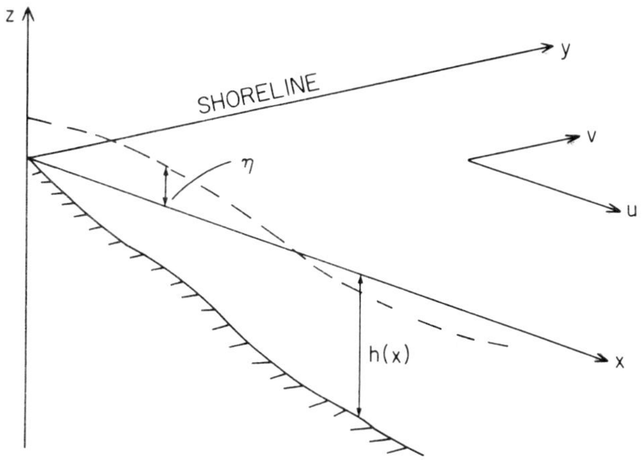

Figure 4.15. Coordinate definition diagram for edge wave theory (for symbols used see Notation).

This energy has been attributed to the presence of an 'edge wave' on the steep beach (Huntley and Bowen, 1973). Before discussing the evidence for this it is necessary to begin with a brief discussion of the theory of edge waves.

Edge waves are free modes of water motion trapped against a sloping boundary by refraction. Ursell (1952) solved the irrotational wave equations in three dimensions to obtain edge waves on a beach of constant slope; but the solutions are complicated and difficult to use. However, he found that the depth dependence of the full solutions is small, so that, except for the longest period and highest mode number edge waves, the velocities can be assumed to be independent of depth and the shallow water wave equations provide a very satisfactory approximation (Eckart, 1951).

Using the notation in Figure 4.15 the shallow water equations on a sloping beach, ignoring dissipation, are

$$\frac{\partial u}{\partial t} = -g\frac{\partial \eta}{\partial x}; \qquad \frac{\partial v}{\partial t} = -g\frac{\partial \eta}{\partial y} \tag{4.9}$$

$$\frac{\partial \eta}{\partial t} + \frac{\partial}{\partial x}(h(x)u) + h(x)\frac{\partial v}{\partial y} = 0 \tag{4.10}$$

where the depth h is a function only of the offshore distance x. These equations can be solved for a velocity potential ϕ, where

$$(u, v) = \nabla\phi \qquad \text{and} \qquad \eta = -\frac{1}{g}\frac{\partial \phi}{\partial \sigma} \tag{4.11}$$

Edge wave type solutions will then be of the form

$$\phi(x, y, t) = \phi(x)\sin(ky - \sigma t) \qquad \text{progressive}$$

or (4.12)

$$\phi(x, y, t) = \phi(x)\cos ky \cos \sigma t \qquad \text{standing}$$

where, for an edge wave, $k = 2\pi/L$ with L the longshore wavelength, and $\sigma = 2\pi/T$ with T the wave period.

The offshore dependent part will then be a solution of

$$h(x)\frac{d^2\phi}{dx^2} + \frac{dh}{dx}\frac{d\phi}{dx} - \left(k^2h - \frac{\sigma^2}{g}\right)\phi = 0 \tag{4.13}$$

Analytic solutions of this equation exist for a few simple forms of $h(x)$, notably for

$$h = x \tan\beta \qquad \text{a linear slope} \tag{4.14}$$

and

$$h = h_0(1 - \mathrm{e}^{-\alpha x}) \text{ an exponential slope} \tag{4.15}$$

If we consider, as an example, a linear beach slope, solutions take the form

$$\phi = \frac{ga}{\sigma} L_n(2kx)\,\mathrm{e}^{-kx} \sin(ky - \sigma t) \qquad \text{progressive} \tag{4.16}$$

$$\phi = \frac{ga}{\sigma} L_n(2kx)\,\mathrm{e}^{-kx} \cos ky \cos \sigma t \qquad \text{standing} \tag{4.17}$$

where L_n is the Laguerre polynomial of mode n, with the first three modes,

$$\begin{aligned} n &= 0 \qquad L_0(2kx) = 1 \\ n &= 1 \qquad L_1(2kx) = 1 - 2kx \\ n &= 2 \qquad L_2(2kx) = 1 - 4kx + 2(kx)^2 \end{aligned} \tag{4.18}$$

and where the frequency σ and longshore wave number k are related by a dispersion relation:

$$\sigma^2 = gk(2n + 1)\tan\beta \tag{4.19}$$

This solution serves to illustrate the main features of edge waves. The amplitude of the edge wave elevation and currents, related to the velocity potential by Equations 4.11, varies sinusoidally along the shoreline. The term $\exp(-kx)$ in Equations 4.16 and 4.17 ensures that the velocity potential dies away offshore while the mode number n gives the number of zero crossings of the function in the offshore direction; the zero order mode therefore decays as a simple exponential but the higher modes are more oscillatory and their amplitude dies away more slowly. Figure 4.16 illustrates this behaviour for the onshore and offshore components of the edge wave velocity field for modes 1 and 2.

The dispersion relation, Equation 4.19, provides a family of dispersion curves, a separate curve for each mode. Figure 4.17 shows the family of dispersion curves relevant to Slapton beach. This shows that, for a given period, each mode has a predicted wavelength, but that several modes with different wavelengths may occur with the same wave period. These curves for Slapton have in fact been obtained from the edge wave solution for an exponential beach slope fitting Equation 4.15 (Ball, 1967), with the exponential parameters chosen to be a best fit to the surveyed beach profile at Slapton. The essential features of the

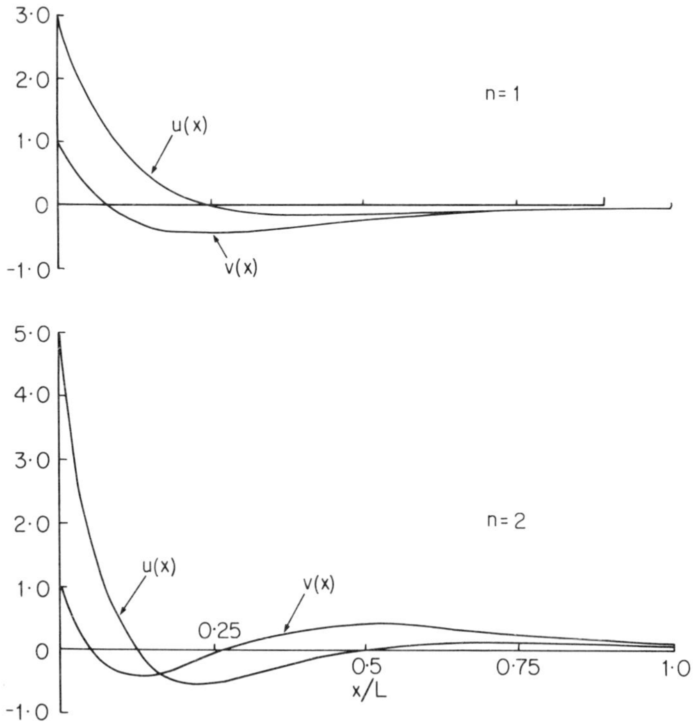

Figure 4.16. Offshore dependence of edge wave onshore $u(x)$ and longshore $v(x)$ currents for modes $n = 1$ and $n = 2$. The amplitudes are normalized to $v(x) = 1$ at $x = 0$.

edge waves for the exponential profile are similar to those for a linear slope, with the notable exception of the existence of a cut-off period for each mode beyond which no trapped edge wave solutions exist. This cut-off period is shown for each mode in Figure 4.17.

The validity of an edgewave interpretation of the 0·1 Hz energy peak at Slapton depends upon relating the observed offshore decay of current amplitudes to the decay predicted by solutions of Equation 4.13 for the exponential beach profile at Slapton. The data presented in Figure 4.14 in fact are found to fit an exponential decay from the shoreline to well within experimental error, with the same decay constant for both longshore and onshore current components. Such an exponential decay is predicted by the theory for the $n = 0$ mode of edge wave on this beach and the decay constant depends on the longshore wavelength of the edge wave, as in the linear beach slope case. From the observed decay constant for the data of Figure 4.14, the wavelength of the edge wave is thus predicted to be 34 $\pm$ 6 m. This can be compared to the wavelength predicted by the dispersion curve for an $n = 0$ mode edge wave of about 10 seconds period; on Slapton beach this wavelength is 32 $\pm$ 3 m (Figure 4.17). These two wavelength estimates agree to well within experimental error and give firm support to the

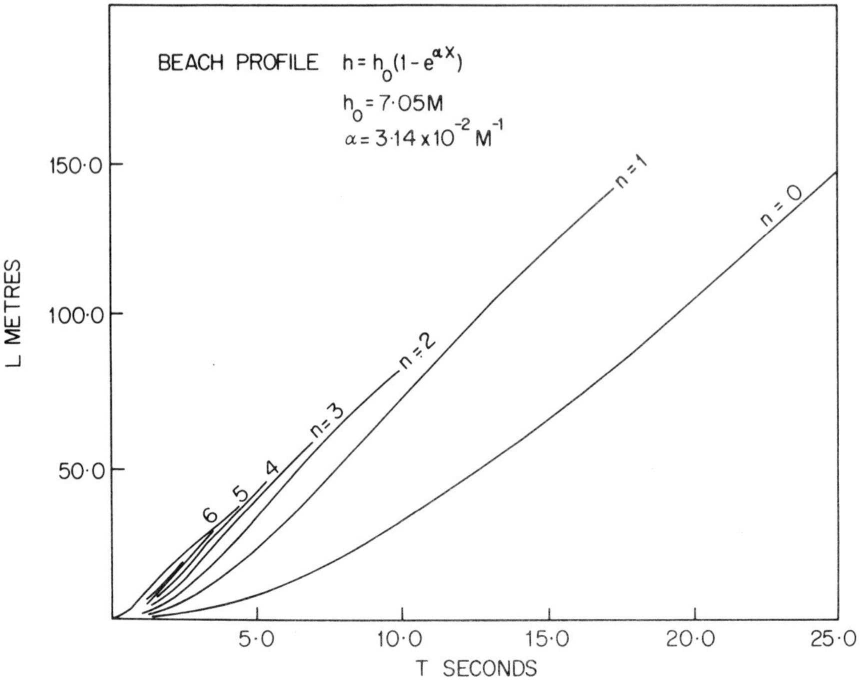

Figure 4.17. Edge wave dispersion curves for the steep beach using the theory of Ball (1967).

edge wave interpretation of this peak in energy at Slapton. A careful consideration of other possible explanations for this energy peak has shown that only the edge wave can explain all the features of the observed motion satisfactorily (Huntley and Bowen, 1973; Huntley and Bowen, 1975).

It is significant that an edge wave of this prominence should occur on a steep beach, while no comparable amplitude of edge waves seems to have occurred on the shallow beach. The swash interaction effects on this beach discussed in Section 4.2 provide a possible generation mechanism for these edge waves at the first subharmonic of the incident wave period. Another possible generation mechanism, also favouring steep beaches, is through coupling between the edge waves and the incident waves reflected at the shoreline (Birchfield and Galvin, 1973; Guza and Davis, 1974).

The influence of edge waves on nearshore sediment at Slapton and elsewhere is discussed in Huntley and Bowen (1975). It is clear that edge waves may provide an explanation for the existence of many nearshore sedimentary features possessing regular longshore variations. Our observations suggest that subharmonic edge waves may be particularly significant for sediment motion on steep beaches.

Spectra from the shallow beach

A characteristic feature of spectra from the shallow beach is the presence of a broad spectrum of energy below the incident wave frequency (Figure 4.13(a)).

The analysis techniques used to plot spectra were too insensitive at low frequencies to be satisfactory for studying this low frequency energy. A low-pass filter, with a cut-off at 8·5 seconds period, was therefore applied to the original data to provide filtered time series showing only the wave motion at frequencies lower than the predominant incident wave frequencies.

Approximate values of the mean amplitude and period of the filtered wave motion were then calculated using Draper's technique (Draper, 1963). The results are shown in Table 4.2. In this table ε is a measure of the spectral width of the wave motion, tending to zero as the spectrum becomes narrow; a common value for a wind wave spectrum is 0.8.

Table 4.2. Saunton Sands. Low frequency oscillations (analysis after Draper (1963))

Run identification	Shoreline distance m	Periods in seconds: Zero crossing T_z	Periods in seconds: Crests T_c	Spectral width ε	U_{rms} cm/s
FFC 7	121	41·0	26·5	0·76	11·2
FFC 6	116	37·5	24·0	0·77	11·9
FFC 8	115	41·0	24·0	0·81	11·5
FFC 5	95	48·0	24·0	0·87	12·1
FFD 1	73	39·0	24·0	0·79	12·2
FFD 2(I)	55	45·0	25·0	0·83	15·6
FFD 2(II)	44	44·0	25·0	0·83	16·0

As expected, these results also suggest a steady increase in the mean energy at low frequencies as the shoreline is approached; but the change is small and, in view of the uncertain accuracy of the analysis technique in this situation, must still be considered tentative. Certainly it has not been possible to relate the slow changes satisfactorily to the predicted changes of amplitude with distance from the shore, for either a reflected incident wave or an edge wave of the average period of either zero crossings or crests. In the latter case, only a second mode edge wave would be reasonably consistent with the observations. Equation 4.19 shows that the wavelengths of a given period of edge wave motion vary as $\tan \beta$ and will therefore be very short on this shallow beach in comparison to the values on the steep beach. For the observed periods, the wavelength of an edge wave on the shallow beach would be at most 70 m, a value which is very small by comparison with the scale of the beach and the surf zone and is therefore unlikely to occur.

The lack of coherence between breaker amplitude and run-up has previously been remarked upon. Not surprisingly a similar lack of coherence was found between the filtered time series at 121 m offshore and the run-up, though the average crest period in the filtered series shown in Table 4.2 compares reasonably with an interval between times of high run-up commonly of approximately 30 seconds. If it is correct to attribute this apparently spatially incoherent energy to breaker interaction, the 25–30-second periodicity may represent a weakly preferred periodicity for the interacting breakers in the surf zone.

At the limit of resolution of the filtered time series there was some evidence for a very low frequency oscillation of about 400 seconds period. It is instructive to compare this frequency with that expected for an edge wave which could be trapped in the bay at Saunton. Saunton beach forms part of a bay approximately 8·6 km long terminated by fairly well defined cliff boundaries. The beach itself forms the northern of two straight beaches separated by an estuary mouth near the middle of the bay. An edge wave trapped in this bay must have longshore current nodes at the boundaries and must therefore have an integral number of half wavelengths across the bay, i.e. if we call the length of the bay B, and the wavelength of the edge wave L, then $b(L/2) = B$, where b is an integer. For Saunton beach, the calculated periods for different values of b and edge wave mode number n are shown in Table 4.3 for a linear slope best fitting the hydrographic data at Saunton to about 3 km offshore. The period of the very low frequency oscillations observed cannot be determined from the data with sufficient accuracy to fix unambiguously the relevant values of b and n, and $n = 0$, $b = 7$ and 8, and $n = 2$, $b = 1$ are equally probable. Possibly, however, there is an additional constraint which can be put on the relevant wavelength of the edge wave. The presence of an estuary mouth approximately at the centre of the bay suggests that an edge wave should have a longshore current node and onshore current antinode at the centre of the bay. This implies that there should be an integral number of full wavelengths across the bay and therefore that b should be an even number. Hence the most probable mode is an $n = 0$ edge wave with four wavelengths across the bay. Since the amplitude of a zero order edge wave decays to a negligible amplitude at a distance from the shoreline equal to the longshore wavelength, this edge wave motion at 400 seconds might be expected to have decayed to a negligible amplitude about 2·2 km from the shore. This is also consistent with a wave trapped between the cliff boundaries of the bay since these boundaries each extend seawards about 1·6 km.

Table 4.3. Saunton Sands. Edge wave periods (seconds) for edge wave mode number n and bay mode number b

n \ b	1	2	3	4	5	6	7	8	9	10
0	1120	794	648	561	502	458	425	397	374	355
1	648	458	374	324	290	265	245	230	216	205
2	502	365	290	251	224	205	190			
3	425	300	245	212	190					
4	374									

4.6 High frequency energy

As previously mentioned, a feature of the spectra from within the surf zone which is common to both Slapton and Saunton beaches is the approximately exponential decay of energy with increasing frequency in the range 0·35–1·0 Hz (Figure

4.13). This is a previously unobserved feature of the surf zone velocity field and is as yet unexplained. However, the high frequency energy seaward of the surf zone will be discussed first.

Outside the surf zone the energy spectrum of the velocity field on the high frequency side of the wind wave peak has previously been found, both theoretically and experimentally, to decay as (frequency, σ)$^{-5}$ for at least some frequency range (Phillips, 1966; Shonting, 1968; Taira, 1971). Other experiments indicate that under steep waves offshore, high frequency turbulence is also generated and can contain more energy than the wind wave orbital velocities above a certain transition frequency (Yefimov and Khristoforov, 1971). At the transition frequency, the frequency dependence of spectral energy is found to change rapidly from σ^{-5} at lower frequencies to σ^{-n} at higher frequencies, where n is between 2 and 3. Our results from seawards of the breakpoint at Slapton show similar frequency dependencies; Figure 4.18 is the spectrum from 9 m offshore plotted

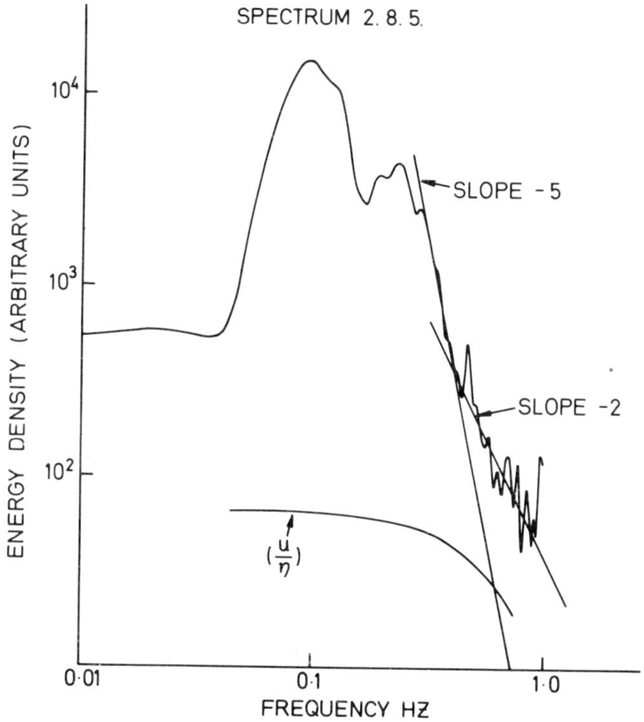

Figure 4.18. An onshore current spectrum outside the surf zone on the steep beach, plotted on a log–log scale. The -5 slope represents the predicted slope for the equilibrium spectrum of deep water wave elevation (Phillips, 1966). The -2 slope represents the empirical slope for wave-induced turbulence found by Yefimov and Khristoforov (1971). The curve (u/η) shows, on an arbitrary scale, the frequency dependence of the ratio of horizontal current u to wave elevation η for this spectrum.

on a log–log scale and shows a region of σ^{-5} dependence changing to approximately σ^{-2} at a transition frequency of about 0·45 Hz, in good agreement with offshore experiments. In direct contrast to the deep sea measurements of Yefimov and Khristoforov, however, the transition frequency is found to decrease with decreasing mean water depth at Slapton, suggesting that the turbulent contribution to the velocity field increases as the wave shoals towards the breakpoint. This may be a local effect or diffusion of vorticity across the breakpoint from the surf zone.

Inside the surf zone the observed decay of the high frequency energy suggests an energy dependence of the form

$$E(\sigma) = E_0 \exp(-p\sigma) \qquad \sigma > \sigma_i \tag{4.20}$$

where σ_i is the incident wave frequency.

For the Saunton data it is possible to compare measured values of the decay constant p at different positions in the surf zone. In Figure 4.19(a) the values of p have been plotted against offshore distance and show that there is a definite trend towards lower values as the local water depth decreases. Dimensional analysis can give some indication of the expected depth dependence of p. Significant parameters controlling p on a linear beach are likely to be the local water depth h, the gravitational acceleration g and the beach slope, $\tan\beta$; these parameters alone will determine, to a good approximation, the local wave energy in the surf zone and the local rate of change of energy. Unfortunately the beach slope $\tan\beta$ is dimensionless and cannot therefore be included in a dimensional analysis. Nevertheless a dimensionless combination of h, g and σ requires that the depth dependent part of $p \sim \sqrt{h/g}$. Unfortunately the data in Figure 4.19(a) are inconsistent with such a dependence on the local depth if we make the physically plausible assumption that p must remain positive or be zero at the shoreline. We must therefore introduce a third parameter. Physical considerations suggest that the amplitude of the incident waves at the breakpoint, a_b, may also be significant to the turbulent intensity at a point in the surf zone. This parameter will determine approximately the incident energy at the breakpoint, the width of the surf zone and the time of travel of the breakers from the breakpoint to the given point in the surf zone, and will therefore include non-local effects controlling the value of p. With this additional parameter dimensional analysis suggests that the depth dependent part of p depends upon

$$p \sim \sqrt{\frac{h}{g}}\left(\frac{h}{a_b}\right)^l$$

where l can be any positive or negative integer.

For Saunton, $l = 2$ is found to give the best fit to the data; the values of p are plotted against $h^{5/2}$ in Figure 4.19(b) and show a reasonable fit to a straight line, within the rather large errors involved in estimating p.

Several recent papers have also described theoretical and experimental work on wave spectra close to, and within, the surf zone. Ijima and coworkers (1970)

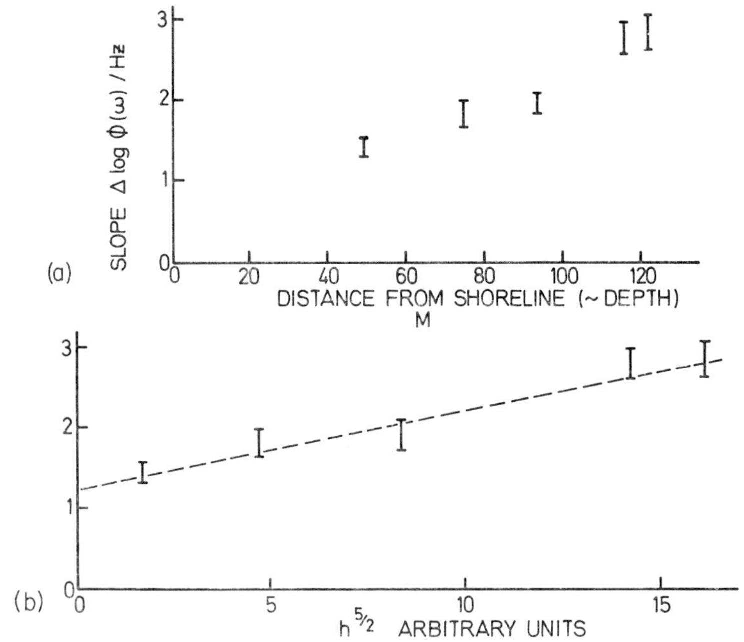

Figure 4.19. (a) High frequency spectral energy decay constant as a function of distance from the shoreline, for onshore currents inside the surf zone on the shallow beach. (b) The same data plotted against $h^{5/2}$, where h is the local water depth.

attempt to calculate an equilibrium wave spectrum using theoretical breaking criteria to specify the maximum wave heights. Results obtained on Japanese beaches show limited agreement with theory. Basinski and St. Massel (1973) describe preliminary investigations of nearshore elevation and current spectra and also observe a secondary peak in the spectra at approximately twice the incident wave frequency. Neither of these papers describe an exponential decay in high frequency spectral energy, however.

It is not yet clear what significance these preliminary results may have for sediment motion close to and within the surf zone, but it seems likely that the turbulent intensity plays an important role in setting the bed material into motion. Further experiments are needed to study this interaction more closely.

4.7 Summary

There are significant differences, both qualitative and quantitative, in the velocity fields occurring on steep and shallow beaches. On a steep beach waves steepen rapidly and, when forming plunging breakers, will collapse and dissipate their energy over a narrow violent surf zone. Observations suggest that within this narrow surf zone the swash plays an important role, interacting with incoming breakers to form undular bores and possibly possessing a natural resonant period which can generate longer period variations in breaker height. The collapse of the breaker itself is a violent process and often involves rebound of the plunging face. On a shallow beach changes of breaker waveform occur much more slowly as the waves traverse a wide surf zone. A steady transformation of the waveform occurs from a wave with slopes of front and back faces approximately equal, characteristic of a breaking solitary wave, to a wave with a steep front face and a relatively gentle back face, characteristic of a hydraulic bore; in this respect wave transformation in the surf zone of a shallow beach resembles the transformation of non-breaking waves on a steep beach as they approach the breakpoint. Observations on a shallow beach also suggest that low frequency oscillations become increasingly prominent in the velocity field as the breakers approach the shoreline and it is suggested that these oscillations, which do not appear to be coherent across the surf zone, are due to weak interaction between the many breakers in the shallow beach surf zone.

Many of these features can be quantified in terms of the moments of the probability distribution of velocities. The square root of the variance of the distributions, the r.m.s. velocities, are of the same order on both beaches but, as expected, the amplitudes vary rapidly on the steep beach while remaining essentially constant on the shallow beach. In fact, on the shallow beach, the spectral energy of the components of wave motion at and above the incident wave frequency is found to decay approximately linearly with water depth towards the zero at the shoreline, and transfers energy into motion at lower frequencies in such a way as to maintain an almost constant total r.m.s. velocity across the surf zone. Higher moments quantify the skewness and kurtosis of the distributions caused by the different forms of wave distortions on the two beaches.

The pattern of mean drift velocities is radically different on the two beaches. On the steep beach the drift velocities were consistent with the existence of nearshore horizontal circulation cells, and the deduced characteristics of the cells are in reasonable agreement with those predicted for the beach. On the shallow beach, on the other hand, a steady longshore current, increasing with distance from the shoreline, was observed in the surf zone and was consistent with a longshore current generated by obliquely incident swell waves.

Observations also suggest that free modes of nearshore oscillation, known as edge waves, can occur on steep beaches, with a period of about twice the incident wave period. Such short period edge waves may be generated on steep beaches either by weak interaction with reflected incident waves or by strong interaction with the swash. In fact these two mechanisms for generating edge waves may be fundamentally the same, since the amplitude and period of the swash is likely to be related to the presence of a reflected wave at the shoreline. Edge wave theory and authors' observations suggest that on a shallow beach edge wave periods, other than at the incident wave period itself, are likely to be much longer than on a steep beach and not clearly related to the incident wave period. Theoretically the wavelength of an edge wave may be determined by the length of the beach and the nature of the end boundaries. This gives a wave motion which is in effect a bay resonance and the authors' observations indicate that such resonances do occur in natural bays.

Within the surf zone the contribution to the velocity field made by motion at frequencies greater than the incident wave frequency is found to have the approximate functional form $E(\sigma) = E_0 \exp(-p\sigma)$ on both steep and shallow beaches. Dimensional analysis suggests that the decay parameter varies as (local depth)$^{5/2}$ but the physical significance of this observation is unclear. The dependence of this effect on beach slope also need furthers investigation.

Relating these hydrodynamic characteristics to sediment dynamics is largely qualitative at present. The onshore drift of sediment expected under the low summer waves on both beaches is consistent with the observed positive values of T_b, the bed load transport parameter suggested by Wells (1967). There are also indications that the transport of material on a steep beach is dependent on the characteristics of the swash; for high waves, dominance of backwash may account for the erosional nature of the waves. The importance of edge waves for near-shore sediment is discussed in Bowen and Inman (1971), Komar (1971, 1972) and Huntley and Bowen (1975). The observations presented here suggest that on a steep beach the short period edge waves may be responsible for relatively small-scale periodic features of beach topography along the shore, such as beach cusps, while on a shallow beach large-scale features, stretching further offshore, are more likely to occur.

Notation

a = wave elevation amplitude
a_b = amplitude of incident waves at the breaker line
b = bay mode number, an integer
c = phase velocity of wave motion
e = the exponential parameter = 2·7182
E_T = the total energy of a wave field
$E_T(\sigma)$ = the total energy of a wave field at the frequency σ
$E(\sigma)$ = the spectral energy of a velocity component at frequency σ, i.e. the mean square of the measured wave velocity in a given direction at the frequency σ
$\langle f(u) \rangle$ = the time average of some function $f(u)$, where u is a function of time
g = the acceleration due to gravity
h = the local water depth, measured from mean sea level
$h(x)$ = the local water depth considered as a function of distance from the shoreline, x, only
h_0 = the water depth at $x = \infty$, for an exponential beach slope, Equation 4.15
k = wave number (rad m^{-1})
l = any positive or negative integer

L = longshore wavelength of an edge wave
L_n = Laguerre polynomial of mode n
n = an integer mode number for L_n and, hence, the edge wave mode number
p = exponential decay constant for high frequency spectral energy, Equation 4.20
$P(u)$ = probability distribution of u
t = time
T = wave period of an edge wave
T_b = bed load transport parameter, Equation 4.8
u = on/offshore component of horizontal velocity
$\bar{u}$ = mean value of u
u_0 = horizontal amplitude of the orbital velocity of a wave
u_w = the instantaneous horizontal velocity in the direction of wave propagation
v = the longshore component of horizontal velocity
(x, y) = horizontal Cartesian coordinates
x = measured seawards from the shoreline
y = measured parallel to the shoreline
z = the vertical coordinate, measured upwards from the mean water level
x_w = the direction of wave propagation
α = exponential parameter for exponential beach slope, Equation 4.15
$\tan \beta$ = linear beach slope, Equation 4.14
γ = the ratio of breaker height to local water depth
ε = a measure of the spectral width of wave motion
η = the vertical displacement of the water surface from the mean sea level, positive upwards
μ_r = the rth moment of a probability distribution
π = 3·142
ρ = the density of seawater
σ = wave frequency (rad s^{-1})
σ_i = incident wave frequency
2τ = the total length, in seconds, of a time series
ϕ = wave velocity potential, Equation 4.11

References

Adeyemo, M. D., 1970, Velocity fields in the wave breaker zone, *Proc. Twelfth Conf. cst. Engng. Washington, D.C.*, Am. Soc. Civ. Engrs., pp. 435–460.

Bagnold, R. A., 1940, Beach formation by waves; some model experiments in a wave tank. *Proc. Inst. civ. Engrs.*, **15**, 17–52.

Bagnold, R. A., 1963, Beach and nearshore processes, Pt. 1. Mechanics of Marine Sedimentation, in Hill (Ed.), *The Sea*, Interscience, New York and London.

Ball, F. K., 1967, Edge waves in an ocean of finite depth, *Deep Sea Res.*, **14**, 79–88.

Basinski, T., and St. Massell, 1973, The coastal research station at Lubiatowo—Investigations, new methods and equipment, *Inter Ocean '73*, Internationaler

Kongress mit Ausstellung für Meeresforschung und Meeresnutzung, Düsseldorf, Vol. 2, pp. 1078–1086.

Binnie, A. M., and J. C. Orkney, 1955, Experiments on the flow of water from a reservoir through an open horizontal channel, II, The formation of hydraulic jumps, *Proc. R. Soc. A.*, **230**, 237–246.

Birchfield, G. E., and C. J. Galvin, 1973, Generation of edge waves through nonlinear subharmonic resonance, *Series in Applied Mathematics: Report no. 73–6*, Northwestern University, Evanston, Illinois.

Bowen, A. J., 1969, Rip Currents, 1. Theoretical investigations, *J. geophys. Res.*, **74**, 5467–5478.

Bowen, A. J., and D. L. Inman, 1969, Rip Currents, 2. Laboratory and field observations, *J. geophys. Res.*, **74**, 5479–5490.

Bowen, A. J., and D. L. Inman, 1971, Edge waves and crescentic bars, *J. geophys. Res.*, **76**, 8662–8671.

Carter, T. G., P. L-F. Lin and C. C. Mei, 1973, Mass transport by waves and offshore sand bedforms, *J. Wat. Ways Harb. Div., Am. Soc. civ. Engrs.*, **99**, HY5, 165–184.

Draper, L., 1963, Derivation of a design wave from instrumental records of sea waves, *Proc. Inst. civ. Engrs.*, **26**, 291–304.

Eagleson, P. S., and R. G. Dean, 1961, Wave-induced motion of bottom sediment particles, *Trans. Amer. Soc. civ. Engrs.*, **126**, 1162–1189.

Eckart, C., 1951, Surface waves in water of variable depth, *Scripps Instn. Oceanogr. Tech. Ser.*, 51–12, 100.

Emery, K. O., and J. F. Gale, 1951, Swash and swash marks. *Trans. Am. geophys. Un.*, **32**, 31–36.

Favre, H., 1935, *Étude théoretique et experimentale des ondes de translation dans les canaux découverts*, Dunrod, Paris.

Gallagher, B., 1972, Some qualitative aspects of non-linear wave radiation in a surf zone, *Geophys. Fluid. Dyn.*, **3**, 347–354.

Galvin, C. J. Jr., 1972, *Waves on beaches and resulting sediment transport*, ed. Meyer, Academic Press, New York and London.

Guza, R. T., and R. E. Davis, 1974, Excitation of edge waves by waves incident on a beach, *J. Geophys. Res.*, **79**, 1285–1291.

Harris, T. F. W., 1964, *A qualitative study of the nearshore circulation off a Natal beach with a submerged longshore sand bar*, M.Sc. thesis, Univ. of Natal, Durban, S. Africa.

Hasselman, K., W. Munk and G. McDonald, 1963, Bispectra of Ocean Waves, Chap. 6, in M. Rosenblatt (Ed.), *Time Series Analysis*, Wiley, New York.

Huntley, D. A., and A. J. Bowen, 1973, Field observations of edge waves, *Nature*, **243**, 160–162.

Huntley, D. A., and A. J. Bowen, 1975, Field observations of edge waves and a discussion of their effect on beach material, *J. geol. Soc. Lond.*, **131**, 69–81.

Ijima, T., T. Matsuo and K. Koga, 1970, Equilibrium range spectra in shoaling water, *Proc. Twelfth Conf. cst. Engng., Washington D.C.*, Am. Soc. civ. Engrs., pp. 137–150.

Inman, D. L., and N. Nasu, 1956, Orbital velocity associated with wave action near the breaker zone, *Beach Erosion Board, Corps of Engineers, Tech. Memo. 79.*

Iwagaki, Y., and H. Noda, 1963, Laboratory study of scale effect in two-dimensional beach processes, *Proc. Eighth Conf. cst. Engng., Berkeley, Calif.*, The Engineering Foundation Council on Wave Research, pp. 194–210.

King, C. A. M., 1972, *Beaches and Coasts*, 2nd ed., Edward Arnold, London.

Kinsman, B., 1960, Surface waves and short fetches and low wind speeds—A field study, Chesapeake Bay Institute, *Technical Report 19.*

Kinsman, B., 1965, *Wind Waves: their generation and propagation on Ocean Surface*, Prentice-Hall, Englewood Cliffs, N.J.

Komar, P. D., 1971, Nearshore cell circulation and the formation of giant cusps, *Bull. geol. Soc. Am.*, **82**, 2643–2650.

Komar, P. D., 1972, Nearshore currents and the equilibrium cuspate shoreline, *Tech. Rep. Dept. Oceanography, Oregon State University, 239.*

Komar, P. D., and M. K. Gaughan, 1972, Airy wave theory and breaker height prediction, *Proc. Thirteenth Conf. cst. Engng., Vancouver, B.C., Canada*, Am. Soc. civ. Engrs., Vol. 1, 405–418.

Lamb, H., 1932, *Hydrodynamics*, 6th ed., Cambridge University Press.

Lau, J., and B. Travis, 1973, Slowly varying Stokes waves and submarine longshore bars, *J. geophys. Res.*, **78**, 4489–4497.

Longuet-Higgins, M. S., 1953, Mass transport in water waves, *Phil. Trans. R. Soc. A*, **245**, 535–581.

Longuet-Higgins, M. S., 1970a, Longshore currents generated by obliquely incident sea waves, Part 1, *J. geophys. Res.*, **75**, 6778–6789.

Longuet-Higgins, M. S., 1970b, Longshore currents generated by obliquely incident sea waves, Part 2, *J. geophys. Res.*, **75**, 6790–6801.

McKenzie, R., 1958, Rip current systems, *J. Geol.*, **66**, 103–113.

Nagata, Y., 1964, Deformation of temporal pattern of orbital wave velocity and sediment transport in shoaling water, in breaker zone and on foreshore, *J. oceanogr. Soc. Japan*, **20**, 7–20.

Patrick, D. A., and R. L. Weigel, 1955, Amphibian tractors in the surf, *First Conf. on Ships and Waves*, The Engineering Foundation Council on Wave Research and the Am. Soc. Naval Architects and Marine Engrs, pp. 397–422; (see also Weigel, 1964).

Phillips, O. M., 1966, *The Dynamics of the Upper Ocean.* Cambridge University Press.

Russell, R. C. H., and J. D. C. Osorio, 1958, An experimental investigation of drift profile in a closed channel, *Proc. Sixth Conf. cst. Engng. Berkeley, Calif.*, The Engineering Foundation Council on Wave Research, pp. 171–93.

Sandover, J. A., and O. C. Zienkiewicz, 1957, Experiments on surge waves, *Water Power*, **9**, 418–424.

Shepard, F. P., K. O. Emery and E. C. Lafond, 1941, Rip currents: A process of geological importance, *J. Geol.*, **49**, 337–369.

Shepard, F. P., and D. L. Inman, 1950, Nearshore circulation related to bottom topography and wave refraction, *Trans. Am. geophys. Un.*, **31**, 555–565.

Shepard, F. P., and D. L. Inman, 1951, Nearshore circulation, *Proc. First Conf. cst. Engng.*, 50–59, University of California, Council on Wave Research.

Shonting, D. H., 1968, Autospectra of observed particle motions in wind waves, *J. mar. Res.*, **26**, 43–65.

Sonu, C. J., 1968, Collective movement of sediment in littoral environment, *Proc. Eleventh Conf. cst. Engng., London.* Am. Soc. civ. Engrs., pp. 373–400.

Sonu, C. J., S. P. Murray, S. A. Hsu, J. N. Suhayda and E. Waddell, 1973, Sea breeze and coastal processes, *Trans. Am. geophys. Un.*, **54**, 820–833.

Taira, K., 1971, Wave particle velocities measured with a Doppler current meter, *J. oceanogr. Soc. Japan*, **27**, 218–232.

Tam, C. K. W., 1973, Dynamics of rip currents, *J. geophys. Res.*, **78**, 1937–1943.

Tucker, M. J., N. D. Smith, F. E. Pierce and E. P. Collins, 1970, A two-component electromagnetic ship's log, *J. Inst. Navig.*, **23**, 302–316.

Ursell, F., 1952, Edge waves on a sloping beach, *Proc. R. Soc. A.*, **214**, 79–97.

Webber, N. B., and G. N. Bullock, 1969, A model study of the distribution of run-up of wind-generated waves on sloping sea walls, *Proc. 11th Conf. cst. Engng., London* Am. Soc. civil Engrs., pp. 870–887.

Weigel, R. L., 1964, *Oceanographical Engineering*, Prentice-Hall, Englewood Cliffs, N.J.

Wells, D. R., 1967, Beach equilibrium and second order wave theory, *J. geophys. Res.*, **72**, 497–504.

Yefimov, V. V., and G. N. Khristoforov, 1971, Spectra and statistical relations between the velocity fluctuations in the upper layer of the sea and surface waves, *Atmos. Oceanic Physics*, 7, 1290–1310.

Zabusky, N. J., and C. J. Galvin Jr., 1970, Shallow water waves. The Korteweg-de-Vries Equation and Solitons, *Bell Telephone Laboratories Report, PCP-70-26.*

Discussion

Dr. P. Holmes, Department of Maritime Engineering, University of Liverpool. Is there a possibility that subharmonic oscillations will result from a linear or non-linear 'forward process', i.e. breaking and run-up, with a 'feedback' process, i.e. back-rush, which modifies the input?

Yes, certainly. This seems to be the kind of interaction process that we have observed in the surf zone. It is perhaps not surprising that the result of this 'feedback' process is an edge wave, since an edge wave is a free mode of oscillation of the nearshore water.

Dr. P. H. Kemp, Department of Civil and Municipal Engineering, University College, London. We made a number of observations of beach cusps on Chesil Beach during a study of the effect of wave characteristics on beach profiles. The cusp length data collected at that time did not seem to correlate with dominant wavelengths predicted by the edge wave equation presented in your paper. The cusp lengths did seem to bear a systematic relationship with the breaker height. Could you comment on whether you observed beach cusps during your measurements and, if so, whether they could be said to originate from edge waves?

At the time of our measurements there was no cusp formation on the beach at Slapton, probably because of a longshore current caused by obliquely incident wind waves.

Previously, we had seen cusps on Slapton beach of considerably shorter wavelength than the observed edge waves, but rough calculation suggests that they would be comparable with the expected wavelength of an edge wave of the incident wave period, not of this subharmonic wave period.

As the dispersion curves show, there are many different edge wave modes for a given wave period, each with a different wavelength. It is possible that an apparent correlation between cusp length and breaker height may be due to the excitation of higher order modes by higher breakers.

Nevertheless, definitive results linking edge wave characteristics to periodic sedimentary features are still missing. We have recently obtained field measurements of the velocity field on a beach during the formation of rip currents and a cuspate shoreline, but these data have not yet been analysed.

W. A. Price, Hydraulics Research Station, Wallingford, Berks. I was interested to hear that maximum uprush does not necessarily tie up with maximum velocities. This observation ties up to some extent with our observations that in testing sea walls for wave overtopping in flumes with *random* seas the highest waves in the wave train do not necessarily produce the most overtopping.

Observations of this kind suggest that realistic wave trains must be used in model studies, rather than monochromatic waves, if the results are to be applicable in the field.

CHAPTER FIVE

Trends in the Application of Research to Solve Coastal Engineering Problems

D. H. WILLIS and W. A. PRICE

Abstract

This paper reviews some of the research that has been done at the Hydraulics Research Station in the coastal engineering field in recent years. The Station, in trying to effect the better design of sea defences, has encouraged certain authorities to solve their coastal erosion problems by beach nourishment rather than by conventional means. Two such schemes are described, one at Bournemouth, England, and the other at Portobello, Scotland.

A mathematical model is described that is capable of predicting changes in the plan shape of a beach following the construction of sea defences or an alteration in the wave climate. The importance of the method in estimating coastal changes due to offshore dredging is mentioned and the future development of the mathematical approach is outlined.

Introduction

At the Hydraulics Research Station, Wallingford, a part of our research effort is directed towards developing better methods of coastal protection. Two lines of research are worth mentioning. These are a method of dealing with coastal erosion problems by the nourishment of an eroding beach from material dredged offshore—a procedure already widely adopted in the United States but only recently applied in Great Britain—and the development of beach mathematical models.

Beach nourishment

In order that a beach should remain stable, the littoral drift of material (the longshore drift that takes place under wave action) should be such that the amount of material leaving a beach section equals that being supplied. A considerable quantity of beach material results from cliff and dune erosion so that the stabilization of these features leads in the long run to beach starvation. This is one of the reasons we have a coastal erosion problem on a large scale.

A common solution is to build a series of groynes backed by a massive sea-wall. The groynes arrest the movement of a certain amount of material that would otherwise travel along the shore. Where the supply of material has been stopped or has been markedly reduced, it follows that groynes are not necessarily a good solution. It is surprising how often this is not understood and, even if it is, how seldom a programme of groyne construction is carried out in association with beach nourishment.

Some years ago Bournemouth Corporation were faced with a beach erosion problem and decided to carry out a pilot beach nourishment scheme. The first task was to find a borrow site where slightly coarser material was present than existed on the beach. Coarse material tends to move in the layers close to the seabed. Here, not only is the drift of water normally towards the shore but the orbital velocities are higher towards the beach than away from it. Hence, coarse material has a greater tendency to move shoreward than finer material which is entrained in the mid-depth layers where the net drift of water is away from the beach.

In the 1970 pilot experiment at Bournemouth 110,000 m^3 of sand were deposited within 200 m of the shoreline by a trailing hopper suction dredger. A smaller dredger then continuously pumped some of this material ashore along a 1·5 km length of beach. Although the beach has since been subjected to considerable storm wave activity, the new beach levels have been well maintained.

Following the success of the pilot experiment at Bournemouth, a more ambitious beach nourishment scheme was carried out at Portobello, near Edinburgh, Scotland. Here 200,000 m^3 of sand were pumped ashore in 1972 and the early results indicate that this scheme, too, is a success (Newman, 1974).

As longer lengths of our shoreline is protected, coastal erosion problems in some other areas will be aggravated. Based on the results of the two beach nourishment schemes described, it would seem realistic to look to the future when perhaps many more of our coastal erosion problems will be resolved by specially designed dredgers carrying out 'topping up' operations of beaches on a national scale. Further information on these schemes may be obtained in Newman (1974).

Beach mathematical models

Price, Tomlinson and Willis (1972) presented a paper to the 13th Conference on Coastal Engineering which described development of a mathematical model for predicting the plan shape of beaches. At that time, the model had a serious limitation in that it could not take account of changes in the wave refraction pattern resulting from changes in the beach. Subsequently, a simplified wave refraction calculation capable of reproducing the effect of beach changes on the waves as well as the effect of waves on beaches was incorporated in the mathematical model. The result was a much more comprehensive model.

Beach plan shape

Plan shape may be defined as the shape, viewed from above, of a beach contour between the breaker line and the swash limit. The derivation presented here deals with the mean sea level contour, but this is by no means the only one which could have been used. The model is thus two-dimensional, since the shape of this contour can be described by an equation in X and Y coordinates only.

The plan shape calculation is essentially a finite difference solution of the continuity equation in the alongshore direction:

$$\frac{\partial Q}{\partial x} + \frac{\partial A}{dt} = 0$$

If a depth D can be assumed, below which beach changes resulting from alongshore transport are negligible, then

$$\partial A = \frac{D}{2} \partial y$$

Wave basin studies indicated that D is approximately equal to twice the breaking wave height, but in practice it is assumed to be constant, that is, independent of wave height, over the model. The continuity equation thus becomes:

$$\frac{\partial Q}{\partial x} + \frac{D}{2}\frac{\partial y}{\partial t} = 0 \tag{5.1}$$

The solution of this equation requires a means of calculating the rate of alongshore sediment transport, Q, given wave and beach properties at an instant in time. Many formulae have been developed for this purpose. The one currently used is the Scripps Institute of Oceanography Equation, as modified by Komar (1969). This was developed for sand beaches on the coast of California, but has also given reasonable results when applied to small-scale beaches at the Hydraulics Research Station.

$$I_L = 0{\cdot}35E(Cg) \sin 2\alpha$$

Converted to a volume rate of transport, this becomes:

$$Q = \frac{0{\cdot}35}{\gamma_s} E(Cg) \sin 2\alpha \tag{5.2}$$

in which γ_s = submerged density of beach material in place. Except for one or two special cases, for example, when wave energies are uniform along the beach, Equations 5.1 and 5.2 are incapable of direct solution for y and must be solved numerically using a digital computer. The differential equations must then be converted to difference equations, with finite steps, denoted by Δ, replacing the infinitely small steps, denoted by ∂. The corresponding system of difference equations is:

$$\frac{Q[n+1, t] - Q[n, t]}{\Delta x} + \frac{Dy[n+\frac{1}{2}, t+\frac{1}{2}] - y[n+\frac{1}{2}, t-\frac{1}{2}]}{2\Delta t} = 0$$

or:

$$y[2+\tfrac{1}{2}, t+\tfrac{1}{2}] = y[n+\tfrac{1}{2}, t-\tfrac{1}{2}] - \frac{2\Delta t}{D\Delta x}(Q[n+1, t] - Q[n, t]) \tag{5.3}$$

$$Q[n, t+1] = \frac{0{\cdot}35\ E[n](Cg}{\gamma_s} \sin 2\alpha[n, t+1] \qquad (5.4)$$

$$\alpha[n, t+1] = \alpha_x[n] - \tan^{-1} \frac{y[n+\frac{1}{2}, t+\frac{1}{2}] - y[n-\frac{1}{2}, t+\frac{1}{2}]}{\Delta x} \qquad (5.5)$$

in which $[n, t]$ refers to the number of Δx and Δt steps, respectively, from the origin, and α_x is the angle between the breaking wave front and the X axis (see Figures 5.1 and 5.2).

The computation procedure is as follows:

(1) Using existing y values, calculate rates of alongshore transport at $t = 0$ from Equations 5.4 and 5.5.
(2) Applying the continuity equation (5.3) claculate new y values between the points of calculated transport rates, half time step forward.
(3) Recalculate rates of alongshore transport from Equations 5.4 and 5.5 at $t = 1$ and return to (2).

The finite difference scheme represents the differential equation only so long as the calculation is 'stable', that is, when errors inherent in the scheme diminish with successive calculation steps. In practice this means that the time step, Δt, must be small enough to ensure that the beach does not oscillate with increasing amplitude about its equilibrium position. With some difference schemes, for example that of Bakker (1968, 1970), the stable time step may be calculated theoretically from the data. The present scheme is rather more empirical than Bakker's and requires that the time step be determined by trial and error. The first step in the beach plan shape calculation is to calculate transport rates at time $t = 1$, using the largest practicable time step. If at any section, $Q[n, 1]$ is the opposite direction and equal to or greater in magnitude than $Q[n, 0]$, then the calculation is unstable. The time step is reduced by half and new trial '$Q[n, 1]$'s are calculated, This is repeated until a stable time step is found. Although this may seem a complicated procedure, its effect is more than compensated for by the speed and accuracy of the programme.

Wave refraction

Clearly the plan shape calculation requires breaking wave heights and directions along the beach. These are obtained from deep water wave conditions by the process of refraction. In deep water, the waves are assumed to be of uniform height and direction. However, as they pass over the seabed travelling towards the shore, they are modified. For the assumption of straight parallel contours, they obey Snell's Law of Optics.

$$\frac{\sin \alpha'}{C} = \text{Constant} \qquad (5.6)$$

in which α' is the angle between the wave front and any depth contour, and C is

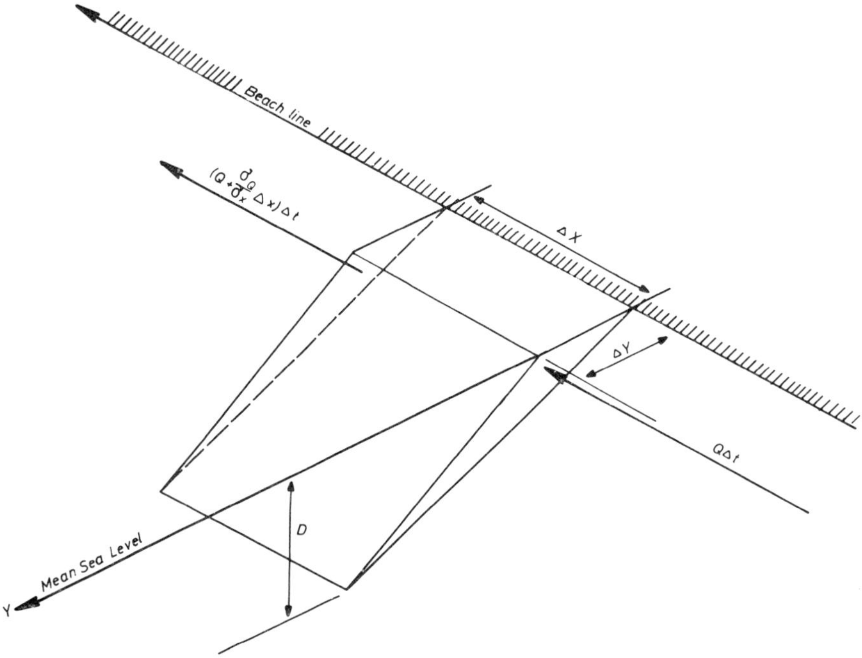

Figure 5.1.

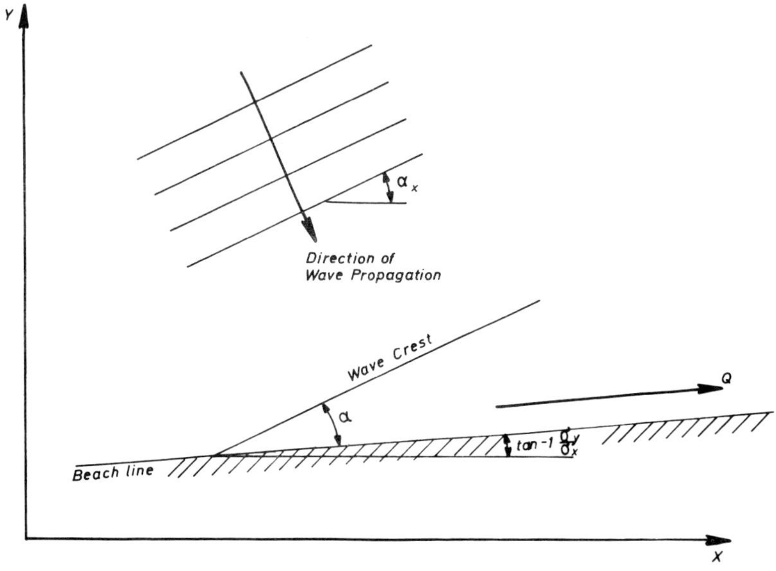

Figure 5.2.

the phase velocity of the waves at this depth. If wave energy can only propagate perpendicular to the wave front (a necessary and reasonable assumption for the use of Snell's Law) then energy will be concentrated in areas where the refracted wave front is concave shoreward, that is, wave heights will be higher here than where the wave front is convex shoreward. In this way uniform deep water wave conditions are converted to non-uniform breaking waves at the beach.

The beach mathematical model includes a simplified version of the Hydraulics Research Station's wave refraction mathematical model developed by Abernethy and Gilbert. Wave orthogonals, lines everywhere perpendicular to the wave fronts, are calculated over an equilateral triangular grid of water depths from deep water to the beach. In each triangle, a plane of wave phase velocity can be defined by the depths at the three apices. Over this plane, that is within the triangle, it can be shown that an orthogonal travels in a circular arc having its centre on the line of zero phase velocity (see Figure 5.3). The model proceeds from triangle to triangle up to the point of breaking, usually when the water depth is approximately equal to the wave height or a little greater, where the direction of the breaking wave, α_x, the wave energy, E, and the group velocity, (Cg), are calculated. From a number of such orthogonals the variation of breaking wave height and direction along the beach is determined.

The model goes through this refraction procedure each time the deep water wave climate is changed. As the beach plan shape is altered, offshore depth contours will also be slightly changed affecting the pattern of refraction waves

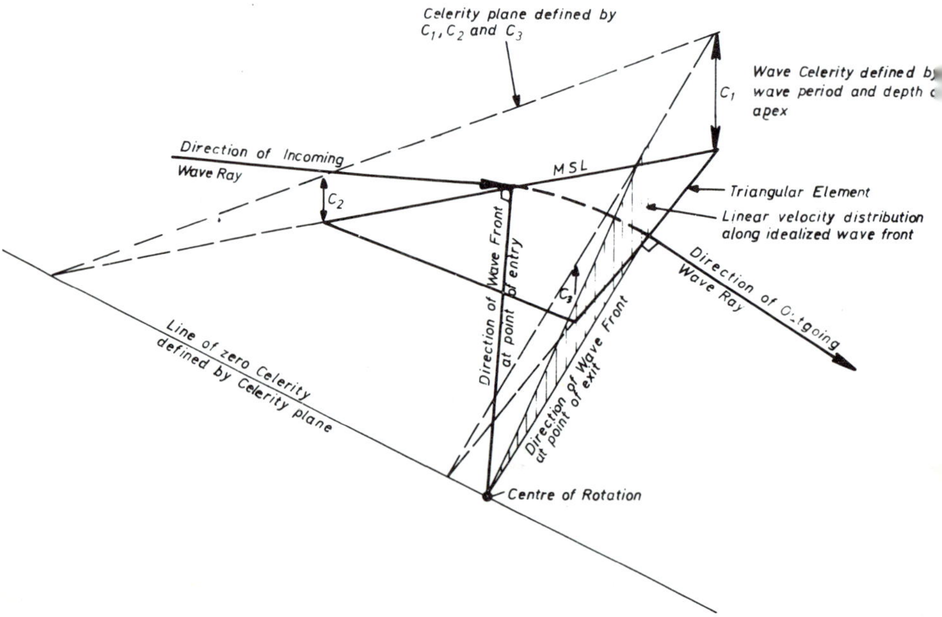

Figure 5.3.

arriving at the beach over the following period. In this manner, waves and beach plan shape tend towards a sort of equilibrium with time in which variations in alongshore sediment transport are minimized along the beach.

Model operation

In principle, the model performs the following operations:

(1) Calculates breaking wave conditions from deep water wave conditions by refraction over the inshore seabed.
(2) Calculates rates of alongshore sand transport, Q, on the beach from the breaking wave conditions, using Equations 5.4 and 5.5.
(3) Determines stable time step, Δt, by trial and error.
(4) Calculates amounts of accretion and erosion, Δy, from the rates of transport, using Equation 5.3.
(5) Distributes accretion and erosion over the inshore seabed, to depth D, as in Figure 5.2.
(6) Recalculates wave refraction and returns to (2).

In addition to specifying deep water wave conditions, the initial beach shape and the initial inshore seabed, it is also necessary to specify two boundary conditions. These may be of the following forms:

(1) *Free boundary*. This approximates to the condition where the non-model side of the boundary is an infinitely long straight beach with uniform breaking wave conditions. The boundary is allowed to erode or accrete and to transmit sediment. Mathematically, this is accomplished simply by forcing the transport rate at the boundary to be equal to that at the first section in from the boundary. This is the most commonly used type of boundary since it can also adequately describe conditions at a river mouth supplying more than the beach requires.
(2) *Fixed boundary*. Any point on the beach can be fixed, that is prevented from accreting or eroding. This may be the case at a rocky outcrop, for example.
(3) *Specified transport boundary*. The rate at which sand enters or leaves the model through a boundary can be specified simply by specifying Q at the boundary. A special case, that of no sand transport, is commonly used to describe the situation at a long groyne or jetty at right angles to the beach. The more general case applies to a river supplying less sand than the beach requires.
(4) *Inerodible boundary*. This is a hybrid of the free and fixed boundaries, useful for describing a beach backed by a seawall or rock cliffs. It behaves as a free boundary, except when it is required to erode landward of a specified point, where it becomes fixed. Just as with a fully-fixed boundary, this may be used at any point along the beach, not only at the ends.

Other boundary conditions can be imagined but are difficult to justify physically. For example, a non-accreting boundary analogous to the inerodible boundary could be put in the model, but could only be described by the situation of a dredger or scraper operating on a beach to remove any accretion that takes place.

The mathematical model is used in the same way as a physical model of the same situation would be. An attempt is first made to prove the model, that is, to reproduce existing conditions. This involves a certain amount of data smoothing and the selection, by a process of trial and error, of deep water wave conditions, boundary conditions, and ratio of breaker depth to breaker height such that the existing beach shape remains relatively stable with time. Once this condition is achieved, the proposed alterations can be put in the model and their effects on the beach plan shape studied.

Further development

The mathematical model was developed for use in studying the possible effects of offshore dredging on adjacent beaches. It is now adequate for that purpose and is being applied on behalf of the Crown Estate Commissioners, the body responsible for licensing the winning of sand and gravel off the coast of the British Isles. However, two aspects are being developed further:

(1) Because the present mathematical model is two-dimensional, it can only describe the behaviour of a single contour in the breaker zone. Clearly there would be some advantage in modelling the entire nearshore seabed, that is, in making the model three-dimensional. Research is therefore being carried out into the transport of sediment in the onshore–offshore direction and the response of beach profiles to wave action.

(2) Abenerthy and Gilbert of the Hydraulics Research Station have recently developed a method for calculating the refraction of an entire directional wave spectrum, rather than a single component. The refraction segment of the beach mathematical model is being modified to make use of this development.

Notation

Δ = beach cross-sectional area
C = phase velocity of the wave at a given depth
(Cg) = group velocity of waves at breaking
D = depth
E = energy density of breaking waves, $\frac{1}{8}\rho g H^2$
g = acceleration due to gravity
H = wave height, trough to crest, at breaking
I_L = submerged weight rate of alongshore sediment transport
Q = volume rate of alongshore sediment transport
t = time
x = distance in the alongshore direction
y = distance normal to the beach

α = angle between breaking wave front and the beach
γ_s = submerged density of beach material in place
α' = angle between wave front and any depth contour
α_x = angle between breaking wave front and X axis
ρ = mass density of water

References

Bakker, W. T., 1968, The dynamics of a coast with a groyne system, *Proc. 11th Conf. cst. Engng., London*, Vol. 1, pp. 492–525.

Bakker, W. T., 1970, The influence of diffraction near a harbour mole on the coastal shape, *Studierapport W W K 70–2*, Rijkswaterstaat, The Netherlands.

Komar, P. D., 1969, *The longshore transport of sand on beaches*, Ph.D. thesis, University of California, San Diego.

Newman, D. E., 1974, A beach restored by artificial nourishment, *Proc. 14th Conf. cst. Engng., Copenhagen.*

Pelnard-Considere, R., 1956, Essai de theorie de l'évolution des formes de rivage en plages de sable et de galets, *Quatriemes Journées de l'Hydraulique, Paris, Les Énergies de la Mer*, Question III, rapport 1, Vol. 1, pp. 289–298.

Discussion

Dr. P. H. T. Beckett, Department of Agricultural Science, University of Oxford. Will you indicate the quantities of material restored to the Bournemouth and Portobello beaches?

120,000 m^3 of sand was put ashore at Bournemouth and 200,000 m^3 at Portobello. The average cost of placing this material on the beach was about £1 per m^3.

E. W. M. Gifford, Gifford and Partners, Southampton. Can Mr. Price confidently predict the depth of water beyond which wave action will not affect gravel movement? Is it not possible that the effect of exceptional gales could be very important?

Using radioactive pebbles placed at 4 depths we observed gravel movements over two winters at Worthing, and during this time the area was subjected to severe storms. At a depth of 18 m no shingle movement was recorded. We feel therefore that from source considerations alone it is safe to allow dredging in depths of water greater than 18 m, for sites with similar wave climates to Worthing.

P. Parkinson, British Gas Council, formerly Nature Conservancy, Newbury, Berkshire. Oil and gas developments on the east coast of Scotland require sand dredging for reclamation schemes. It is possible to predict adequately the effects of offshore dredging, as in the area north of Peterhead?

Yes, it is possible to predict the effect of offshore dredging by using our wave refraction programme linked to the beach mathematical model.

P. Sargent, Nature Conservancy, Newbury, Berkshire. The over-use and misuse of groynes for coast protection purposes is an accepted fact. However, is beach replenishment the complete answer or is it still a relatively short-term solution? For example, how often is it necessary to top-up replenished beaches? Replenishment schemes of the type used at Dungeness would seem to be long-term continuing and expensive commitments.

Obviously beach nourishment is not the solution to a beach erosion problem for ever. However, it is our experience that when a coarse material is used for replenishment purposes the beach drawn-down during the winter is very much less than with the indigenous material. It is not possible to say how often a beach would have to be renourished but even if it had to be done, say, every 10 years, it might still be a more economical solution than groynes backed by a seawall. Even the losses from re-nourished beaches benefit downdrift beaches—the disadvantage to groynes is that they produce downdrift erosion.

E. W. M. Gifford. Do you operate your mathematical model with tides?

No. It is important to understand that we use bulk flow sediment transport equations in our mathematical model. We would have to be able to describe the sediment transport at all points on the beach to be able to build in the effects of tides. This we hope will be possible in the next generation of beach mathematical models.

Dr. A. H. W. Robinson, Department of Geography, University of Leicester. Hallsands provides a classic example of an offshore zone that does not affect the adjacent beach. Even the presence of the Skerries sandbank 1·5 km offshore has not provided sediment for the depleted beach. Hence it has not improved since the artificial removal of shingle at the beginning of the century. It would seem that whatever the influence the Skerries have in relation to wave refractive effects on the beaches of Start Bay, its own changing configuration is due to tidal currents—a further proof of the complexity of nearshore sediment dynamics.

Hallsands is often quoted as a classical example of beach erosion following offshore dredging. However, no-one these days would ever license people to dredge quite so close inshore.

Dr. J. R. Hails, Institute of Oceanographic Sciences, Taunton. With reference to your comments on dredging in Start Bay opposite Hallsands, I would like to add that the extraction of gravel was extended to the intertidal zone. The geophysical and geological work completed by the Institute of Oceanographic Sciences, Taunton, shows that there are three discrete lithological units within the Bay and that gravel extends only a few metres beyond LWM spring tides. There is no conclusive evidence to suggest that material is entering the Bay either alongshore or onshore–offshore. In fact, it might be claimed that Start Bay is a closed system. It is suggested therefore that where such situations exist, offshore dredging of gravel should be limited to the depths you mentioned. In conclusion, I would like to refer to the work that Professor Holmes has completed for us on wave refraction patterns and wave energy dissipation. There seems little doubt that Hallsands village was sited on a sector of the coast which is particularly vulnerable to northeasterly storm wave attack.

I think the problem of offshore dredging is a very serious one. Because it is sand and gravel and hence relatively cheap, there is no pressure to restrict its export—the Dutch are buying huge quantities of it. There is not an infinite supply of it by any means; we should conserve it.

Dr. I. N. McCave, School of Environmental Sciences, University of East Anglia. The beach mathematical model you mentioned is for longshore transport caused by obliquely incident waves. Catastrophic events on coasts producing great losses of sand in a short time must involve sand movement in an onshore–offshore rather than an onshore direction. What is known or understood about this sense of movement, and can it yet be incorporated in mathematical models?

Dr. McCave is right in pointing out that our existing beach mathematical model only deals with sediment transport on a bulk flow basis in the alongshore direction. Before we can attempt to build truly three-dimensional models we need to establish the laws of sediment transport both in waves and currents. We are making a start on this subject ourselves. In connection with our Maplin project in the Thames Estuary a tower has been built in shallow water off the edge of the sands where the effect of waves on sediment transport will be established. Very little has been done to quantify the onshore–offshore transport of sediment—this is an area of research that requires a great deal of effort given to it.

Dr. G. Evans, Department of Geology, Imperial College, London. Can you take account of tides in your calculations—not only moving the breaker zone up and down the shore, but more so the effect of tides flowing alongshore interfering and combining with wave effects as the tide rises. The latter combination seems to make British beaches more difficult to understand compared to the comparatively simple beach regimes of the western USA.

In answer to a previous speaker, I said that we do not include the effects of tides but we realize their importance. If we get some more information on the laws of sediment transport under wave action, especially in the breaker zone, we will be in a very much better position to take fluctuating water levels into account.

CHAPTER SIX

Marine Bars and Nearshore Sedimentary Processes, Kouchibouguac Bay, New Brunswick

BRIAN GREENWOOD and ROBIN G. D. DAVIDSON-ARNOTT

Abstract

Various types of bar topography have been studied and described from a number of different areas. This paper gives a detailed analysis of the morphological and spatial characteristics of an inner and outer systems of bars in Kouchibouguac Bay, New Brunswick, Canada, together with an account of their temporal development. The processes responsible for submarine bar formation in this particular area are discussed in the light of the nearshore sediment budget. Work completed so far shows that, in the case of Kouchibouguac Bay, the direct mechanisms of sediment transport are currents associated with the orbital motion of incident waves and secondary longshore and rip currents.

Introduction

Until relatively recently both theoretical and empirical work on beaches has been restricted to the exposed beach face. However, it seems highly probable that the nearshore submerged topography and the beach face are related dynamically and, if so, the total zone under wave activity should be considered in questions concerning 'beach equilibrium' and 'sediment budget'. Sediment transfer within the zone of wave-induced motion on low angle sandy coasts is, frequently, intimately related to the initiation, growth and migration of large-scale bedforms, known under the general term of subaqueous bars. It is clear that sediment circulation patterns, rates of sediment movement, and the resulting dynamic equilibrium of the beach, are strongly dependent on the morphology and dynamic nature of this irregular bed topography.

Subaqueous bar topography of various kinds has been studied and described from a number of different areas including: the Great Lakes (Evans, 1939, 1940; Davis and McGeary, 1965; Bajorunas and Duane, 1967; Saylor and Hands, 1970; Saylor and Upchurch, 1970; Davis and Fox, 1972); the Baltic (Otto, 1911; Hartnack, 1924; Knaps, 1959); the Black Sea (Egorov, 1951; Zenkovich, 1967);

the North Sea (Bruun, 1954; Bakker, 1968; Dyhr-Nielsen and Sorensen, 1970); the Mediterranean (King and Williams, 1949; Clos-Arceduc, 1962 and 1964); Japan (Hom-ma and Sonu, 1962); the Eastern United States (Sonu and Russell, 1966; Sonu, 1968; Hayes and coworkers, 1969; Niederoda and Tanner, 1970; Niederoda, 1972; Nilsson, 1973; Sonu, 1973); and the Western United States (Shepard, 1950; Bascom, 1953). There has also been some experimental work by Keulegan (1948), Johnson (1949), Nayak (1970), King (1972), Paul and coworkers (1972) and Raman and Earattupuzha (1972), and theoretical work by Clos-Arceduc (1962, 1964), Bowen and Inman (1971, 1972), Bowen (1972), Barcilon and Lau (1973), Lau and Travis (1973) and Sonu (1973) on the origin of certain bar forms.

At present, there is a lack of definition concerning bar types and their relationships to one another, a lack of a strong theoretical framework for explaining bar formation, and only a limited amount of information concerning bar growth, sediment movement and the interrelationships with wave and current patterns. This paper presents a detailed analysis of the morphological and spatial characteristics of a system of bars, together with their temporal development, in Kouchibouguac Bay, New Brunswick, Canada. An assessment of the possible controls on bar morphology and sedimentation is also presented.

Location

Kouchibouguac Bay is located at the western end of Northumberland Strait, New Brunswick (Figure 6.1). A series of barrier spits and barrier islands backed by lagoons and separated by three major inlets fringe the whole bay. These transgressive barriers are marked on the seaward side by a complex marine topography.

The sand-sized material (Table 6.1) present in the Bay is derived from a combination of: (a) cliff and nearshore bedrock erosion near Point Sapin and (b) reworking of the relict Pleistocene sands contained in the transgressive barriers and the associated nearshore zone. Little sand-sized material is presently derived from the small, drowned, catchments draining into the lagoons and most

Table 6.1. Mean and range of size-frequency statistics of sediments from the inner and outer bar systems

	Number of samples	$m\phi$ min.	$m\phi$ av.	$m\phi$ max.	$s\phi$ min.	$s\phi$ av.	$s\phi$ max.	$sk\phi$ min.	$sk\phi$ av.	$sk\phi$ max.
Inner bar–trough system	103	1·61	2·25	2·57	0·24	0·39	0·69	−1.91	−0·55	0·25
Outer bar–trough system	62	0·83	2·12	2·83	0·36	0·53	1·10	−1·18	−0·27	0·42

Note: $m\phi$ = phi mean; $s\phi$ = phi standard deviation; $sk\phi$ = phi skewness; computed using the method of moments.

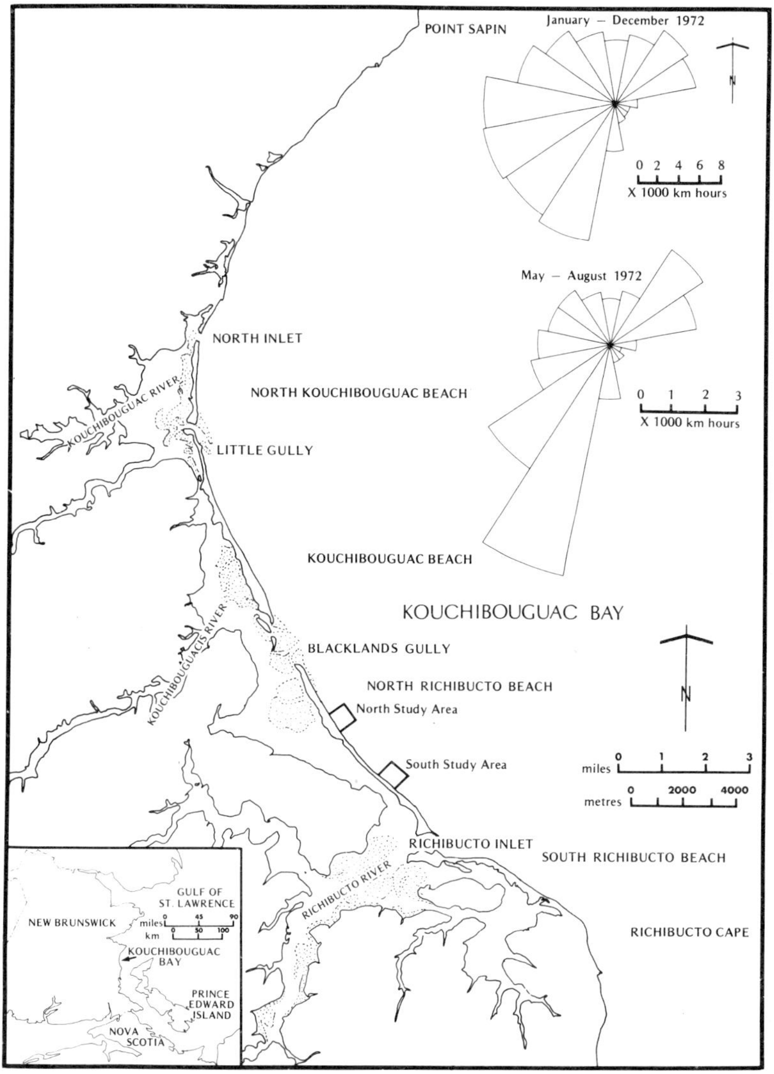

Figure 6.1. Map of Kouchibouguac Bay showing location of study areas. Wind roses are in kilometre hours (sum of average hourly velocities by direction).

of the material represents a relatively thin relict sand body undergoing reworking (Kranck, 1967).

Offshore slopes are approximately 1:1000 from 10 m to 20 m for the whole bay; however, the 0 to 10 m slope is steepest off North and South Richibucto beaches (1:200 compared with 1:500 to 1:330 off North and South Kouchibouguac beaches).

During winter, the nearshore zone is ice-covered for 3–4 months with a well developed ice foot, a factor which clearly may influence the characteristic bar and trough morphology and dynamics at this time.

Tides, winds and waves

The tidal regime in Kouchibouguac Bay is of the mixed, semi-diurnal type and about 80% of the tidal effect in Northumberland Strait is attributable to the lunar semi-diurnal tidal constituent, M_2 (Kranck, 1972). The maximum tidal range is approximately 1·25 m (neap tide range 0·35 m) and fairly strong tidal currents flow parallel to the beach during spring tides (up to 30 cm/s) and in the necks of the tidal inlets (up to $>$1·0 m/s). Kranck (1972) gives the speed of the M_2 tidal constituent (or average maximum tidal current) measured 7 m below the surface in Kouchibouguac Bay ranging between 0·13 m/s and 0·2 m/s. Blackford (1965) documents bottom currents in the Bay moving both northwest and southeast parallel to the coast with a net movement (net water transport) to the southeast and also a distinct onshore component associated with upwelling coincident with offshore winds. Currents flowing southeast had magnitudes 50% greater than those for the northwest component. The general residual current in the Bay is southeasterly (Lauzier, 1965 and 1967) although a flow laterally across the Strait toward the New Brunswick shore has been cited by Kranck (1972) as the means of concentrating sediment near these shores.

The prevailing winds are from the southwest and west (see wind roses Figure 6.1), which are offshore winds occurring for long periods during the summer months. However, the coastal zone in the Bay is affected primarily by waves generated within the Gulf of St. Lawrence (McCann and Bryant, 1972) and by local wave generation within the Strait itself. Waves generated by winds blowing from the northeast (the direction of maximum fetch—approximately 300 km) during the passage of a depression, or refracted through this window, have the most important effect on the nearshore topography. The frequency of strong northeasterlies is shown in the wind roses in Figure 6.1.

The Bay is characterized by short period waves (generally $<$7·0 s) with a steepness of 0·020 or greater. Low energy waves break on the inner bar or at the beach face whilst during storms waves break offshore, and one or more breaker lines are present: maximum wave height measured during storms was 2·8 m with a period not greater than 5–6 seconds. Figure 6.2 documents the wave characteristics for one period in August 1973 together with the coincident wind data. Mean and maximum values of significant wave heights and periods measured each hour are given, together with the significant wave values for the total period. This segment of the wave climate is characteristic of the late summer and early autumn period whilst the late autumn, winter and late spring differs only in the higher frequency of storm waves. Figure 6.3 shows the wind and wave generation associated with the passage of a depression and illustrates a clear correlation between relatively high speed winds from the northeast quadrant and high energy wave conditions.

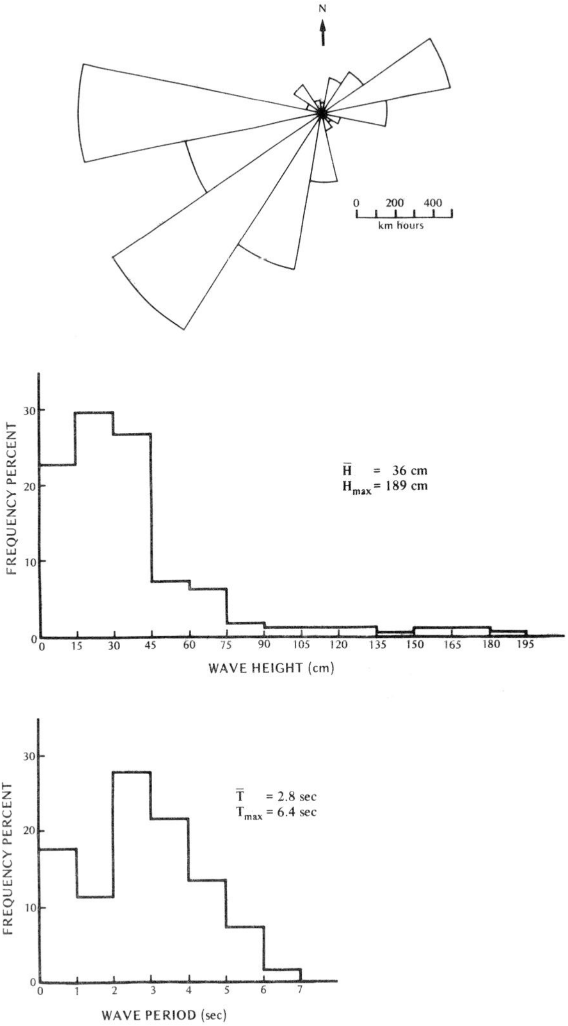

Figure 6.2. Wind and wave characteristics for a period 9–23 August 1973. $\overline{H}$ = average wave height; $H_{\max}$ = maximum wave height; $\overline{T}$ = average wave period; $T_{\max}$ = maximum wave period.

Examination of Figure 6.1 reveals the orientation of the beaches relative to the direction of significant wave generation. The predominant sediment drift in the littoral zone is to the southeast under the influence of waves generated by winds from the northeast. This is supported by data on the net water transport and by shoal orientation, particularly near the inlets (Blackford, 1965). The short-term changes in the nearshore topography to be documented here follow this pattern closely.

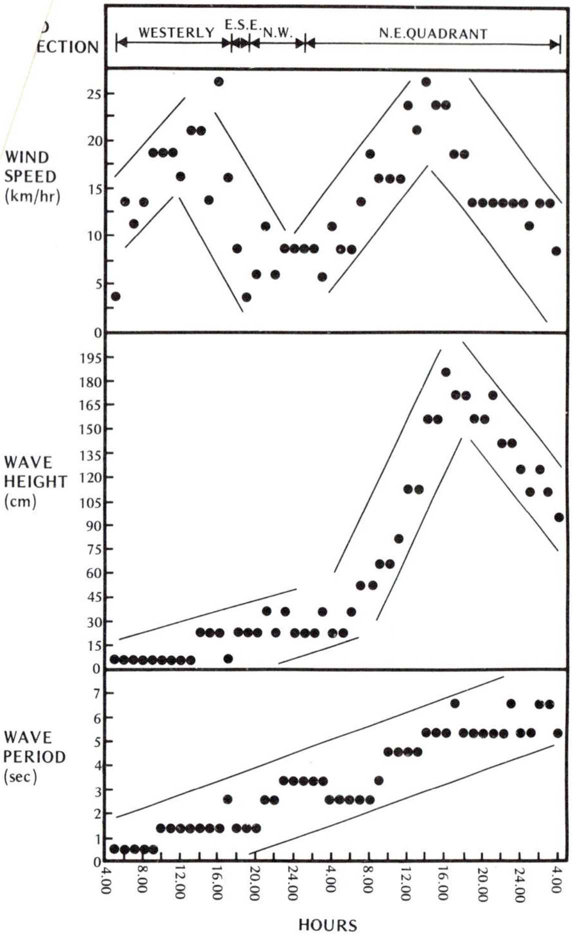

Figure 6.3. Storm wind and wave characteristics, Kouchibouguac Bay, 0500 hours 12 August 1973 to 0400 hours 14 August 1973. Values for grouped data are plotted.

Therefore Kouchibouguac Bay physically satisfies the characteristics common to barred coastlines and frequently quoted as necessary for their development, namely: (a) gentle offshore slopes (1:200 to 1:50), (b) availability of sand-sized material, (c) small tidal range and (d) limited fetch or absence of long period swell.

Field and laboratory procedures

Analysis of aerial photographs (1959, 1965, 1969, 1970, 1971, 1973) and direct aerial reconnaissance were used to determine the major bar patterns along the Bay. Repeated surveys and observations were carried out in two areas, 450 m in length, on North Richibucto Beach and are designated the north and south study areas respectively (see Figure 6.1).

Topographic changes in the inner bar system were monitored by repeated surveys of transects normal to the shoreline and spaced 30 m apart. The profiles were surveyed using an engineer's level and stadia rod and were extended to the limit of wading on the seaward side of the inner bar. Measurement beyond the limit of wading was carried out along alternate lines (60 m spacing) using a Heath Company digital depth sounder with a resolution of 0·3 m and an accuracy of 2·0%. Operation in less than 1·0 m of water was possible. The boat was kept on course by lining up two (1·5 m by 0·15 m) fluorescent orange markers placed on the beach foreshore, and all lines were run from deep water towards the shore. The position of the boat on the line was fixed every 30 seconds by sighting from a theodolite or level on the beach and communication was by two-way radios. A local tide gauge was used during echo sounding to correct for water level changes.

Contour maps of the inner and outer bars were drawn separately from the surveying and echo sounding data and then combined. Because the echo sounding was carried out primarily at high tide and surveying at low tide there was always an area of overlap between the two measurement techniques. Depth measurements were tied to a permanent bench mark on the backshore which was used as the datum for all maps and profiles and is 0·5 m above higher high water. Contours are given in metres below this datum.

Current vectors in the nearshore were determined by tracking drogues from two positions on the beach. Only discrete wave measurements were made during the 1971 and 1972 field season. In 1973 wave height and period were monitored using a continuous resistance wave wire mounted on a tower in 7 m depth of water, operating for a 3-minute period every hour, and output recorded on a Hewlett-Packard 15 cm strip chart recorder. Hourly wind data were obtained from the meteorological station at Chatham (40 km west-northwest of Point Sapin) and correlated with discrete measurements in the Bay itself.

Spatial and morphological characteristics of bar topography

General patterns

Bar topography is a permanent feature of the nearshore bathymetry in Kouchibouguac Bay and although considerable variation in detailed form occurs within short periods of time, each major beach unit appears to have a characteristic pattern which is maintained from year to year.

Broadly speaking two distinct systems of bars are present in any area and it is possible to define an *outer* and *inner* bar system separated by a well developed trough (Sonu, 1973). The latter, is found in water depths less than 2 m. There are considerable morphological and dynamic differences between the two bar systems. Form is far more variable and complex in the inner system and is subject to more rapid change. Furthermore, the inner bar is frequently connected to the beach face by giant cusps. The amplitude of relief is also much smaller in the inner system.

South Kouchibouguac and North Richibucto beaches have well defined inner and outer bar systems, the latter being characteristically crescentic in form

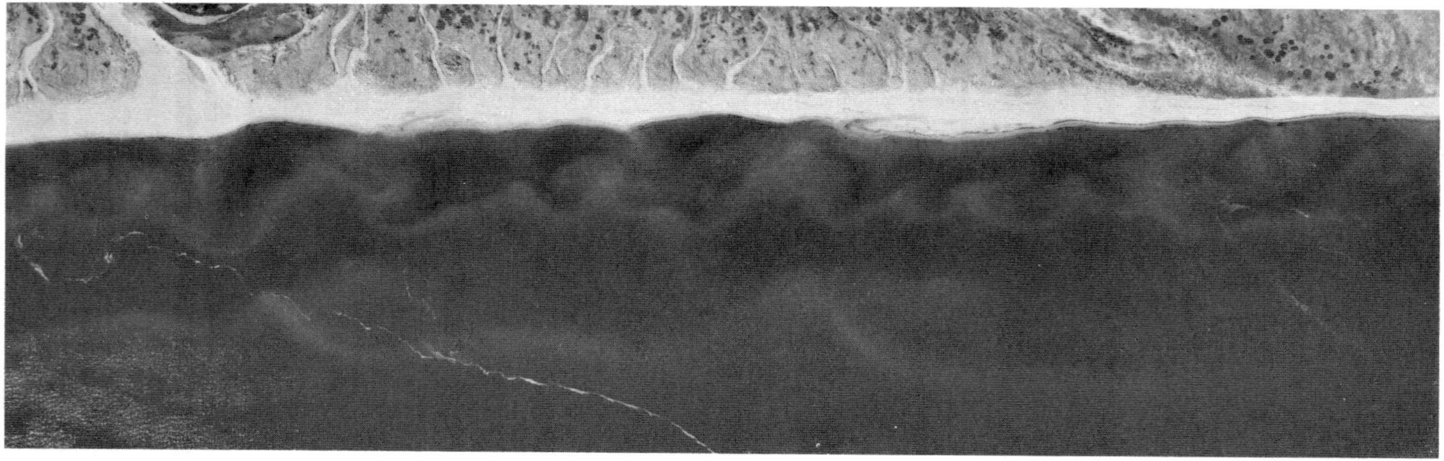

Figure 6.4. Vertical aerial photograph (18 July 1973) of a portion of Kouchibouguac Bay showing complex inner bars and associated rip cells together with a crescentic outer bar, North Richibucto Beach. The photograph includes the north study area and its location is shown in Figure 6.1. (Original photo supplied by National Air Photo Library, Department of Energy, Mines and Resources; ref. A234 35–10.)

(Figure 6.4). The longshore development of the bar systems continues onto the shoal areas at the tidal inlets on the updrift (northern) side and is disrupted for some distance downdrift. However, this simple pattern does not occur at North Kouchibouguac and South Richibucto beaches. These two are characterized by irregular bar development with the former exhibiting numerous gravel bars. Exposed bedrock shelves at shallow depth are common to both locations.

Outer bar system

General characteristics. The outer bars along South Kouchibouguac and North Richibucto beaches generally form a rhythmic pattern of crescents which extends for most of the length of the barrier islands. The bars are similar to those described by Clos-Arceduc (1962, 1964), Hom-ma and Sonu (1962) and Sonu (1973), in that the crescent horns point shoreward and there is a broad shoal area where two crescents join. Unlike the inner system, there is only one outer bar which is usually separated from the inner system, by a deep trough (Figures 6.4, 6.5, 6.6). It is orientated essentially parallel to the shoreline and topographically is not as complex as the inner system. There is considerable variation in crescent size and regularity. While some approach an ideal crescent shape, others may be skewed downdrift, having a wave-like form between shoal areas, or may be elongated with curved ends where they join the next crescent. In many instances the crescents overlap each other where they join at the shoal (Figures 6.4 and 6.6), with the shoal itself extending in a direction opposite to the dominant drift.

There does not appear to be any direct spatial relationship between the outer bar and the inner bar, and beach face (cf. Sonu, 1973). The wavelength of the outer crescent is also normally several times larger than the rip cell spacing of the inner system. Measurements of the wavelength, amplitude and distance offshore of the outer bar crescents along North Richibucto Beach taken from vertical aerial photographs, flown in 1965 and 1971, are presented in Table 6.2.

The general characteristics of one crescent from the north study area in August 1972 and June 1973 are shown in Figures 6.5 and 6.6. The wavelength in 1972 was approximately 500 m and the amplitude 45 m. Water depth over the bar crest was shallowest over the shoal area and deepest near the middle of the bar, the difference being approximately 0·8 m. The actual depth of water over the shoal area at spring low tides was approximately 1·8 m. Unlike the bar described by Sonu and Russell (1966) and Sonu (1973) the trough shoreward of the bar was as deep or deeper near the shoal area as it was opposite the central part of the crescent, and thus the bar relief was greatest at this point, being 2·75 m. The shoreward face of the bar also tended to be steepest at the shoal and the bar profile more asymmetrical. These general relationships appear to hold true for the more complex form of the bar found in 1973 (Figure 6.6) and for other crescents along the beach.

Morphological changes. Examination of aerial photographs of North Richibucto and South Kouchibouguac beaches indicates that although there are changes

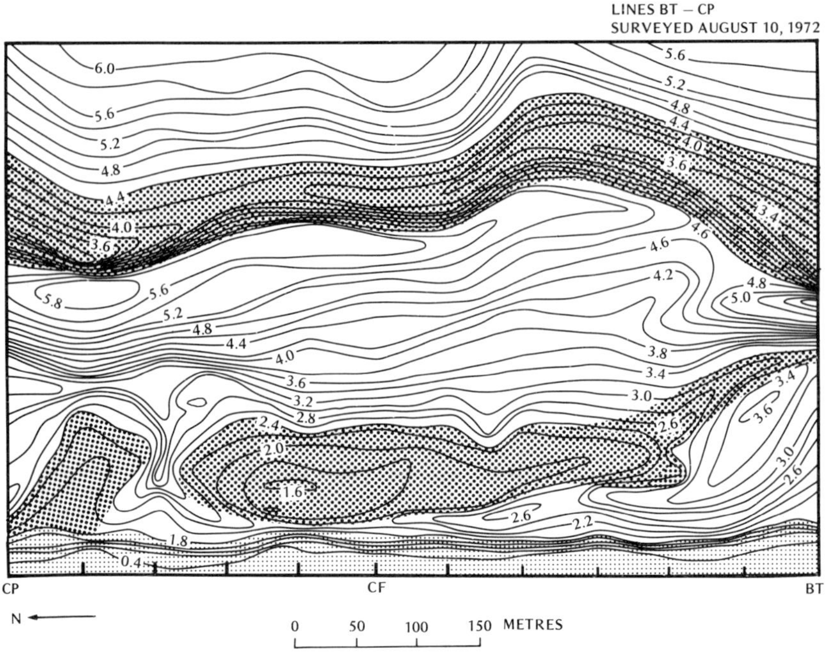

Figure 6.5. Map of the north study area surveyed on 10 August 1972, showing inner and outer bar systems. The beach above the lowest low water mark is indicated by light stippling and the general bar form by dark stippling; heights are in metres below a fixed datum on the beach and contour interval is 0·2 m (similar stippling and contouring are used in all other maps).

Table 6.2. Measurements of wavelength (distance between crescent horns), amplitude (1/2 maximum seaward displacement of centre of crescent from a line joining the two shoal areas) and distance offshore (measured to centre of crescent) of outer bars on North Richibucto Beach.

	Mean (m)	Range (m)	Observations
(a) 1965			
Wavelength	608	288–1232	9
Amplitude	42	28–68	9
Distance offshore	249	200–320	9
(b) 1971			
Wavelength	497	288–738	11
Amplitude	28·5	18–45	11
Distance offshore	260	189–351	11

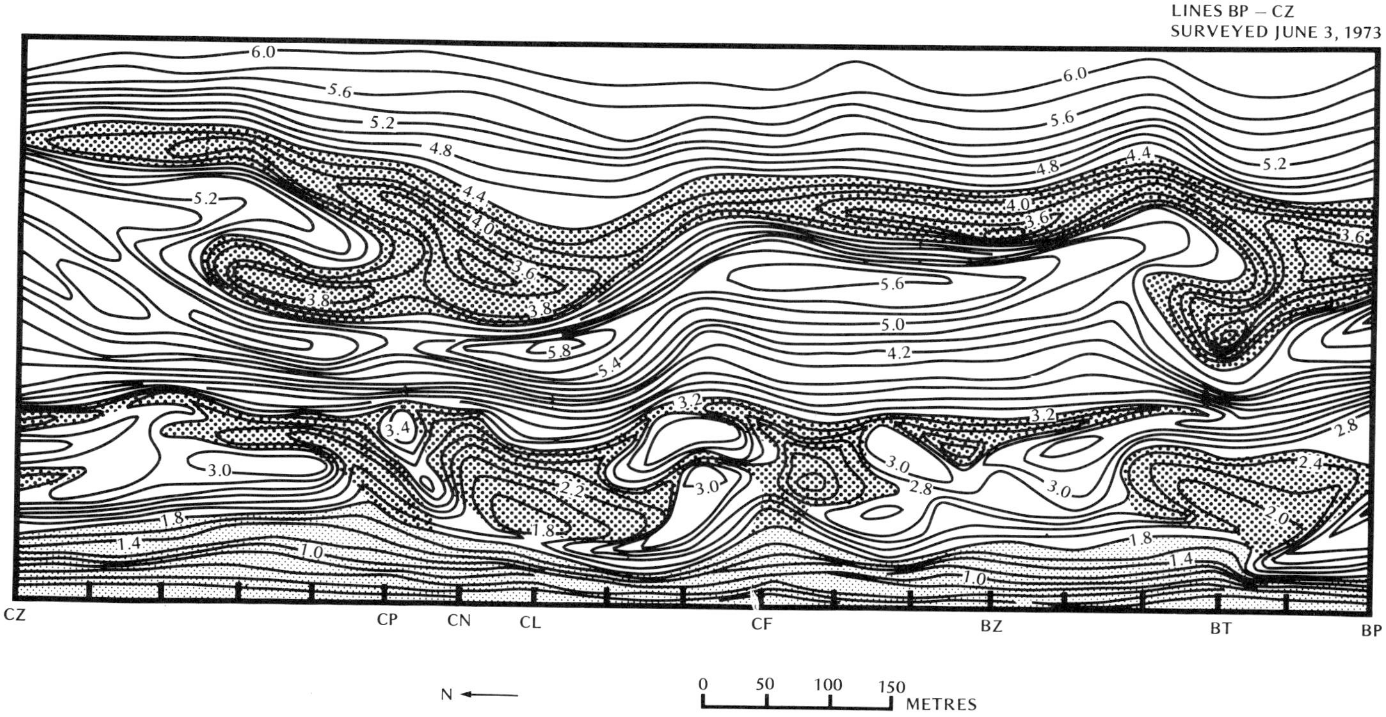

Figure 6.6. Map of the north study area surveyed on 3 June 1973.

in the numbers of crescents along the beach and in the shape of individual crescents, the outer system remains essentially similar in form and position from year to year. This stability of form, compared with the rapid changes that occur in the inner bar systems, has already been noted by previous investigators elsewhere (e.g. Sonu, 1973). The outer bar system does not appear to be greatly altered or destroyed during the winter months when the entire area is ice covered. A comparison of profiles (Figure 6.7) along two lines surveyed in August and November 1972, and April and June 1973 shows that although changes in bar form and position occurred, the bar itself remained as an easily identifiable unit over the period. The survey in November was carried out during the initial stages of freeze-up and the survey in April during ice break-up when there was ice foot extending 50–100 m offshore. Investigation of bars in Lake Michigan (Davis, 1973) and at Richibucto Cape (E. H. Owens, personal communication) support the view that seasonal ice cover restricts normal wave and current processes but does not cause any significant changes in the pre-freeze-up configuration.

Repeated surveys of the outer bar system, however, show that significant movement and alteration of the bar form does occur during the ice-free months.

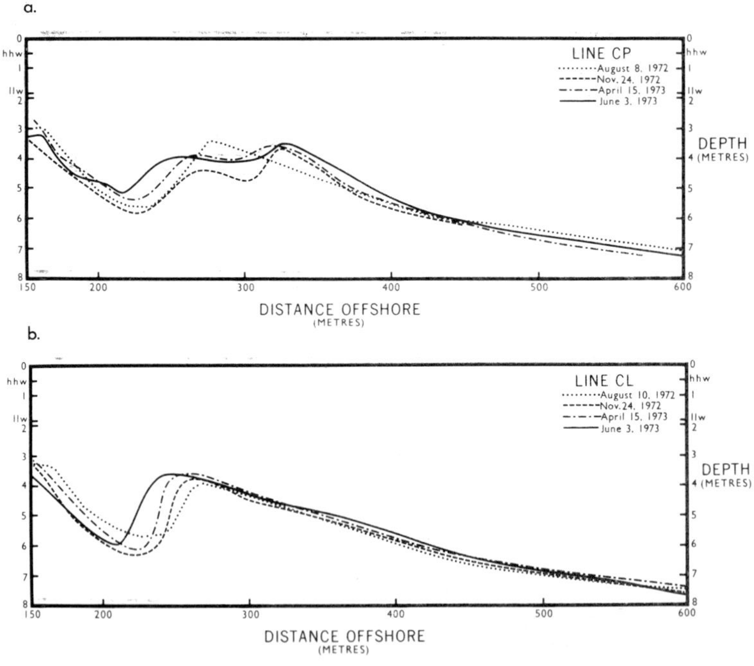

Figure 6.7. Profiles across the outer bar (north study area) during the period 10 August 1972 to 3 June 1973: (a) profile along line CP. Note development of 'double bar' profile between 10 August and 24 November 1972; (b) profile along line CL. Note progressive shoreward displacement of bar crest and increased relief as shoal area moves southward. Location of the profiles is shown in Figure 6.5 and Figure 6.6.

Two types of movement have been identified. The first involves changes in the location of the shoal areas where the bar crescents join. An examination of Figure 6.6 shows the presence of well defined 'tails' on the north side of the shoal area in 1973 which were not present in 1972 (Figure 6.5). A comparison of oblique aerial photographs taken in July 1972 and July 1973 shows the presence of similar 'tails' at several other shoal areas in 1973, which were not present in 1972. The 'tails' appear as the result of a southward shift in the bar such that each crescent overlaps the next crescent southward, forming a new shoal area and leaving the previous shoal as a remnant form. This produces a double bar profile at this point (Figures 6.7(a) and 6.8(a)). The appearance of this feature between August and November 1972 along line CP (Figure 6.7(a)) suggests that it results from movement during one or two major storms, which occur most frequently during the autumn period. Once formed, the extended 'tail' section appears to persist for a considerable period of time. However, surveys of the north study area during 1973 indicated a gradual reduction in height and extent over the summer period.

A second type of movement of the outer bar occurs more regularly over shorter time periods and involves landward or seaward displacement of the bar crest between the shoal areas. There is, however, no net movement of the bar in either direction and, thus, along any one profile the bar crest and form appear to oscillate landward and seaward. Similar oscillations of an outer bar system in Lake Michigan have been noted by Saylor and Upchurch (1970). Displacement of the crest does not take place uniformly along the whole crescent and as a result the crescentic form is frequently distorted (Figures 6.4 and 6.8(c)). In general, the larger the crescent the greater the number of irregularities in the bar crest. This 'wave-like' form and the displacement of the outer bar crest are clearly visible in Figure 6.8(c). Displacement of the outer bar crest results in a shift of the whole bar profile (Figure 6.8(b)). Landward movement of the crest is accompanied by erosion of the seaward slope (and vice versa for seaward movement), thus maintaining an equilibrium profile. Short-term changes in the position of the bar crest are confined primarily to the crescent areas with the shoals remaining relatively stable. Thus, in Figure 6.8(c), the shoal areas appear to act as fixed nodes with 'wave-like' oscillations of the crest between them. Along individual profiles, net amounts of erosion and deposition on the seaward slope of the bar are generally much less across the shoals than the centre of the crescent (Figures 6.8(a) and 6.8(b)).

Comparison of maps of the outer system over a summer period (1973) suggest that there may also be a slow southerly migration, of the order of 10 m per month, of the complete crescent and shoal form. Such movement might be expected under the influence of northeast storms, but the rate of migration tends to be masked by more rapid changes in form and crest position over the period of observation. The rate of movement measured was that for the summer period and it might be reasonable to expect an increase with storm frequency in the autumn (assuming uniformity of movement per storm). Nevertheless, the rhythmic outer bars in Kouchibouguac Bay do not appear to migrate at the high

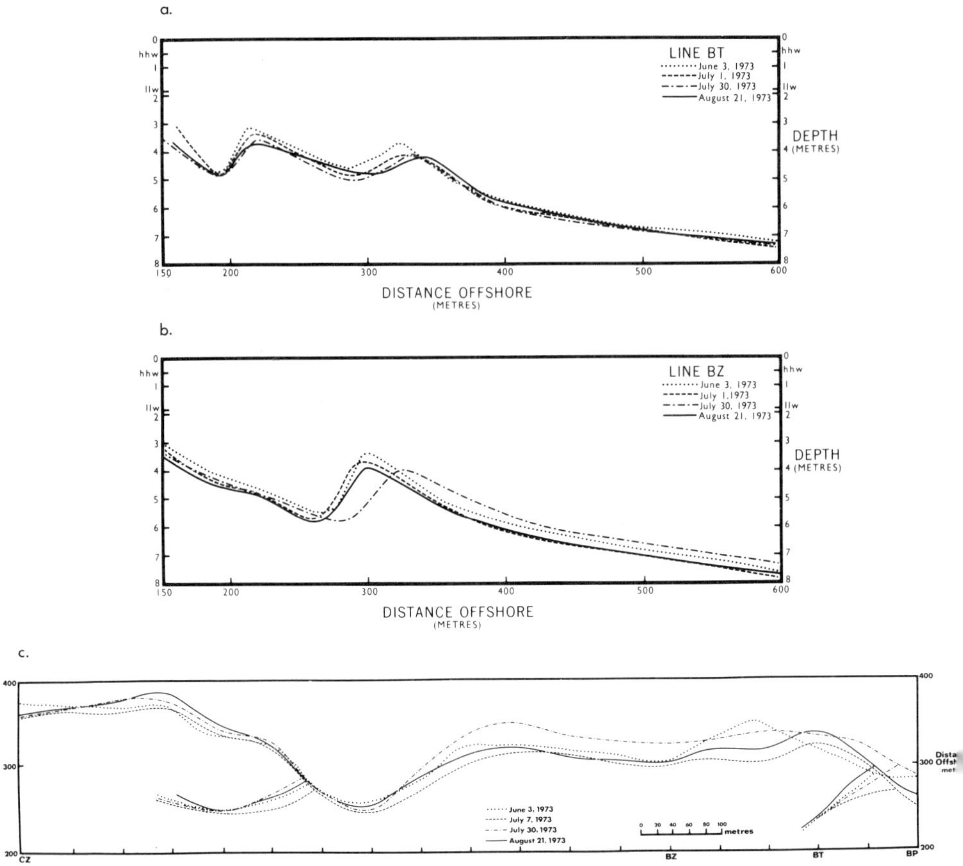

Figure 6.8. Changes in the outer bar system during the period 3 June to 21 August 1973: (a) profile across a shoal area, line BT; (b) profile across centre of a crescent, line BZ; (c) changes in the position of the outer bar crest. Locations of profiles in (a) and (b) are shown in Figure 6.5 and Figure 6.6.

rate noted by Sonu (1961) of 200 m per month off the Niigata coast, Japan, for features of similar size.

Inner bar system

General characteristics. The inner bar system in Kouchibouguac Bay, particularly along South Kouchibouguac and North Richibucto beaches, is more complex and variable than the outer bar system and consists of one, two and occasionally three bars and their associated troughs (see Figure 6.4). The bars are asymmetrical in profile with a steep face on the shoreward side and are normally permanently submerged, but those close to the shoreline may be exposed for short periods during spring low tides. Heights of the bars above the trough seldom exceed 1·0–1·5 m compared to 2·0–3·0 m for the outer bar. During periods with wave heights greater than 0·5 m, waves break on the bar at all stages of the tide and generate longshore currents in the trough. These flow seaward as rip currents, usually in well defined channels. As reported by previous workers (Evans, 1940; Hom-ma and Sonu, 1962; Davis and Fox, 1972; Sonu, 1973) the inner bar rip cell system is usually associated with shoreline undulations or giant cusps (Komar, 1971) and the bars are often attached to the beach face at the cusp horns. Cusp wavelengths measured on North Richibucto Beach ranged from 90 m to 300 m with average rip cell spacing of 243 m. These can be compared to values of 150–1000 m obtained by Dolan (1971) a mean of 90 m by Evans (1939), a mean of 182 m by Davis and Fox (1972) and values of 100–300 m by Sonu (1973).

The bars may be parallel, transverse, crescentic or of intermediate shapes. There is considerable variation in bar form, both spatially and temporally, and all morphological types may occur within a short stretch of coastline. Figure 6.9(a) shows part of a 900 m section of the shoreline mapped in 1971, including the north study area. The bars here were primarily transverse or spit-like in form. They appear similar to forms described by Evans (1940), Miller and Ziegler (1964) and Sonu (1968, 1973), but changed very little over the summer months despite the occurrence of several storms, indicating a much smaller degree of mobility. The same area in 1972, in contrast (Figure 6.5), was dominated by an essentially parallel bar and trough system which was joined to a crescentic bar at the southern end, indicating more significant topographical changes in the autumn and winter period.

Part of a 2 km length of shoreline mapped in 1971 including the south study area is shown in Figure 6.9(b). Two parallel bars are present in the area shown, as well as one spit-like shoal at the end of a cusp horn. The two bar forms remained fairly stable during the summer period despite several storms and migrated shoreward a distance of only 10 m over a period of about 10 weeks. The same area in 1972 was characterized by crescentic rather than straight bars (Figure 6.13), again suggesting considerable changes on an annual basis.

The inner crescentic bars are similar in many respects to those of the outer bar. The horns of the crescents join at a distinct shoal area and water depths are greatest near the centre of the bar. However, they tend to be less regular and continuous than the outer crescentic bars and occur in groups of two or three. The wavelength of the inner crescents is similar to that of the shoreline cusps with the shoal areas located opposite the cusp horns (Sonu, 1973). Amplitudes range from 10 m to 20 m. The shoal area may be fairly small and almost circular in form or may extend as a straight parallel bar for some distance. Single crescentic bars may link two parallel bars and smaller crescents may extend across the seaward end of a rip channel. In almost all cases, the shoal areas are separated from the shoreline by a distinct trough.

Morphological changes. In order to monitor growth and movement of the inner bar system, two study areas on North Richibucto Beach were surveyed at intervals of one to two weeks during the summer of 1972. Observations of

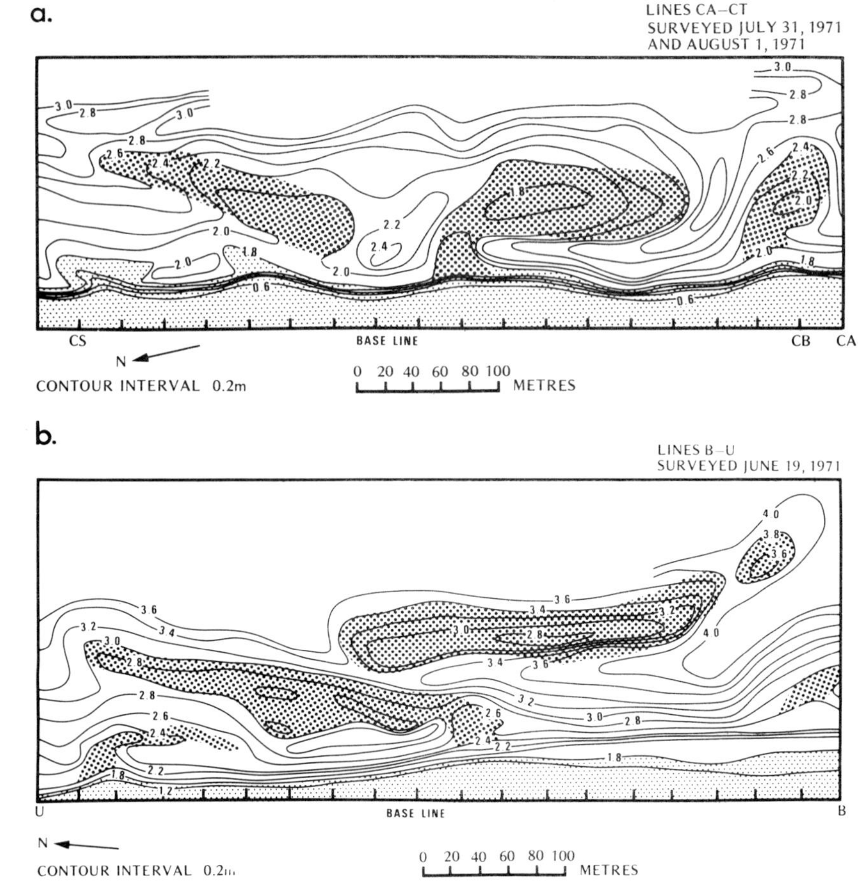

Figure 6.9. Inner bar topography surveyed in 1971: (a) north study area; (b) south study area.

current speeds were also undertaken. The bars in the two study areas, approximately 2 km apart, were unique in form one to the other and although southward migration occurred in both areas during the summer period, the manner by which this took place appeared to be quite different.

(1) *North study area*: This area was dominated by the growth and migration of a straight, parallel bar. Four of the ten surveys over a 3-month period are presented in Figure 6.10 to show the major changes in the bar form and position that occurred. The principal changes were: (a) erosion of a new rip channel through the bar at the point of attachment to the cusp horn (there was no deepening of the trough parallel to the shore, see below); (b) simultaneous infilling of the previous rip channel and establishment of a new rip current cell; (c) migration of the bar front onshore and attachment to the shoreline again at a cusp horn, approximately 180 m south of the original point of attachment.

Infilling of the rip channel at the south end of the study area and excavation of a new channel at the north end did not occur during a single storm period.

Initial changes occurred primarily during two storms and the new rip channel was considerably widened and deepened during a subsequent storm. This indicates the competency of currents in the rip neck. Observations when waves approximately 1 m in height were breaking on the bar, showed the main longshore current still flowing south along the trough and seaward through the rip channel near the south end of the area. Farther south, a current flowing northward also turned seaward through the same channel. However, the circulation pattern after the new rip channel was excavated was somewhat different. Current speeds and directions measured with waves of 0·8 m breaking on the bar are shown in Figure 6.11. From near the centre of the study area one current flowed northward and then seaward through the new rip channel while a second current flowed southward and eventually seaward through a rip channel about 100 m south of the study area. In the central area currents were weak and speeds increased away from the divide.

During July and August the bar front migrated shoreward and eventually became attached to a cusp horn which developed in the area where longshore currents were weakest. Examination of two profiles, normal to the shoreline, CC and CL, at four points in time (Figure 6.12(a) and (b)), gives some indication of the movement of the bar face, the build-up of sediment in the bar form and the extent of erosion in the new rip channel. Over the study area landward migration of the bar face averaged 25–30 m and there was aggradation ranging from 20 cm to 65 cm over the study period. Only minor changes in the profiles on the seaward side of the bar occurred and thus considerable accumulation of sediment in the bar form is indicated. However, this does not appear to occur at the expense of erosion of the beach face. There must, therefore, be considerable net transport of sediment alongshore or onshore, or both, to account for this sediment build-up. Bar front migration is as an avalanche slip face producing medium-scale, tabular, units of cross-bedding, approximately as thick as the bar height (Davidson-Arnott and Greenwood, 1974). Steepening of the bar face is coincident with shoreward migration. Trough depth tends to remain fairly constant initially, but as the bar face approaches the shoreline, there is a gradual infilling of the trough, particularly at the point of eventual attachment to the shoreline.

(2) *South study area*: This area was characterized by the presence of small shoal areas joined by crescentic bars with general features similar to those already described. Three of the six surveys over the period June 14 to August 12, are shown in Figure 6.13. Initial surveys carried out prior to June 14 revealed no distinct bar system and the topography mapped on June 14 appears to have developed, or at least moved shoreward and grown in size, during a storm from June 10 to June 12 (breaker heights greater than 1 m).

The major changes that occurred over the study period were: (a) growth and development of the shoals and bars and establishment of a cellular rip system; (b) longshore and onshore migration of the shoals and bars. The shoal at the north end of the study area migrated 34 m shoreward and 30 m southward during this period while the shoal near the south end migrated 34 m shoreward and 42 m along the coast. There was, initially, an increase in the height of the shoal areas of 20–50 cm, but the height remained fairly constant or decreased slightly

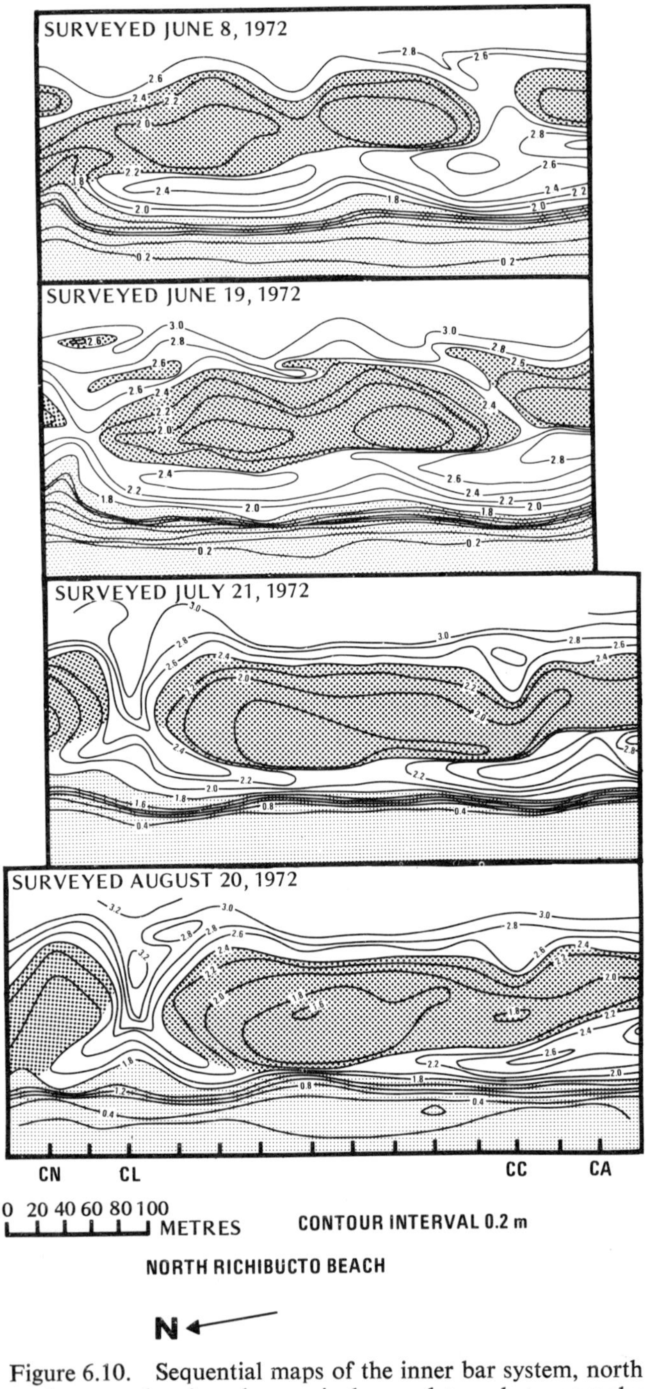

Figure 6.10. Sequential maps of the inner bar system, north study area, showing changes in bar and trough topography over the period 8 June to 20 August 1972.

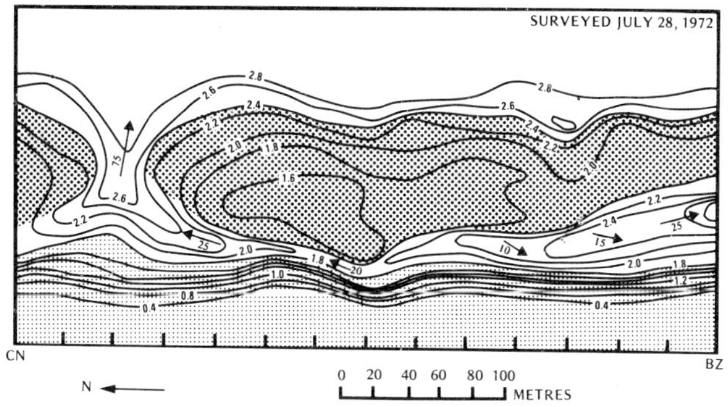

Figure 6.11. Current vectors, north study area, measured on 28 July 1972. Figures next to the arrows are speeds in cm/s.

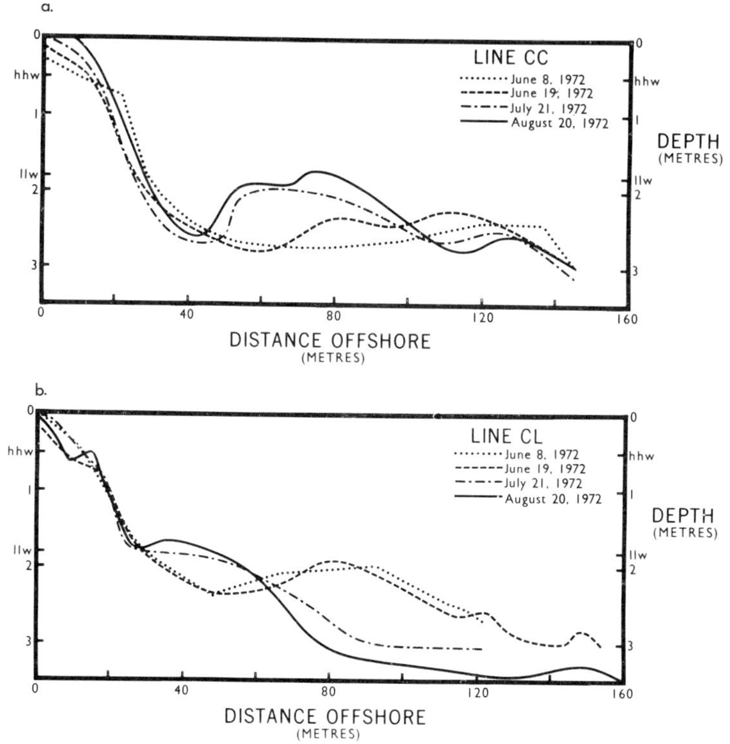

Figure 6.12. Changes in inner bar topography along selected profiles, north study area, over the period 8 June to 20 August 1972. Location of the profiles is shown in Figure 6.10.

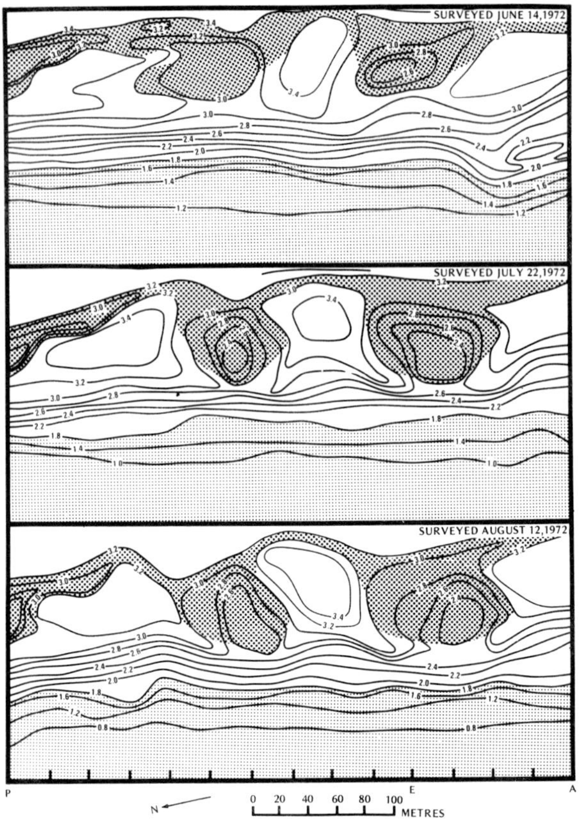

Figure 6.13. Sequential maps of the inner bar system, south study area, showing changes in bar and trough topography over the period 14 June to 12 August 1972.

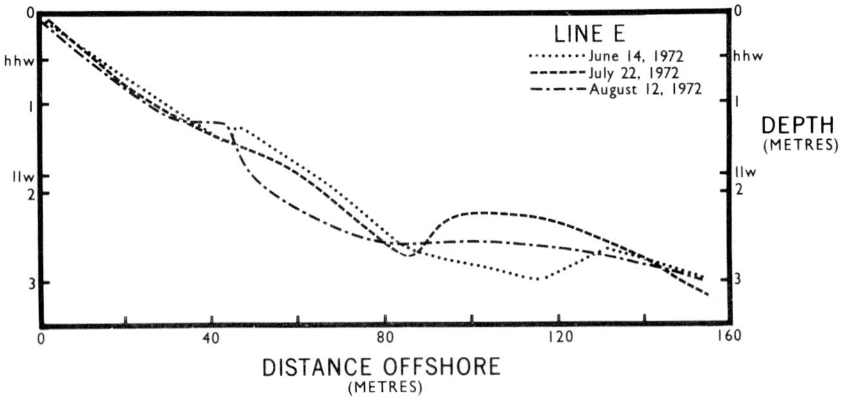

Figure 6.14. Changes in the inner bar topography along one profile, line E, south study area. 14 June to 12 August 1972. Location of the profile is shown in Figure 6.13.

after June 30. Profiles along line E over the time period illustrate the changes associated with migration of the shoal area through that section (see Figure 6.14). The bar front at the shoal areas is very steep compared to the slope along the crescent and there is a well defined trough between the shoal and the shoreline at all times.

Evaluation

While bar genesis and migration are extremely important in the nearshore sediment balance, there is no comprehensive quantitative model presently available to explain the diverse character of the features. Indeed, the complexity of the fluid dynamics within the nearshore zone makes such a treatment very difficult at present. Most authors have stressed that the bar form represents a modification of the offshore slope, which is in dynamic equilibrium with a particular set of wave and current conditions (Johnson and Eagleson, 1966; Zenkovich, 1967), and it has been argued that measurements of bar spacing, depth of trough to depth of crest ratio (Shepard, 1950) etc., provide evidence of the existence of such an equilibrium. Measurements of depth to crest, depth to trough, and the ratio between them for the inner and outer system in Kouchibouguac Bay are presented in Table 6.3, together with published values from other areas. However, the variability of calculated ratios together with other aspects of bar form indicates the complexity of the interaction of factors controlling this equilibrium.

Model and field studies correlate bar location with the position of breaking waves, and this is certainly the case in Kouchibouguac Bay. The depth of water measured over the outer bar crest (see Table 6.3) means that waves approximately 1·5 m in height, or greater, are necessary for breaking to occur. Although infrequent, these waves do occur regularly during storms throughout the year.

Johnson (1949) related bar formation to conditions of wave steepness greater than 0·025 and Paul and coworkers (1972) found experimentally that bars formed under similar wave characteristics. Rector (1954) and Watts (1955) both suggested wave steepness values greater than 0·016 were necessary for bar formation. Bivariate plots of wave steepness versus the ratio between wave height and median sand diameter have been used in an effort to define the controlling parameters (Iwagaki and Noda, 1963) and Nayak (1970) summarized several different criteria for the generation of offshore bars. Wave steepness for the largest storm wave recorded in Kouchibouguac Bay was 0·066 and most waves have steepness values greater than 0·02. The ratio between wave height for storm waves and median sediment diameter is of the order of 10^3 and this, together with the wave steepness values, falls in the field of values for bar formation put forward by Iwagaki and Noda (1963).

The nature of bar sediment accumulation in Kouchibouguac Bay is not associated with excavation of material in the trough and seaward movement of sediment under the breaking wave, as has been suggested by Keulegan (1948), Shepard (1950), Herbich (1970) and King (1972). Neither are the bars formed of

Table 6.3. Form characteristics of the outer and inner bar systems in Kouchibouguac Bay together with published values from experimental and other field studies.

	Average trough depth (H_t) (m)	Average crest depth (H_c) (m)	Average ratio H_t/H_c
Outer system	4·43 (3·8 to 5·2)	2·72 (2·0 to 3·3)	1·63 (1·3 to 2·2)
Inner system			
(South study area)	2·06 (1·3 to 2·6)	1.56 (1·0 to 2·1)	1·32
(North study area)	1·58 (1·1 to 2·1)	0·87 (0·5 to 1·1)	1·82
Keulegan (1948) Experimental			1·69
Shepard (1950) California			1·16
Shepard (1950) Cape Cod			1·34
Evans (1940) Lake Michigan			1·42–1·55
Herbich (1970) Texas Coast			1·25–1·35
Otto (1911) Pomeranian Coast			1·56–1·87

Note: For Kouchibouguac Bay data averages determined from 20 profiles at two discrete time intervals. Figures in parenthesis indicate range of values.

the coarsest material, which is supposed to accumulate at the breakpoint (Fox and coworkers, 1966; Wood, 1969). In fact, waves normally break on the seaward side of the bar, not on the crest or in the trough, and the bar materials are finer than in the succeeding trough (Greenwood and Davidson-Arnott, 1972). Although longshore currents generated in the trough by breaking waves are not responsible for excavation of the trough (in contrast to the suggestion by Knaps, 1959, quoted in Zenkovich, 1967), they are undoubtedly important in transporting and preventing the deposition of sediment entrained by the breaking waves. This longshore and eventual seaward transport of sediment is probably responsible for preventing the trough from infilling and preserving the steep, landward-facing slope of the bar. Examination of the internal structure of inner bars (Davidson-Arnott and Greenwood, 1974) and preliminary work on the outer bar system indicates that accumulation of sediment in the bar form results primarily from landward transport of sediment and that seaward dipping structures are restricted to thin, narrow units in rip current channels. The bars are, therefore, accumulative, rather than scour, features resulting from the energy loss and reduction in transporting competency associated with the breaking waves.

The two bar systems present in Kouchibouguac Bay are both formed by the same basic set of processes associated with breaking waves, but there are clearly major differences between the two systems resulting from differences in wave energy, wave characteristics, and the secondary currents generated. Because of the restricted fetch, waves generated by winds blowing from directions other than the northeast quadrant are generally too small to initiate sediment movement on the outer bar. This, together with the low frequency of southeast and east winds, means that the outer system responds only to a limited spectrum of waves. The inner bars are also strongly affected by the northeast storm waves and major changes, such as the excavation of new rip channels, occur during these high energy periods; but unlike the outer system, modification and migration of the bar form does occur under all but the lowest energy conditions. The wide range of wave heights (0·5–3·0 m) and directions of wave approach affecting the inner bar system are partly responsible for the variability in bar form. The inner system is clearly related to the nearshore circulation pattern of rip cells and to shoreline cusps but there are strong feedback mechanisms between the bar form and the circulation pattern, which makes the process–response relationships complex. Bar shape appears to be related to, if not controlled by, the length of feeder currents (and thus rip current spacing) and the orientation of the rip channels.

Crescentic inner bars frequently occur across the area of flow expansion beyond the necks of rips flowing perpendicularly to the shoreline. The crescentic shape mirrors the shape of the rip heads and suggests sediment deposition where net landward bed-velocities generated by waves are matched by seaward bed-velocities generated by the dispersing rips. These rip cells are characterized by very short feeder currents generated by waves breaking on the shoal areas at the ends of the crescents. Transverse bars are associated with rip currents flowing obliquely to the shoreline with the feeder currents again being generated by waves breaking on the bar. Parallel bars are associated with longshore currents of considerable length as the surf bores generated across the bar tend to extend the length of the feeder channel until a particular storm surge creates a new rip channel spacing.

Nearshore current circulation is undoubtedly controlled by wave 'set-down' and 'set-up' (Bowen and coworkers, 1968), which in turn reflects the submarine bathymetry (Sonu, 1972a). Changes in the circulation pattern should therefore reflect changes in wave characteristics and direction of wave approach and, in turn, result in modification of the bars. The transverse, parallel and crescentic bars all migrate, both normal and parallel to the shore and have a 'sand-wave' character (Sonu, 1968) induced by the strong lateral wave incidence particularly during the northeast storms. Shoreward migration may result in the eventual merging of bars with the beach face but bar growth does not appear correlated with erosion of the beach face landward of the bar and seaward shift of this sediment. There is no evidence in Kouchibouguac Bay to support the suggestion of Bernard and coworkers (1959), Davis and Fox (1972) and Coakley and coworkers (1973) that the inner bars are destroyed during storms and regenerated as wave activity decreases, but clearly the greater size of features in this study area is an important consideration. Nevertheless storm waves cause significant changes in form through the development of new rip channels and the associated sediment redistribution.

The landward and seaward displacement of the outer bar crescent poses the problem of accounting for the net seaward movement of sediment during high energy storm conditions. Observation and tracking of current drogues released in the inner bar system during storms showed no apparent link between the inner and outer circulation patterns. Currents in rip channels generally flow seaward for only a short distance beyond the inner bar and then either dissipate rapidly or merge with flow parallel to the shoreline. While net shoreward motion associated with shoaling waves can account for shoreward transport of sediment, the seaward displacement of the bar crescent between the shoal areas indicates also the presence of seaward flowing currents. Preliminary study of internal structures in the outer bar has revealed the presence of medium-scale seaward-dipping, cross-stratified units on the bar crescent (Figure 6.15) indicating that seaward movement of bed load in migrating megaripples can occur. The stability of the shoal areas and the fact that seaward movement of the bar form occur only over the crescent area suggests that water motion and sediment transport on the shoals is primarily in a landward direction. This is also indicated by current drogue measurements which show a shoreward movement across the shoal area and then movement parallel to the bar along the trough under deep water wave heights of approximately 2·0 m and current speeds of 5–10 cm/s. These factors suggest that the circulation pattern associated with the outer bars is similar to that associated with crescentic inner bars; that is, shoreward flowing currents across the shoal area with longshore currents in the trough and seaward flowing 'rip' currents across the centre of the bar crescent. The wave-like form of the bar crest and its landward and seaward displacement may, therefore, reflect variations in the seaward return flow both temporally and spatially. Confirmation of this will require further current measurements during storms and investigation of sedimentary structures.

Bowen and Inman (1969) have suggested the existence of edge waves as a controlling factor in rip current spacing and, more specifically, as a cause of crescentic bar formation (Bowen and Inman, 1971). While such a theoretical model may provide a possible explanation for sediment transport patterns in the nearshore zone, it is difficult at present to establish its relevance in the prototype (Sonu, 1972b). It is evident from study in Kouchibouguac Bay that the direct mechanisms of sediment transport are currents associated with the orbital motion of incident waves and secondary longshore and rip currents. Hopefully, more detailed knowledge of sediment movement patterns, reflected in structural and form changes, together with wave and current characteristics, will assist in establishing the most significant controls in bar sedimentation.

Acknowledgements

Research for this paper was supported by a grant to one of us (B.G.) from the National Research Council of Canada (NRC A7956). R.D-A. acknowledges receipt of a postgraduate scholarship from the same body. Miss P. Keay and Mr. P. Hale gave able field assistance and invaluable support was provided by the Technical Services staff and Graphics Department of Scarborough College, University of Toronto. Finally we would like to thank the reviewers and editors for their helpful suggestions.

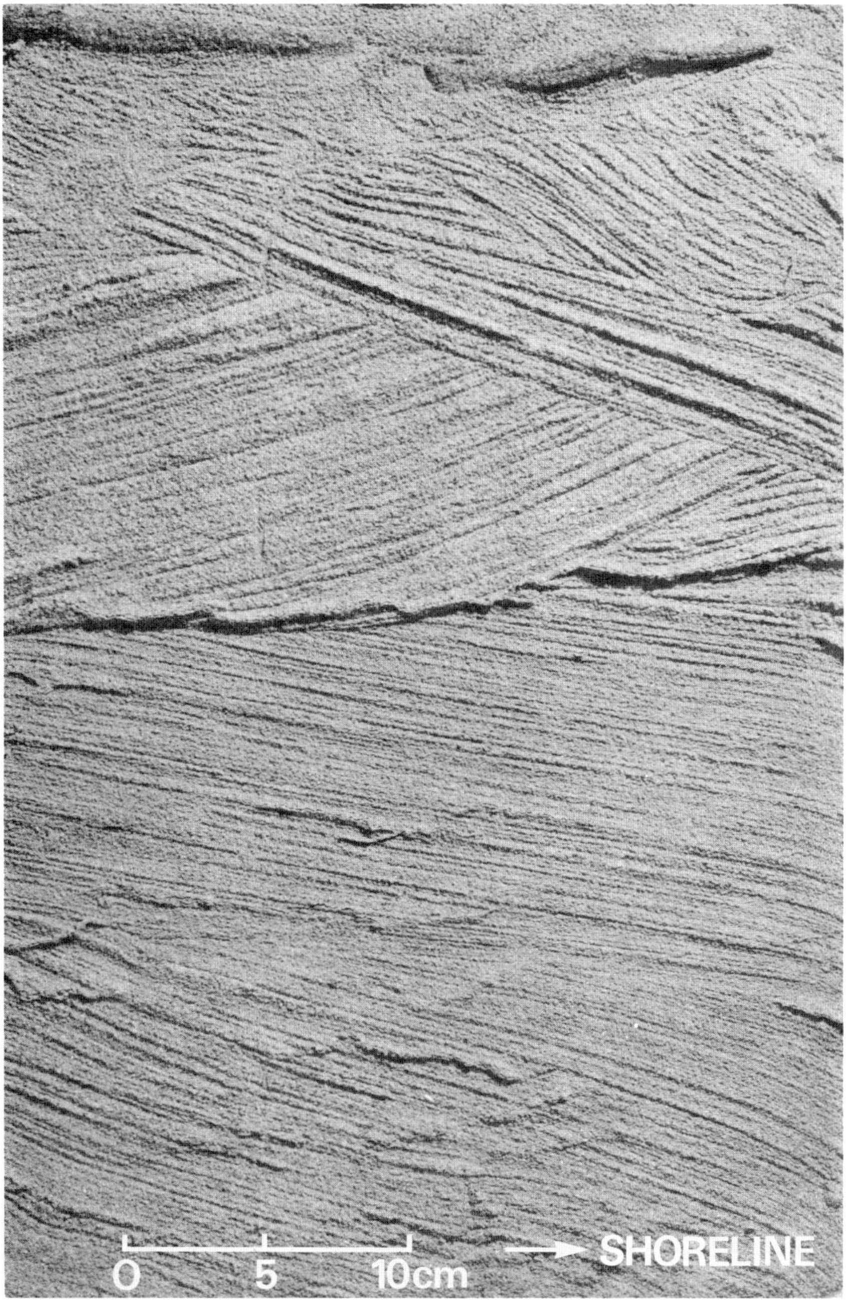

Figure 6.15. Peel made from a box core taken on the crescent area of the outer bar system.

References

Bajorunas, L., and D. B. Duane, 1967, Shifting offshore bars and harbour shoaling, *J. geophys. Res.*, **72**, 6195–6205.

Bakker, W. T., 1968, A mathematical theory about sand waves and its application on the Dutch Wadden Isle of Vlieland, *Shore and Beach*, **36**, 5–14.

Barcilon, A. I., and J. P. Lau, 1973, A model for the formation of transverse bars, *J. geophys. Res.*, **18**, 2656–2664.

Bascom, W. N., 1953, Characteristics of natural beaches, *Proc. 4th Conf. cst. Engng., Chicago, Ill.*, pp. 163–180.

Bernard, H. A., C. F. Major Jr. and B. S. Parrott, 1959, The Galveston barrier island and environs: a model for predicting reservoir occurrence and trend, *Trans. Gulf-Cst. Ass. geol. Socs.*, **9**, 221–224.

Blackford, B. L., 1965, Results from a current meter moored in Kouchibouguac Bay, June and July 1964, Fish. Res. Bd. Can., *M.S. Rept. Ser.* (*Oceanog. Limnol.*) *No. 196*.

Bowen, A. J., 1969, Rip currents: 1. Theoretical investigations, *J. geophys. Res.*, **74**, 5467–5478.

Bowen, A. J., 1972, Edge waves and the littoral environment, *Proc. 13th Conf. cst. Engng., Vancouver, Canada*, Vol. 2, pp. 1313–1320.

Bowen, A. J., and D. L. Inman, 1969, Rip currents: 2. Laboratory and field observations, *J. geophys. Res.*, **25**, 5479–5490.

Bowen, A. J., and D. L. Inman, 1971, Edge waves and crescentic bars, *J. geophys. Res.*, **76**, 8662–8671.

Bowen, A. J., and D. L. Inman, 1972, Reply (to comment by C. F. Sonu, on Bowen and Inman, 1971), *J. geophys. Res.*, **77**, 6632–6633.

Bowen, A. J., D. L. Inman and V. P. Simmons, 1968, Wave set-down and set-up, *J. geophys. Res.*, **73**, 2569–2577.

Bruun, P. M., 1954, Migrating sand waves and sand humps with special reference to investigations carried out on the Danish North Sea coast, *Proc. 5th Conf. cst. Engng., Grenoble, France*, pp. 269–295.

Clos-Arceduc, A., 1962, Étude sur les vues aériennes, des alluvions littorales d'allure périodique, cordons littoraux et festons, *Bull. Soc. photogramm. Fr.*, **4**, 13–21.

Clos-Arceduc, A., 1964, La photographie aérienne et l'étude des dépots prélittoraux, *Etude de photo-interpretation No. 1*, Institut Geographique National, Paris.

Coakley, J. P., W. Haras and N. Freeman, 1973, The effect of storm surge on beach erosion, Point Pelee, *Proc. 16th Conf. Great Lakes Res.*, pp. 377–389.

Davidson-Arnott, R. G. D., and B. Greenwood, 1974, Bedforms and structures associated with bar topography in the shallow-water wave environment, *J. geophys. Res.*, **44**, 698–704.

Davis, R. A. Jr., 1973, Coastal ice formation and its effect on beach sedimentation, *Shore and Beach*, **41**, 3–9.

Davis, R. A. Jr., and D. F. R. McGeary, 1965, Stability in nearshore bottom topography and sediment distribution, southeastern Lake Michigan, *Proc. 8th Conf. Great Lake Res.*, pp. 222–231.

Davis, R. A. Jr., and W. T. Fox, 1972, Coastal processes and nearshore sand bars, *J. sedim. Petrol.*, **42**, 401–412.

Dolan, R., 1971, Coastal landforms, crescentic and rhythmic, *Bull. geol. Soc. Am.*, **82**, 177–180.

Dyhr-Nielsen, M., and T. Sorensen, 1970, Some sand transport phenomena on coasts with bars, *Proc. 12th Conf. cst. Engng., Washington, D.C.*, Vol. 2, pp. 855–865.

Egorov, E. N., 1951. Observations of the dynamics of submarine sand ridges, *Trudy Inst. Okeanol. Akad. Nauk. SSSR*, **6**, (quoted in Zenkovich, 1967).

Evans, O. F., 1938, The classification and origin of beach cusps, *J. Geol.*, **46**, 615–627.

Evans, O. F., 1939, Mass transportation of sediments on subaqueous terraces, *J. Geol.*, **47**, 325–334.

Evans, O. F., 1940, The low and ball of the east shore of Lake Michigan, *J. Geol.*, **48**, 467–511.

Fox, W. T., J. W. Ladd and M. K. Martin, 1966, A profile of four moment measures perpendicular to a shore line, South Haven, Michigan, *J. sedim Petrol.*, **36**, 1126–1130.

Greenwood, B., and R. G. D. Davidson-Arnott, 1972, Textural variation in the sub-environments of the shallow-water wave zone, Kouchibouguac Bay, New Brunswick, *Can. J. earth Sci.*, **9**, 679–688.

Hartnack, W., 1924, Uber Sandriffe, *J. b. Geogt. Ges. Greifswald*, 40–42.

Hayes, M. O., and coworkers, 1969, Coastal environments, northeastern Massachusetts and New Hampshire, Coastal Research Group, Univ. of Massachusetts, *Contrib. No. 1—CRG*, 462 pp.

Herbich, J. B., 1970, Comparison of model and beach sand patterns, *Proc. 12th Conf. cst. Engng., Washington, D.C.*, Vol. 2, pp. 1281–1300.

Hom-ma, M., and C. J. Sonu, 1962, Rhythmic patterns of long shore bars related to sediment characteristics, *Proc. 8th Conf. cst. Engng., Mexico City*, pp. 248–278.

Iwagaki, Y., and H. Noda, 1962, Laboratory study of scale effects in two-dimensional beach processes, *Proc. 8th Conf. cst. Engng., Mexico City*, pp. 194–210.

Johnson, J. W., 1949, Scale effects in hydraulic models involving wave motion, *Trans. Am. geophys. Un.*, **30**, 517–525.

Johnson, J. W. and P. S. Eagleson, 1966, Coastal processes, in A. T. Ippen (Ed.), *Estuary and Coastline Hydrodynamics*, McGraw-Hill, New York.

Keulegan, G. H., 1948, An experimental study of submarine sand bars, *B.E.B. Tech. Rept. 3*, 40 pp.

King, C. A. M., 1972, *Beaches and coasts*, 2nd ed., Edward Arnold, London, 570 pp.

King, C. A. M., and W. W. Williams, 1949, The formation and movement of sand bars by wave action, *Geog. J.*, **112**, 70–85.

Knaps, R. Ya., 1959, Some features of the development of submarine sand ridges, *Trudy Okeanol. Akad. Nauk. SSSR*, **2** (quoted in Zenkovich, 1967).

Komar, P. D., 1971, Nearshore cell circulation and the formation of giant cusps, *Bull. geol. Soc. Am.*, **80**, 2643–2650.

Kranck, K., 1967, Bedrock and sediments of Kouchibouguac Bay, New Brunswick, *J. Fish. Res. Bd. Can.*, **24**, 2241–2265.

Kranck, K., 1972, Tidal current control of sediment distribution in Northumberland Strait, Maritime Provinces, *J. sedim. Petrol.*, **42**, 596–601.

Lau, J., and B. Travis, 1973, Slowly varying Stokes Waves and submarine longshore bars, *J. geophys. Res.*, **78**, 4489–4497.

Lauzier, L. M., 1965, Drift bottle observations in Northumberland Strait, Gulf of St. Lawrence, *J. Fish. Res. Bd. Can.*, **22**, 353–368.

Lauzier, L. M., 1967, Bottom residual drift on the continental shelf area of the Canadian Atlantic coast, *J. Fish. Res. Bd. Can.*, **24**, 1845–1859.

McCann, S. B., and E. A. Bryant, 1972, Beach changes and wave conditions, New Brunswick, *Proc. 13th Conf. cst. Engng., Vancouver, Canada*, Vol. 2, pp. 1293–1304.

Miller, R. L., and J. M. Ziegler, 1964, A study of sediment distribution in the zone of shoaling waves over complicated topography, in R. L. Miller (ed.), *Papers in Marine Geology: Shepard Commemorative Volume*, Macmillan, New York, 133–153.

Nayak, I. V., 1970, Equilibrium profiles of model beaches, *Proc. 12th Conf. cst. Engng., Washington, D.C.*, Vol. 2, pp. 1321–1340.

Niederoda, A. W., 1972, Waves, currents, sediments and sand bars associated with low energy coastal environments, *Trans. Gulf-Cst. Ass. geol. Socs.*, **22**, 229–240.

Niederoda, A. W., and W. F. Tanner, 1970, Preliminary study of transverse bars, *Mar. Geol.*, **9**, 41–62.

Nilsson, H. D., 1973, Multiple parallel sand bars of southeastern Cape Cod Bay, in D. C. Coates (Ed.), *Coastal Geomorphology*, Publications in Geomorphology, State Univ. of New York, 99–102.

Otto, T., 1911–12, Der Darss und Zingst, *Jber, geogr. Ges. Greifswald*, **13**, 235–485.

Paul, M. J., J. W. Kamphuis, and A. Brebner, 1972, Similarity of equilibrium beach profiles, *Proc. 13th Conf. cst. Engng., Vancouver, Canada*, Vol. 2, pp. 1217–1236.

Raman, H., and J. J. Earattupuzha, 1972, Equilibrium conditions in beach wave interaction, *Proc. 13th Conf. cst. Engng., Vancouver, Canada*, Vol. 2, pp. 1237–1256.

Rector, R. L., 1954, Laboratory study of the equilibrium profiles of beaches, *Beach Eros. Bd. U.S. Technical Memo. 41*, 38 pp.

Saylor, J. H., and E. B. Hands, 1970, Properties of longshore bars in the Great Lakes, *Proc. 12th Conf. cst. Engng., Washington, D.C.*, Vol. 2, pp. 839–853.

Saylor, J. H., and S. B. Upchurch, 1970, Bottom stability and sedimentary processes at Little Lake Barbor, Lake Superior, U.S. Army, Corps of Engineers, Lake Survey District, Detroit, Michigan, *Research Rept. 2-1*, 18 pp.

Shepard, F. P., 1950, Longshore bars and longshore troughs, *B.E.B. Tech. Memo. 15*, 32 pp.

Sonu, C. J., 1961, A treatise of shore processes and their engineering significances, Ph.D. thesis (unpublished), University of Tokyo.

Sonu, C. J., 1968, Collective movement of sediment in nearshore environments, *Proc. 11th Conf. cst. Engng., London*, Vol. 1, pp. 373–400.

Sonu, C. J., 1972a, Field observations of nearshore circulation and meandering currents, *J. geophys. Res.*, **77**, 3232–3247.

Sonu, C. J., 1972b, Comment on paper by A. J. Bowen and D. L. Inman, Edgewaves and crescentic bars, *J. Geophys. Res.*, **77**, 2569–2577.

Sonu, C. J., 1973, Three-dimensional beach changes, *J. Geol.*, **81**, 42–64.

Sonu, C. J., and R. J. Russell, 1966, Topographic changes in the surf zone profile, *Proc. 10th Conf. cst. Engng., Tokyo, Japan*, Vol. 1, pp. 502–524.

Watts, G. M., 1955, Laboratory study on the effect of varying wave periods on beach profiles, *Beach Eros. Bd. U.S. Technical Memo. 53*, 19 pp.

Wood, A. M. M., 1969, *Coastal hydraulics*, Macmillan, London, 187 pp.

Zenkovich, V. P., 1967, *Processes of coastal development*, ed. J. A. Steers, Oliver and Boyd, Edinburgh and London.

CHAPTER SEVEN

Sediment Mobility and Erosion on a Multibarred Foreshore (Southwest Lancashire, U.K.)*

W. R. PARKER

Abstract

The stability of an alluvial coast relies on the dynamic equilibrium between the supply of sediment to the foreshore and the removal of sediment therefrom. A diminution in supply may result in erosion of the foreshore and dunes.

The nearshore subtidal and intertidal areas of southwest Lancashire have undergone large bathymetric changes. The most recent phase of intertidal erosion, affecting the coastline since 1900, is manifest in erosion of the foreshore, the dunes and the surface of Holocene deposits underlying the foreshore.

Intertidal sand transport has been examined in an attempt to relate sand mobility to coastal erosion. Although the interrelationships between energy and transport are well documented, it is difficult to prove mobility and continuity of transport from supposed source areas to particular sites, especially on foreshores with complex patterns of sand transport.

In this context, inhibition of sand movement is seen as being of particular importance in the processes of natural beach maintenance. The presence of intertidal and subtidal mud areas and permanently wet areas of the foreshore has a dominant influence on mobility and, combined with the transport patterns in the intertidal zone, on the stability of the coast.

Introduction

Formby Point is situated 17 km northwest of Liverpool on the east side of the Irish Sea (Figure 7.1). Liverpool Bay consists of a series of channels and sandbanks which extend west, southwest and south of Formby Point. The gently westward sloping floor of the Irish Sea is to the northwest (Figure 7.2) while to the north the long expanses of sandy foreshore and sand dune systems lead to the Ribble Estuary (Figure 7.3).

The Holocene evolution of this coastline, which consists of a barrier complex situated in the vicinity of the present coastline with an intertidal and subsequently supratidal marsh to the east, has been outlined by Greswell (1953) and discussed by Sly (1966).

* Titles and areas as existing before the local government reorganization of April 1974.

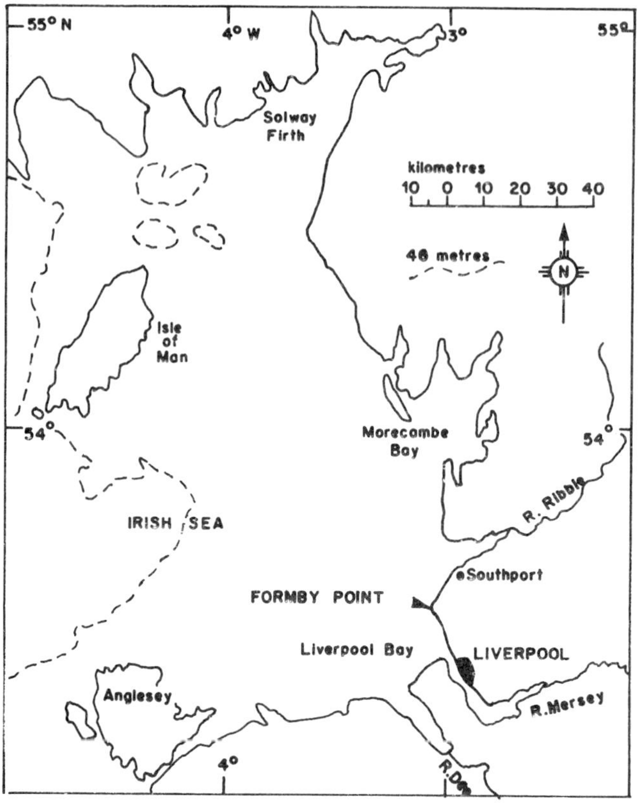

Figure 7.1. Location of Formby Point.

As is typical of alluvial coasts, historical documentary evidence suggests cycles of coastal accretion and erosion, the present retreat of high water mark commencing between 1900 and 1906. Recession at Formby Point has been continuous since then.

Regional and local environments

Formby Point is significantly positioned on the margin of the eastern Irish Sea, where littoral drift divides north towards the Ribble and south towards the Mersey. In the offshore areas west of Jordan's Spit (Figure 7.2) major tidal flood streams are thought to diverge, northeast towards the Ribble and southeast towards the Mersey, and ebb streams converge. This is the area of a corresponding divergence of bed load transport suggested by Sly (1966) and a divergence of near-bed residual water drift shown by Ramster and Hill (1969). The locale has a high tidal range (8·2 m mean spring range), is subject to large surge contributions to high water (Lennon, 1963a and 1963b) and suffers maximum local wave conditions with strong winds from the southwest, through west to

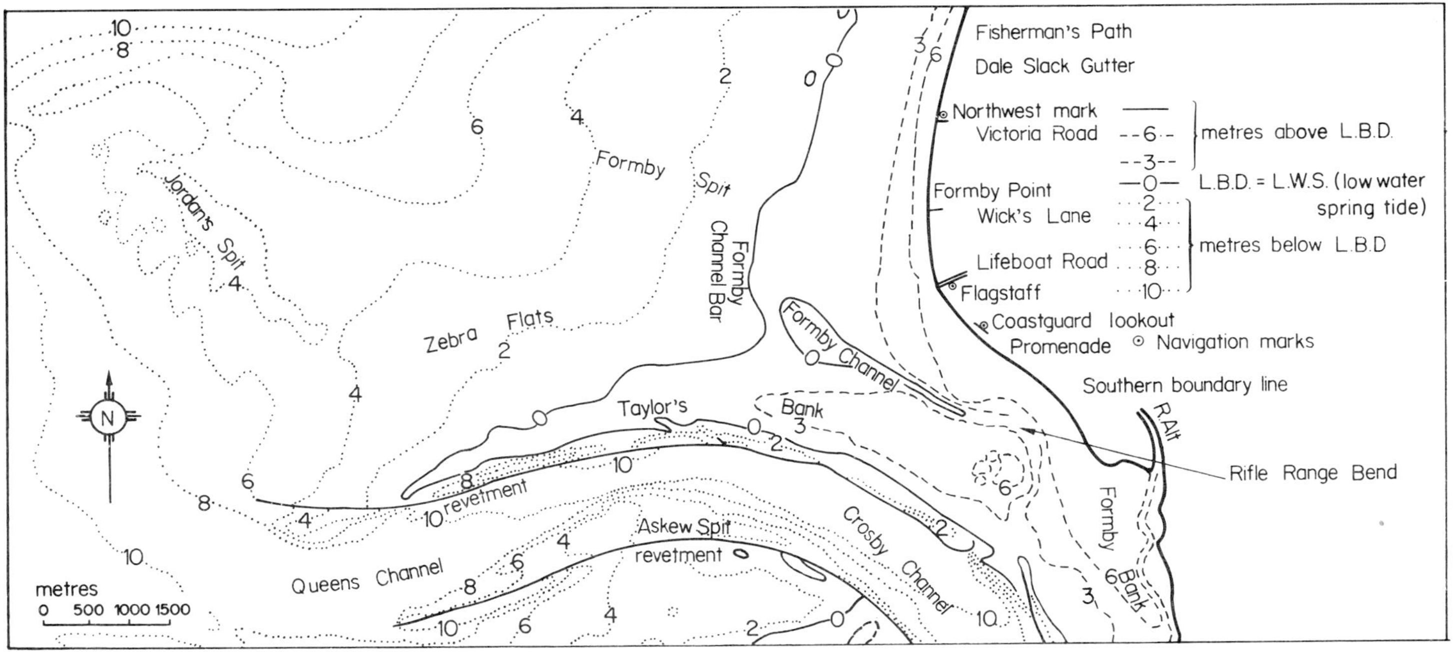

Figure 7.2. Local geography and bathymetry. L.B.D. = Liverpool Board Datum, which is approximately equal to low water spring tides (−4·42 m O.D.).

Figure 7.3. View south over Formby Point, towards Taylor's Bank (large arrow) from Fisherman's Path. Note ridges, runnels and rip channels, coastal facet junctions (C F), wet areas on the upper foreshore (W). Note transition from cliffed facet (1) to uncliffed though retreating coastline (2) and dune systems. D = Dale Slack Gutter, V = Victoria Road and Northwest Mark, L = Lifeboat Road. (Oblique from 140 m; courtesy John Mills Ltd.)

northwest. Reports by Darbyshire (1958) and Murthy and Cook (1962) both suggest that wave spectra at Formby are saturated except from the southerly or easterly quadrants. Recent analyses of local wave data (Draper and Blakey, 1969) show little seasonal variation in either the wave period or spectral width parameter. The most common wave conditions were those with significant wave height of between 0·6 m and 1·0 m and a zero-crossing period of between 4·0 and 4·5 seconds (6·1% of the time). The highest measured wave reported was H_1 = 9·4 m (Tucker, 1963; Draper, 1966) associated with a zero-crossing period of 8·7 seconds. As would be expected, wave heights are generally higher in the autumn and winter months. The regional oceanographic setting has been documented by Bowden (1955, 1960), Bowden and Sharaf El Din (1966a and 1966b) and Sharaf El Din (1970). Regional sedimentology has been described by Sly (1966), Cronan (1969) and Wright and coworkers (1971).

The floor of the southeastern Irish Sea is predominantly of fine to medium sand, with local areas of gravel or mud. Sly (1966) and Belderson and Stride

Figure 7.4. View from off Wick's Lane looking south over Formby Channel, Taylors Bank (T B) and Queen's Channel (Q C). Old Lifeboat house at Lifeboat Road arrowed (cf. Figure 7.6). Note continuity and offshore trend of runnels; narrow intertidal flats in near ground widening southwards; wet areas (W) on upper foreshore and the large platform area (P) south of Lifeboat Road. (Oblique from 140 m; courtesy John Mills Ltd.)

(1969) have used sand-waves, supplemented by some hydrodynamic evidence, to interpret sediment transport directions. They suggest convergence of bed load transport on Liverpool Bay. The model results of Price and Kendrick (1963) and the field data reported by Best (1972), Best and coworkers (1973), and Halliwell (1973) support this suggestion.

Foreshore morphology (Figure 7.5)

The nearshore sublittoral area is characterized by a series of symmetrical sand ridges. These are 0·5–1·0 m high with a wavelength of 300–500 m, and are aligned sub-parallel to the intertidal beach ridges. The area is separated from the lower foreshore sand flats by the subtidal slope which terminates on the seaward edge of the intertidal zone in a low water berm.

The foreshore is convex westwards (Figures 7.2 and 7.3). The lower part comprises a typical intertidal muddy sandflat, much wider in the south of the

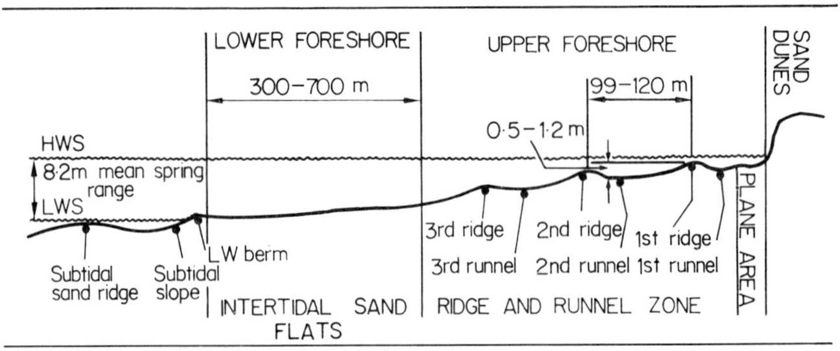

Figure 7.5. Coastal profile (diagrammatic).

area than the north (Figures 7.3 and 7.4), while the upper portion comprises a ridge and runnel zone with as many as four ridges and an upper foreshore plane area. The latter is defined here by the author as 'a planar seaward-sloping zone lying between the most landward runnel and the sand dunes at high water mark'.

The ridges and runnels lie at a tangent to the curve of high water mark such that northwards or southwards from the apex of the point they trend away from high water mark (Figures 7.3 and 7.4). About the apex of the point the discontinuous ridges are commonly broken by rip channels (Figure 7.3) whereas to the north and south the runnels form unbroken channels (Figure 7.4). South of Lifeboat Road (Figures 7.2 and 7.4) a large platform occurs landward of the ridge and runnel zone. This platform is not backed by a 'plane area' on its inshore side but passes directly into the complex of sand dunes which occur along this coastline. Much of the complex system of dune ridges has been destroyed but the most recent groups survive in the north of the area (Figure 7.3).

Stability on alluvial coasts

If, for a sandy alluvial coast, a dynamic balance between onshore sand transport and littoral drift is a valid model of the mechanism of coastal stability, then the erosion of the sand dune cliffs is but a symptom of the budgetary relationship of the foreshore and adjacent offshore areas. The erosion of the dunes may be interpreted as indicating that there is a deficit in supply across the foreshore. An explanation of coastal erosion must be sought in the mechanisms and rates of landward sand transport from the offshore areas across the foreshore to the frontal dunes. Although the evidence supporting the probability of sublittoral sand transport towards the Lancashire coast is considerable, no direct evidence of sand movement in the area of Zebra Flats is available. Samples from this shallow nearshore area comprise interbedded mud and rippled sand layers suggesting at least periodic sand mobility. Large emphemeral mud areas occur widely over much of the otherwise sandy (or muddy sand) seabed in this region (Parker, 1971).

More information concerning sand transport on the foreshore is available to provide comparisons with other authors (e.g. Reineck, 1960 and 1963) and it is the purpose of this paper to discuss this aspect in some detail. However, before sand movement, together with possible resulting patterns of migration, on a multibarred foreshore is considered, the processes influencing dune formation and destruction will be examined since these are relevant to the other processes which influence the form of this coastline.

Aeolian sand transport

Of all the agents credited with landward movement of sand to supply the frontal dunes, wind is the only demonstrably successful one in this particular area but its effectiveness is greatly influenced by two factors:

(1) The relative location of the potential sand supply areas and the wet areas along the foreshore that act as complete barriers to aeolian transport.
(2) The low efficiency of the present frontal dune cliffs and deflated remnants in trapping that sand which is blown onto them.

Taking the foreshore as a whole it is the ridges which dry most rapidly and apparently therefore lend themselves to aeolian sand supply. However, the runnel between the furthermost inshore ridge and the upper foreshore plane area (Figure 7.5) remains wet at all times and completely blocks landward aeolian sand movement from the foreshore ridge and runnel zone.

The upper foreshore plane area is, therefore, the only potential aeolian source of sand for the dunes but, because of the presence of the Holocene sediments underlying the foreshore, it is frequently wet where the water table associated with the frontal dunes and inland slacks outcrops. Despite this moisture some sand is still blown to the cliffs but it does not appear to accumulate along the eroding sector of the coast. Small (1·0–1·5 m) cliffs become buried by sand but at the base of higher cliffs a small triangular sand body develops upwind of the 'roller' vortex which persists at the cliff toe. Once this deposit is formed, sand is fed along the cliff toe to escape landward through blow-through paths.*

In those conditions when the wind is not directly onshore, much of the sand blown from the upper foreshore plane area to the cliffs is directed along the cliff to the first blow-through whence it escapes inland.

Processes of sand cliff erosion

Combined marine and wind erosion are thus responsible for cliff retreat at Formby Point and the wind blows sand from the cliff face, cliff top and the sand areas immediately behind the cliff top. Much of the dune system at Formby is denuded of vegetation as a result of trampling by pedestrians. Consequently these dunes are sculpted to form deep gullies through their eroded remnants.

* The development of the usual hemispheroidal blowout extends to the water table. The down-wind face continues to retreat producing elongate, flat-bottomed, sinuous gullies, often bifurcating landwards, here termed 'blow-through paths'. These gullies provide routes through which sand is blown inland.

The blow-through paths have two important functions. First, they facilitate rapid marine erosion through otherwise relatively resistant dune ridges (Figure 7.6) and, second, they provide a path whereby such sand as is blown from the foreshore to the dunes passes through the front of the dune system to the vegetated areas landward of the deflated zones.

Marine erosion of the sand cliff is achieved by one or both of two processes which attack the base of the cliff, undercutting and soaking. At high water, under normal undisturbed spring tide conditions, waves break approximately 3–5 m from the cliff toe. The swash approaches the cliff obliquely and runs along the cliff toe. The current thus produced cuts a slot in the cliff foot which may penetrate to a depth of 20–30 cm before the unsupported cliff above slides down. The slide debris is eroded in the same way. If surge conditions occur at spring tide high water then local still-water level reaches an altitude such that water stands against the base of the cliff. As a result the cliff becomes saturated, fails, and large masses of sand slide down into the water to become rapidly dispersed thereafter. The extent of erosion has been measured by driving a series of 4 m long, 5 cm bore, steel pipes 2 m into the beach near the foot of the cliff at 14 sites along the eroding section of coastline. During the period September 1966 to April 1968, the position of the cliff base, relative to each of these 14 pipes, was recorded before and after each high water predicted to reach +8·6 m Liverpool Bay Datum (L.B.D.). Liverpool Bay Datum is 4·42 m below Ordnance Datum (O.D.): it approximately corresponds with low water spring tides, while Ordnance Datum is approximately equivalent to mean sea level. Tide levels were estimated at the posts and compared both with estimates from Mersey Dock and Harbour Board tide staffs at Formby Point (off Lifeboat Road), and recorded data for Princes Landing Stage, Liverpool, supplied by Mr. G. W. Lennon of the Institute of Oceanographic Sciences, Bidston (then the Institute of Coastal Oceanography and Tides).

The relationship between erosion per tide and tide height is shown in Figure 7.8. The data show a gradual increase in cliff erosion per tide followed by a rapid rise in erosion beyond a still-water level of +9·6 m L.B.D. Erosion per tide rises from initiation at approximately +8·5 L.B.D. to 0·75 m per tide at a still-water level of +9·4 m. At still-water levels of +9·6 m erosion is of the order of 1·5 m per tide and when levels of +9·75 m are attained erosion is between 4·5 m and 6 m of cliff, irrespective of cliff height or degree of vegetation. The change in gradient in the curve marks a change from tide levels where only undercutting occurs to tide levels where soaking occurs as well.

Local tide levels above +9·6 m L.B.D. are only rarely produced by undisturbed astronomical tides (tide levels greater than +9·6 m may occur once in 3 years (Lennon 1963b)). Thus, during the period of the study described here, the greater part of the marine erosion of the cliff was achieved by tides where the still-water level was raised by meteorological surge contributions to high water.

Surge contributions to high water levels increase the rate of erosion not only by adding soaking to wave attack but also by prolonging wave attack through maintaining water levels at or above a normal high water elevation longer than would otherwise occur.

Figure 7.6. Dune ridges eroded into deep 'blow-through' gullies with isolated dune remnants between. Near Wick's Lane. (Photography by P. Clark, Lancashire County Planning Department.)

Figure 7.7. Marine erosion has exploited a blow-through in a well-vegetated dune ridge. Blow-through follows unofficial footpath. Note angle between erosion facet of cliffs and dune ridge formed in 1920. Compare with oblique of same area in Figure 7.3. Nature reserve on right of picture lies below high water spring tides.

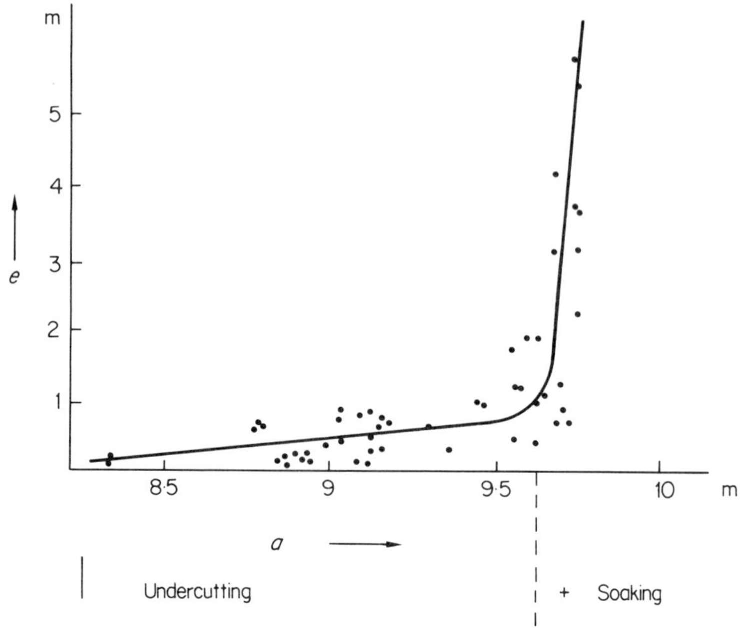

Figure 7.8. Dune cliff erosion, *e*, v. Tide height, *a*. Tide height relates to still-water level above Liverpool Board Datum (−4·42 m O.D.).

The relationship between erosion and tide height (still-water level) shown by these field observations contrasts with the model results of Edelman (1968) who reported gradually increasing erosion with increasing tide height (similar to curves where only undercutting occurs). Apart from discrepancies between Edelman's model and this natural beach (longshore currents are ignored in sand removal in the model) the difference arises from the fact that, as these field observations suggest, at more elevated tide levels two processes operate, their combination leading to a rapid increase in rates of erosion.

Similar comparisons between field data and the experimental work of Meulen and Gourlay (1968) also reveal significant differences in results. Their model tests show rapid erosion of the dune foot at the start of the tests followed by a decrease in erosion rate as the tests proceed. Like Edelman, Meulen and Gourlay conclude, as would be expected, that dune height inversely affects erosion. Their most interesting point, however, is their demonstration of the initial rapid recession of the dune foot. It is suggested that it is these data, from the first 2–3 hours of their tests, which are most significant; they represent the conditions which might occur as a high water with a meteorological surge contribution approaches. Such high waters in tidal seas are more likely to stand at the cliff for 2–3 hours rather than for up to 20 hours as in the various tests undertaken by Edelman and Meulen and Gourlay.

The difference between this field prototype and the model tests is considered to be most important. As Edelman (1968) suggests, during storms most of the upper part of the beach is 'planed off' and may not adjust to the incident wave spectrum before the tide ebbs. The effectiveness of repeated short savage attacks over high water, as opposed to the prolonged attack simulated in the model tests, may render the estimates of erosion derived from such tests inapplicable to field situations especially in areas of significant tidal range and subject to large surge contributions to high water.

The eroding cliff-line is not a smooth curve as maps of this area might suggest. It is comprised of a series of facets which meet at quite distinct 'corners' (Figure 7.3). The positions of the 'corners' can be seen to move irregularly and facets generated at the centre of the section of eroding coast (around Wick's Lane in Figure 7.2) spread north and south. As they do so the next facet along the coast is also spreading and, towards the northerly limit of the present erosion (Figure 7.3), there is a difference of 16° between the azimuth of the first erosion facet (015° N) and the dune ridges (Figure 7.7) formed about 1920 (Blanchard, 1953) along high water mark.

Subsurface erosion of the foreshore

It is in the northern area of transition between non-cliffed (though retreating) and cliffed coastline that the substantial degree of subsurface foreshore erosion is revealed. The sub-beach Holocene silts and clays are frequently and widely exposed at the edge of the runnel which is nearest to high water mark (Figure 7.9). They are exposed in all the runnels down to low water mark ordinary tides off Wick's Lane and Dale Slack Gutter (Figure 7.2).

On examining the thickness of sand over the surface of these Holocene deposits, three types of profiles have been found (Figure 7.11). In the north of the area where there is a region of balanced foreshore budget the typical profile is that of Figure 7.11(a). In the centre, off Wick's Lane or south of the area off Lifeboat Road, the profile is typical of Figure 7.11(e). At Dale Slack Gutter (Figure 7.2) a transition between balanced (to the north) and deficient sand budgets has been demonstrated by the Hydraulics Research Station, Wallingford (see H.R.S. (1969), Figure 9). Between here and just north of Victoria Road (Figure 7.2) a transition between the two sorts of Holocene surface morphology and beach structure can be observed (Figure 7.9). The following sequence of events is suggested to explain the observed profiles.

As the foreshore sand deficit spreads north (to be followed by the spread of cliff erosion) the beach profile is reduced down to a Holocene platform with piles of sand resting on it in the sites of the beach ridges (Figures 7.9, 7.11(a) and (b)). The exposed platform is eroded by the formation of large pits while the platform margins in the runnels are destroyed by retreat of a small cliff (in the case of the site shown in Figure 7.9 the cliff retreats northwards). This produces the profile shown in Figure 7.11(c). The movement of the beach ridges breaks down the cores of Holocene sediment by exposing them to wave attack and the profile Figure 7.11(e) results. Landward retreat of the whole intertidal area frequently exposes the edges or large parts of the steps in the profile of Figure 7.11(e).

Figure 7.9. Exposure of Holocene in runnels (small arrows) and on the edge of the intertidal flats (large arrow) between Victoria Road and Dale Slack Gutter. Group of three small arrows marks cliff in Holocene retreating northwards. Note wet areas on upper-foreshore plane area. (Oblique from 140 m, looking south; courtesy John Mills Ltd.)

Figure 7.10. Thin mud layer on sand. Cohesive surface deformed to symmetrical wave-generated ripples. Mud layer breaking up along its margin by water draining from within. Survey staff in Imperial Units; intermediate divisions 0·1 foot (3 cm).

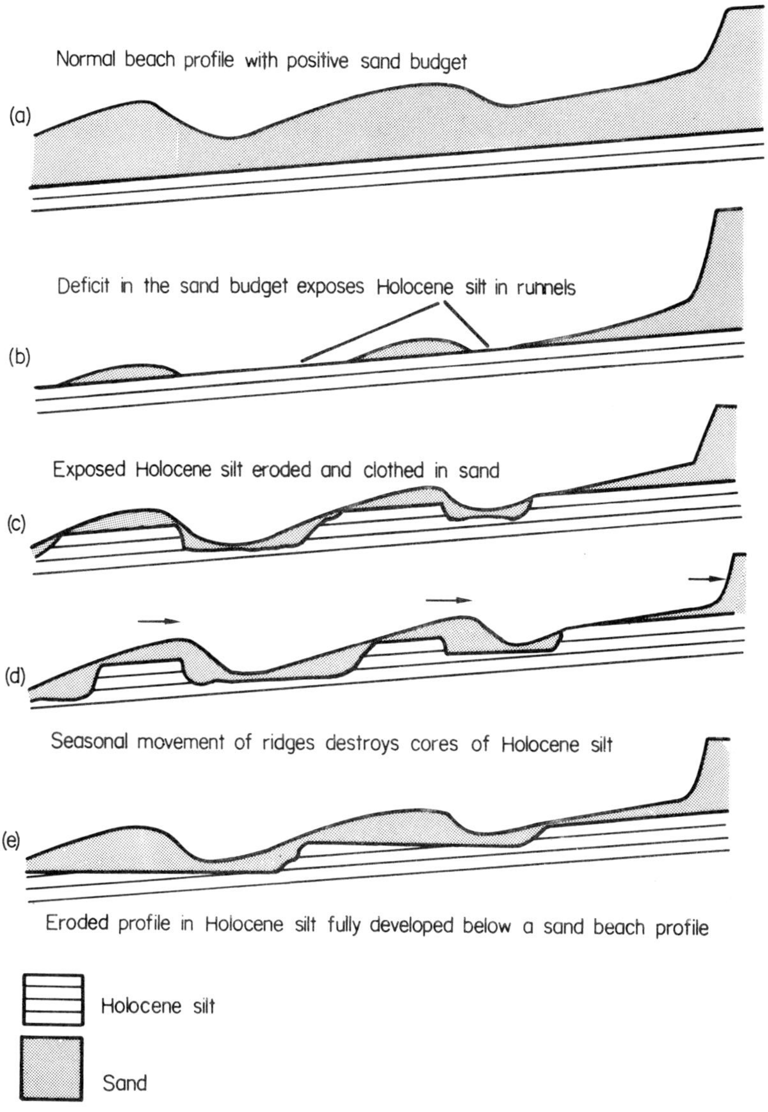
Normal beach profile with positive sand budget
(a)
Deficit in the sand budget exposes Holocene silt in runnels
(b)
Exposed Holocene silt eroded and clothed in sand
(c)
(d)
Seasonal movement of ridges destroys cores of Holocene silt
(e)
Eroded profile in Holocene silt fully developed below a sand beach profile
Holocene silt
Sand

Figure 7.11. Subsurface erosion.

Thus the whole of the coastal profile between low and high water mark is being eroded and is retreating eastwards. The disequilibrium between the present coastline, the foreshore and the offshore morphology, and their wave regime, as shown by these field observations, is paralleled in numerical model studies by the Hydraulics Research Station which show that for most wave conditions the angle between the breakers and the shore is large, indicating that the present coastline is far from being in equilibrium with its environment (H.R.S., 1969).

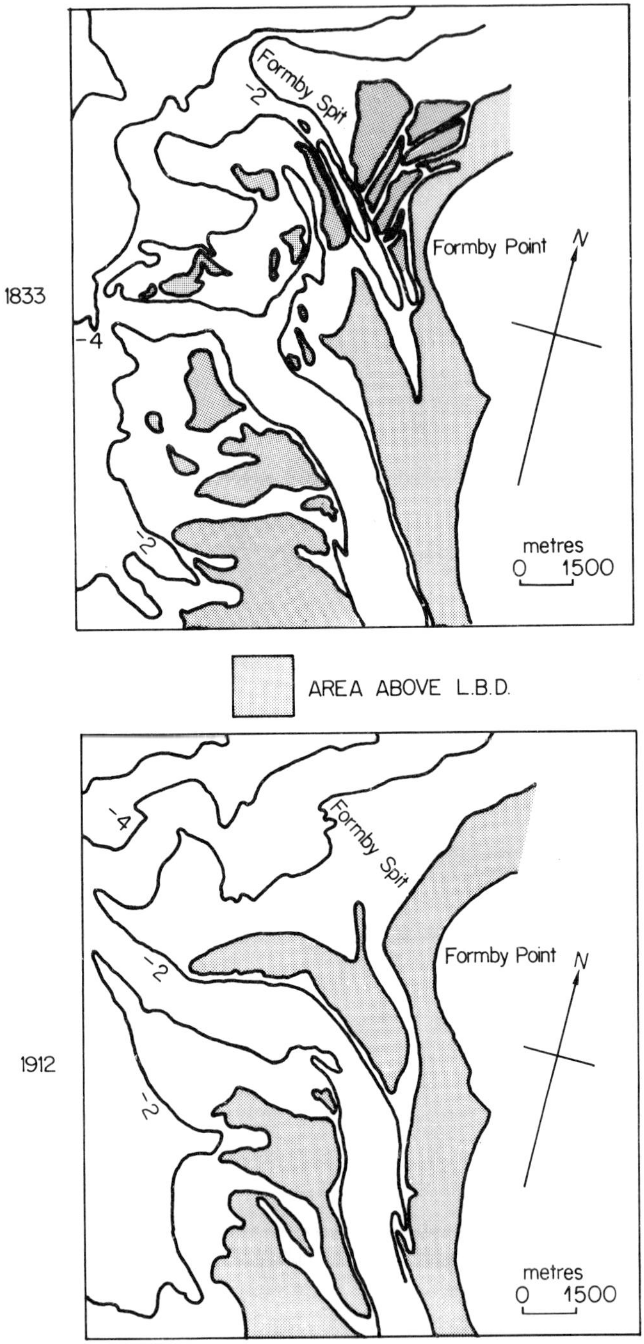

Figure 7.12. Changes in area above Liverpool Board Datum (−4·42 m O.D.), 1833–1912 (contours are in fathoms). Prepared from Mersey Dock and Harbour Board charts.

Figure 7.12, drawn from surveys conducted by the Mersey Dock and Harbour Board, shows large changes in the form of the intertidal area. The observations of erosion between 1964 and 1970 demonstrate the speed of changes in the position of high water mark. The transition from stable to eroding profiles, illustrated in Figure 7.11, indicates that it is necessary to define the depth below the sediment surface to which erosion is effective as well as observing the behaviour of the surface of the foreshore.

It is apparent, therefore, that the long-term erosion of the sand cliffs is achieved by processes whose short-term rates are closely linked to the combined effects of spring tides and meteorological surges. That these processes produce erosion is a result of the mechanisms influencing sediment movement both subtidally and intertidally.

Processes of sand movement on a multibarred foreshore

Within the intertidal zone waves and tidal currents are the two sources of energy available for movement of sediment (Figure 7.5). Waves move sediment either directly by their interaction with the bed or indirectly by their interaction with the foreshore topography inducing the development of 'secondary' currents such as longshore currents (Bruun, 1962, 1963; Chiu and Bruun, 1964), rip currents or other types of recirculation (Sonu and coworkers, 1966; Sonu and Russell, 1966). Tidal currents (including currents produced by water flooding into or draining out of runnels) may predominate over wave generated currents in determining directions of transport, especially on the lower parts of the intertidal area. Two types of hydrodynamic environment may be distinguished on the intertidal area. On topographic highs (ridges) breakers are dominant in the sediment transporting agencies. In topographically low areas wave generated currents (such as longshore currents) or tidal currents dominate. This division is reflected in the factors determining the directions of net transport and in the distribution of sediment types.

Sediment transport on ridges

On the seaward faces of the beach ridges breaker and swash/backwash action dominate sand movement processes with the predictable directional dependence on wave approach. On the crests of the ridges breaker and swash action dominate and net transport is landward. When the angle between the breakers and the ridges is large, dune bedforms develop on the crests of ridges (Simons and Richardson, 1961) (Figure 7.13). The dunes face along the ridge crest suggesting high bed load discharge along the ridge rather than landwards. However, such occurrences are not common at Formby.

The fate of the sand swept landward over the ridge is central to the sand circulation pattern on the foreshore and is determined by the timing of movement.

The evolution of hydrodynamic conditions on the crest of a ridge during one tide broadly follows a simple cycle. As the tide rises, the 'upper flow regime'

Figure 7.13. Dune bedforms on ridge off Fisherman's Path. View from first ridge towards upper foreshore plane area. The dunes have been substantially eroded by ebb breakers. Two thin metal rods are 2 metres apart. Dunes face north along ridge crest.

environment of the swash/backwash zone is replaced by the 'lower flow regime' of the breaker zone (Clifton and coworkers, 1971). Sediment movement on the ridge crest is predominantly landward into the next runnel. During high water, lower flow regime conditions persist over the ridge and the runnel to landward. Wave-induced transport is landward into the next runnel. During the ebb tide, lower flow regime conditions on the ridge are replaced by upper flow regime conditions, again with landward transport.

On both the flood and the ebb the most intensive sand transport occurs when swash water sweeps sand landward, over the crest of the ridge, into the next runnel. On other parts of the ridge at these times transport is either alongshore or seawards into the runnel seaward of the ridge.

The most intense landward movements of sand from the ridges are associated with those times on the ebb or flood when breakers are active over, or just seaward of, the ridge. At both these times longshore currents are active in the runnels landward of the ridge. On the flood maximum sand movement precedes the full development of the runnel currents; on the ebb the sand is fed into the waning longshore currents. The detailed sequence of events is shown in Table 7.1.

Thus the sand of the ridges is either swept landward or seaward into runnels where the currents move the sediment along the runnel, or it is moved along the seaward faces of the ridges. Landward transport paths are short.

It is apparent that the runnels and their associated longshore and rip current systems play a central role in the sand circulation pattern on this multibarred foreshore. The intensity and efficacy of sand transport in the runnels may be gauged from two lines of evidence. The runnels are the sites of widespread and persistent development of dune bedforms indicating longshore movement of

Table 7.1. Sequence of sand transport and energy

Low water	Currents draining runnel move sand alongshore.
Flood tide	Sand swept landward by swash over ridge crest into runnel filling with water. Breaker zone over ridge crest. Sand moved landwards into runnel which is site of longshore current. Development of dune bedforms in runnel. Breaker zone decays. Longshore currents decline. Low sand movement.
High water	Low currents, little sediment movement.
Ebb tide	Breaker zone develops over ridge. Longshore currents generated. Sand moved along runnel. Sand swept landward over ridge into longshore current. Sand swept landward by swash into runnel draining alongshore. Breaker zone develops over next ridge to seaward. Ridge exposed. Runnel drains. Currents move sand alongshore.

sediment. South of Formby Point the runnels discharge sand into Formby Channel. This sand feed has maintained two large 'deltas' at the exits from the runnels. As the ridges and runnels move so do the 'deltas'. The parts not being fed with sand are rapidly removed by the strong tidal currents in this channel but the active 'deltas' are well maintained.

It has been suggested elsewhere that the movement of ridges on a ridge and runnel beach is significant in the context of estimates of sand volume supplies across the foreshore. The three-dimensional nature of water flow and sand movement in the runnels ensures exchange of sand from one side of the runnel to another side even when the dominant movement is alongshore. It may be more realistic to view the ridges as forms through which sediment passes in exchange from one runnel to another.

The likelihood of rapid direct landward movement of sand across such efficient sediment traps as runnels appears small. This conclusion is supported by the work of Seibold (1963) and Reineck (1960 and 1963).

Intertidal sand transport pattern

For any given sediment, the rates and modes of transport depend upon the energy, steadiness and dimensions of the flow. The net transport of sediment in flows of varying or reversing direction will depend upon the energy budget related to particular flow directions.

In many marine situations the variability of bedforms is considerable and their significance as indicators of the direction of sediment movement must be examined carefully. The bedforms related to high energy but intermittent flows (swash or backwash) are different from those related to lower energy but more prolonged flows (tidal or longshore currents) and are of similarly varying significance.

Field observations in this area have shown that within the ridge and runnel zone of the foreshore, currents run in the same directions, on the flood and the ebb. This is because the currents are generated by the interaction of waves with topography and it is only in very exceptional circumstances that the wave approach direction varies from flood to ebb. The relative orientation of topography and waves remains virtually the same.

The consistency of flow patterns has been explored by hydrological experiments in both the shallow subtidal and intertidal zones using small boats and scaffolding towers (during storms) from which the relevant instruments were deployed. Various arrangements were attempted to measure currents, temperature, salinity and suspended solids. The use of optical suspended sediment meters was hampered by the high (500–5000 p.p.m.) concentrations of 'mud' in suspension. Samples collected with a horizontal bottle sampler suggested that, except within the breaker zones and very close to the bed, little sand moved in suspension (on the whole less than 300 p.p.m.). It was thus concluded that the dominant mode of sand transport was as bed load and that, bearing in mind the temporal consistency of flow directions, the patterns of dunes, ripples and other bedforms might be used to examine the patterns of bed-load transport on the foreshore.

Parker (1971) has demonstrated that the temporally predominant conditions of any part of this foreshore, except the upper foreshore plane area, are those of Clifton's lower flow regime (Clifton and coworkers, 1971). The sediment moves predominantly in ripples and dunes. Even within the upper foreshore plane area, during the period of tidal immersion, dune bedforms are frequently developed but are destroyed by the brief period of swash/backwash action of the receding tide. Dune bedforms (called *transverse-megaripples* by Straaten (1953) and *gross-rippelen* by Reineck (1960)) emerge as useful tools for following the patterns of sediment transport as demonstrated both by the field experiments on this foreshore, as well as those of Reineck (1960, 1961, 1963).

The sand transport pattern on this intertidal area will be considered in two parts. On the lower foreshore (Figure 7.5) tidal currents dominate the processes of sediment movement. The sense of transport based on dune bedforms (Figure 7.14) is northwards north of Wick's Lane and southwards south of Wick's Lane.

Within the upper foreshore the runnels are sites of predominantly longshore transport whereas the ridges are dominantly sites of sediment movement along vectors normal to the coast.

The pattern shown in Figure 7.15 is based on field experiments reported elsewhere (Parker, 1971) and the evidence from dune bedforms. The essence of the pattern is the divergence of bed load transport away from the centre of the erosion front, Wick's Lane. This mode of transport is concentrated in the runnels. Currents in the runnels actually move the sediment away from high water mark because of the relative orientation of the coast and the ridges. This offshore component of movement is reinforced by seaward discharges through rip channels, (Figures 7.3 and 7.4). Only the transport route along the upper foreshore plane area maintains a supply along the coastline which does not predominantly move sand offshore.

The geographical relationship between beach structure and cliff erosion has already been mentioned. The relationship between cliff erosion and sediment

Figure 7.14. Dune field on intertidal flats. Bedforms partly eroded by ebb breakers. Leeside pools rapidly fill with mud.

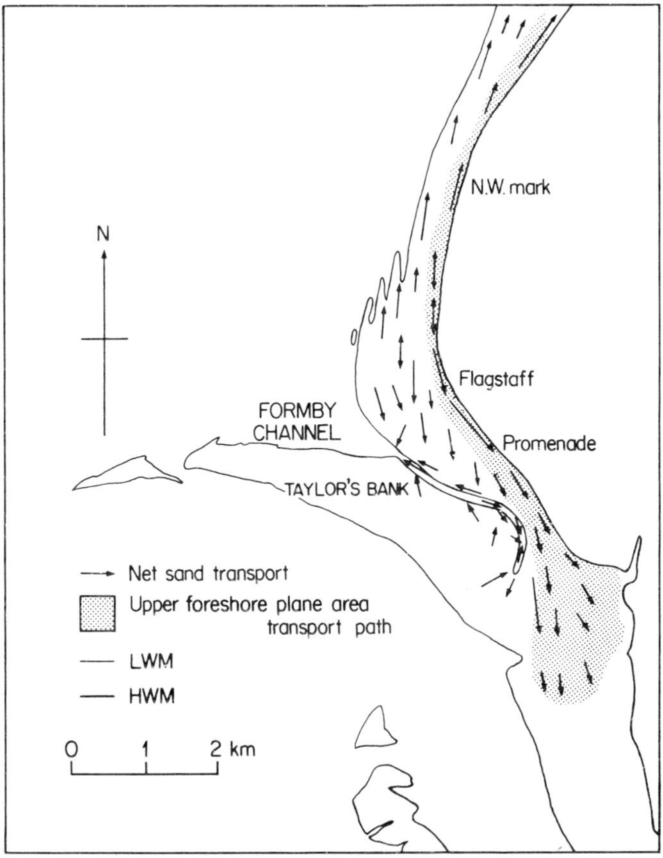

Figure 7.15. Intertidal sand transport.

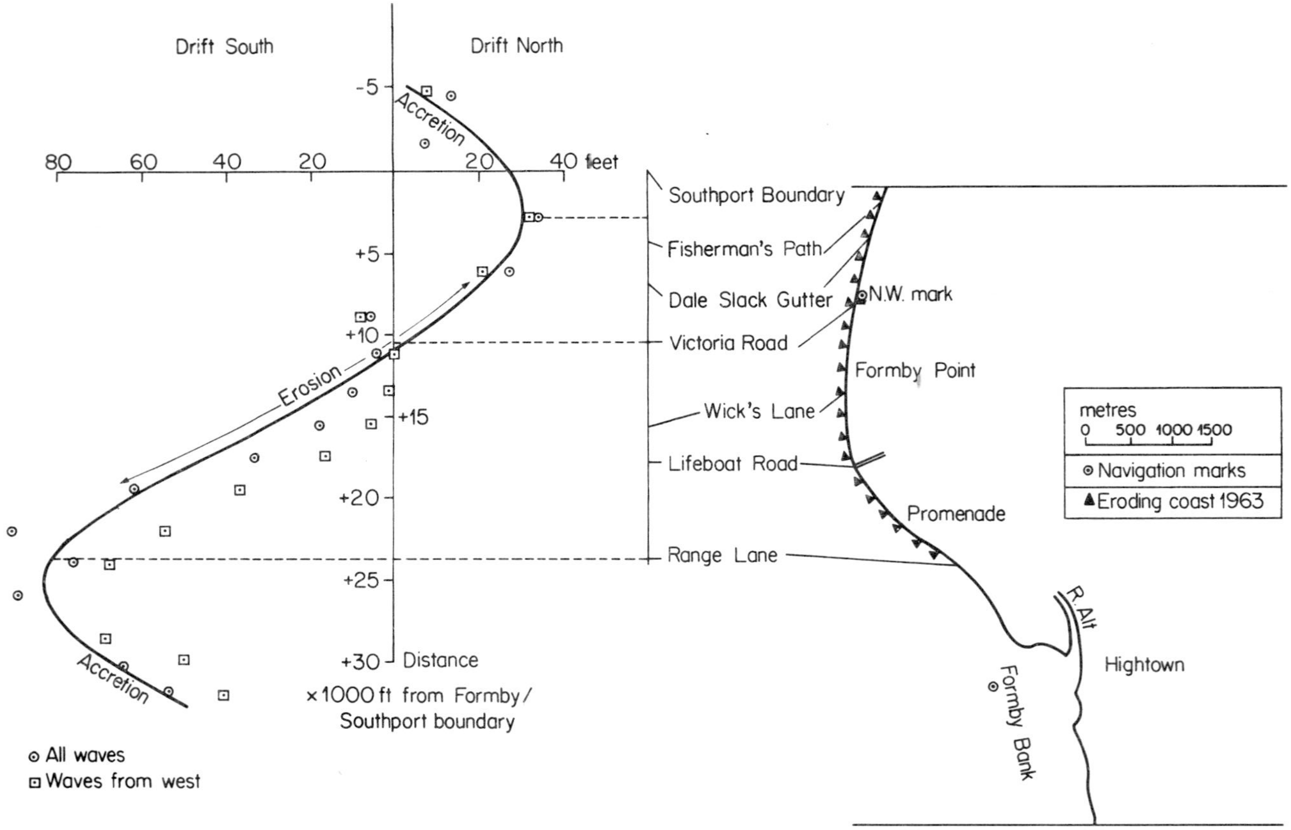

Figure 7.16. Computed littoral sediment transport, 1963, in arbitrary units. Increasing littoral transport produces erosion. Accretion starts where rate of movement of this sand decreases. Reproduced from Figure 14, H.R.S. (1969). Crown Copyright, reproduced by permission Controller HMSO, courtesy Hydraulics Research Station, Wallingford, England.

transport is of greater importance. The only attempt to estimate littoral sediment discharge along this coast has been made by the Hydraulics Research Station (H.R.S., 1969) which computed longshore mass sediment transport by wave refraction calculations based on wave energy data and bathymetry. The shape of the graphs of computed rates of littoral drift (Figure 7.16) correspond to the limits of cliff erosion. The H.R.S. data refer only to the upper foreshore plane area. This agreement between theory and the field situation is comforting, but somewhat perplexing at first sight. However if the sediment circulation on this foreshore is as suggested, the correspondence between the H.R.S. results for the extent of erosion and the field situation is to be expected since sediment transport in the runnels would be almost irrelevant to cliff erosion, except for the exchange of sand between runnels and the upper foreshore plane area whence the cliffs are nourished.

The correspondence between the H.R.S. analytical results and the field situation is interpreted as supporting the contention that runnel sand transport may not be directly relevant to the maintenance of the cliffs in present circumstances.

Inhibition of sediment movement

All the processes involved in the sand movement on this foreshore appear to act in concert to prevent sand reaching the eroding dunes or to remove such sand as may reach them. The actual efficacy of these sand transporting processes is also influenced by another process, perhaps not normally associated with sandy beaches, namely mud sedimentation. The importance of the mobility of the sediment is suggested by the sensitivity of this coastline to the littoral sediment flux (H.R.S., 1969). Thus factors which tend to minimize the effects of applied forces are not only sources of error in theoretical considerations but are also of great significance in the field situation.

Mud sedimentation increases the natural cohesion, and hence stability, of sand substrates by physically binding grains and by decreasing the permeability and porosity of the bed, thereby reducing lift induced by pressure differences across the sediment/water interface (Martin, 1972; Müller and coworkers, 1971 and 1972; Posey, 1969, 1971a and 1971b; Sleath, 1970; Watters and Rao, 1971).

Although field studies of mud erosion or settlement are comparatively few, the dynamics of the erosion of cohesive beds and some aspects of their sedimentation have been reported by Abdel-Rahman (1963), Einstein and Krone (1961, 1962), McCave (1970), Mignot (1968), Moore and Masch (1962), Partheniades (1964), Partheniades and Kennedy (1966), Pierce and coworkers (1970), Postma (1967), Terwindt (1967) and Terwindt and coworkers (1968).

There seems general agreement amongst these authors that clay additions or clay laminae in sandy sediments increase the resistance of the sediment to erosion. The stabilization of the sand effectively removes it from the budgetary system of the area.

Mud sedimentation

The fine sediment in suspension in the water at Formby is flocculated and occurs in concentrations as great as 5000 p.p.m. (in the shallow waters at low tide on the seaward edge of the intertidal flats) but within the intertidal zone the concentration of suspended fine sediment rarely exceeds 2000 p.p.m. Einstein and Krone (1962) have found that suspended sediment concentrations and flocculation kinetics are dominant parameters in mud sedimentation. They suggest that at high concentrations (20,000 p.p.m.) 'fluid mud' deposition may occur at low velocities. At suspension concentrations of below 10,000 p.p.m. a flocculent suspension may deposit some material whilst bed material is re-suspended, depending on bed shear stresses and the internal shear of the flow. DeWork and Kohler (1940), Einstein and Krone (1962), Mignot (1968) and Owen (1970) amongst many, have examined the consolidation of flocculated sediment. Initially, there is the formation of 'fluid mud' with concentrations above 20,000 p.p.m. Once concentrations have risen to the order of 100,000 p.p.m. settling is hindered because de-watering is controlled by the rate of escape of inter-flock water. In this period there are two well defined phases each of decreasing rate of consolidation but increasing concentration and cohesion. According to Einstein and Krone (1962) the end of the first stage of hindered settling marks the high limiting concentration for 'fluid mud'. The transition to the second stage of hindered settling takes place at the bottom of the mud layer before the surface has passed from the fluid mud stage. Again, according to Einstein and Krone, this transition period may be as short as two hours. Of particular relevance in this context of foreshore sedimentation is the work of Owen (1970) who has demonstrated the influence of mud thickness on consolidation rates, thicker layers consolidating more slowly than thinner ones.

Mud erosion

Laboratory studies of cohesive-bed erosion have been detailed by Abdel-Rahman (1963), Mignot (1968), Partheniades (1964) and Partheniades and Kennedy (1966).

The susceptibility of the mud bed to erosion depends upon the degree of de-watering it has undergone. In the field, the layers of still fluid mud are most susceptible to re-suspension and the topography of the transition between the first and second stages of hindered settling will have an important influence on the morphology of the mud surfaces eroded during the tidal flow period following deposition.

At Formby, the intertidally deposited mud quickly forms a coherent surface. De-watering is assisted by drainage and evaporation. Within a single tide mud layers in the runnels acquire, during low water, sufficient surface coherence to withstand currents of 30–40 cm s^{-1} in water 30 cm deep, yet they have such low shear strengths as to allow deformation by wave orbit currents to symmetrical

Figure 7.17. Dune bedforms in runnel of Lifeboat Road blanketed with mud. Mud and dunes eroded in runnel thalweg.

wave-generated ripples and then re-deformation of these ripples by currents draining the runnel. Mud with wave-generated ripples which when drained prove too weak to sustain the ripple shape is common in this area (Figure 7.10). Although one might expect that the survival of mud is determined by the proportion of the layer which has reached an advanced stage of de-watering, the parallel development of a surface coherence in intertidal muds must be borne in mind.

Figure 7.18. View along runnel. The small ridge (25 cm high, large arrow) provides protection for mud survival across the whole of the runnel (between small arrows). First runnel off Lifeboat Road looking south.

Mud survival

The amount of mud, especially in layers, found in the sediment depends upon three factors:

(a) Mud availability.
(b) Hydrodynamic environment, as reflected by
  (i) development of surface coherence, and the proportion of any settled layer that de-waters to a cohesion sufficient for survival;
  (ii) the proportion of the mud layer which is re-suspended on the next tide;
  (iii) the nature of the site as it affects local erosion or survival of mud.
(c) The degree of mechanical or biological mixing which stirs mud into the sand, thus affecting retention or re-suspension.

The supply of mud to this foreshore from the turbid waters of Liverpool Bay (Halliwell and O'Connor, 1966) ensures frequent mud sedimentation. The predominant influence on intertidal mud sedimentation and survival is the topography of the intertidal zone. Topographic highs on both the intertidal flats and the ridge and runnel zone suffer relatively high energy conditions during the ebb tide and consequently their sediments have a low mud content. Topographic lows are sites of mud sedimentation, mud occurring mainly as layers in the sand, as a blanket of soft mud over the sand, or mixed in as a muddy sand. Only in a very few areas is there no mud. Although mud sedimentation may be widespread over the intertidal area at Formby, especially in quiet weather conditions, mud survival is restricted to those sites protected from ebb breakers. This leads to the development of large mud patches on the lower parts of the intertidal flats or in areas where pools in the lee of dune bedforms fill with mud.

In the ridge and runnel zone of the foreshore, mud survival is most persistent in the runnels where protection is afforded by a ridge to seaward (Figures 7.17 and 7.18). Scouring, which removes mud where the water drains between dunes in the bottom of the runnel is common but often the whole of the sheltered zone of the runnel, and its bedforms, are clothed in a layer of mud (Figure 7.17). Mud layers in this situation frequently become stable within the 3 or 4 hours they are exposed over low water. A small protecting topographic high is necessary to induce a large area of mud sedimentation, as shown in Figure 7.18.

Conclusion

Observations on the multibarred foreshore at Formby Point, Lancashire, suggest that the depth to which erosion may occur within the beach is as significant a phenomenon as the behaviour of the sediment surface. It may, in this case, point to the profound nature of the readjustment of the foreshore and indicate the degrees of change to be considered in the design of any protective engineering works.

The relationship between the morphology of the foreshore and sediment transport requires further research. It is not thought that sediment transport that is predominantly alongshore and concentrated in the runnels is unique to this eroding area. It does, however, render the problems of identifying sediment source areas for particular coastal sections less tractable. If the upper foreshore plane area has the best potential for supplying the frontal dunes (and its importance is demonstrated (Figure 7.16) by the close correlation between the theoretical longshore sand flux and cliff erosion), what precisely is the relevance of subtidal and intertidal sediment transport? A quantitative understanding of the influence of mud on the mobility of non-cohesive sediment has yet to be attained. The effects of mud sedimentation can only reduce the accuracy of theoretical calculations of sediment flux. Subtidal and intertidal mud sedimentation must be borne in mind in planning any nourishment programmes and in recreational exploitation. What is clear is that there is still no substantial understanding of the mechanisms of alluvial coastal balance other than in the most general terms.

Acknowledgement

This study was undertaken during the tenure of a Lancashire County Council Postgraduate Studentship at the Department of Geology, Liverpool University. The assistance of members of that Department, especially Joe Lynch, the guidance of Dr. R. G. Bathurst, the cooperation of the Mersey Dock and Harbour Board and the award of the studentship are gratefully acknowledged.

References

Abdel-Rahman, N. M., 1963, The effect of flowing water on cohesive beds, *Mitt. Vers. Anst. Wass. Bau Erdbau, Zurich*, **56**, 1–114.

Belderson, R. H., and A. H. Stride, 1969, Tidal currents and sand wave profiles in the northeastern Irish Sea, *Nature, Lond.*, **222**, 74–75.

Best, R. (Ed.), 1972, *Out of sight, out of mind*, Report of a working party on sludge disposal in Liverpool Bay, H.M.S.O., London, 2 vols.

Best, R., G. Ainsworth, P. C. Wood and J. E. James, 1973, Effects of sewage sludge on the marine environment: a case study in Liverpool Bay, *Proc. Instn. civ. Engrs.*, **55**, 43–66; Discussion: **55**, 755–765.

Blanchard, B. A., 1953, The ecology of the sand dune systems of southwest Lancashire, Ph.D. Thesis, Liverpool University.

Bowden, K. F., 1955, Physical oceanography of the Irish Sea, *Minist. Agriculture, Fish and Food Fishery Invest.*, Series 2, Vol. 18, No. 8.

Bowden, K. F., 1960, Circulation and mixing in the Mersey Estuary, *Int. Assoc. Scient. Hydrol Surf. Wat.*, Publs. 51, 352–360.

Bowden, K. F., and S. H. Sharaf El Din, 1966a, Circulation, salinity and river discharge in the Mersey Estuary, *Geophys. J.R. astr. Soc.*, **10**, 383–400.

Bowden, K. F., and S. H. Sharaf El Din, 1966a, Circulation, salinity and river the Liverpool Bay area of the Irish Sea, *Geophys. J.R. astr. Soc.*, **11**, 279–292.

Bruun, P., 1962, Longshore Currents in one and multi-bar profiles relation to littoral drift, *Proc. 8th Conf. cst. Engng., Mexico City*, pp. 211–247.

Bruun, P., 1963, Longshore currents and longshore troughs, *J. geophys. Res.*, **68**, 1065–1078.

Chiu, T.-Y., and P. Bruun, 1964, Computation of longshore currents by breaking waves, *Engng. Prog. Univ. Fla., Tech. Paper No. 279.*

Clifton, H. E., R. E. Hunter and R. L. Phillips, 1971, Depositional structure and processes in the non-barred high-energy nearshore, *J. sedim. Petrol*, **41**, 651–670.

Cronan, D. S., 1969, Recent sedimentation in the central north eastern Irish Sea. *Rep. No. 69/8*, Inst. Geol. Sci., 10 pp.

Darbyshire, M., 1958, Waves in the Irish Sea, *Dock Harb. Auth.*, **39**, 245–248.

Draper, L., 1966, The analysis and presentation of wave data—a plea for uniformity, *Proc. 10th Conf. cst. Engng., Tokyo*, Vol. 1, pp. 1–11.

Draper, L., and A. Blakey, 1969, Waves at the Mersey Bar light vessel, *N.I.O. Internal Report A37*, pp. 1–4.

Edelman, T., 1968, Dune erosion during storm conditions. *Proc. of 11th Conf. cst. Engng., London*, Vol. 2, pp. 719–722.

Einstein, H. A., and R. B. Krone, 1961, Estuarial sediment transport patterns, *J. Hydraul. Div. Am. Soc. Civil Engrs.*, **87**, 51–59.

Einstein, H. A., and R. B. Krone, 1962, Experiments to determine modes of cohesive sediment transport in salt water, *J. geophys. Res.*, **67**, 1451–1461; Discussion: **67**, 1463–1464.

Greswell, R. K., 1953, *Sandy shores in South Lancashire*, Liverpool University Press 181 pp.

Halliwell, A. R., 1973, Residual drift near the sea bed in Liverpool Bay: an observational study, *Geophys. J.R. astr. Soc.*, **32**, 439–458.

Halliwell, A. R., and B. A. O'Connor, 1966, Suspended sediment in a tidal estuary (Mersey), *Proc. 10th Conf. cst. Engng., Tokyo*, Vol. 1, pp. 687–706.

Hydraulics Research Station, 1969, The south west Lancashire coastline, a report of the sea defences, *Report No. EX 450*, 43 pp.

Lennon, G. W., 1963a, The identification of weather conditions associated with the generation of major storm surges along the west coast of the British Isles, *Q.J.R. met. Soc.*, **89**, 381–394.

Lennon, G. W., 1963b, A frequency investigation of abnormally high tidal levels at certain West Coast ports, *Proc. Instn. civ. Engrs.*, **25**, 451–484.

McCave, I. N., 1970, Deposition of fine-grained suspended sediment from tidal currents, *J. geophys. Res.*, **75**, 4151–4159.

Martin, C. S., 1972, Hydrodynamic effects of seepage on bed particles. Discussion of Paper by Watters and Rao, 1971, *J. Hydraul. Div. Am. Soc. civ. Engrs.*, **98**, 276–279.

Meulen, T. van der, and M. R. Gourlay, 1968, Beach and dune erosion tests, *Proc. 11th Conf. cst. Engng., London*, Vol. 1, pp. 701–707.

Mignot, C., 1968, Étude des propriétés physiques de différents sediments très fins et de leur comportement sous des actions hydrodynamiques, *Houille blanche*, **7**, 591–620.

Moore, W. L., and F. D. Masch, 1962, Experiments on the scour resistance of cohesive sediments, *J. geophys. Res.*, **67**, 1437–1446.

Müller, A., A. Gyr and T. Dracos, 1971, Interaction of rotating elements on the boundary layer with grains of a bed; a contribution to the problem of the threshold of sediment transportation, *J. hydraul. Res.*, **9**, 373–411.

Müller, A., A. Gyr and T. Dracos, 1972, Reply to comment of C. J. Posey, *J. hydraul. Res.*, **10**, 351–355.

Murthy, T. K. S., and J. Cook, 1962, Maximum wave heights in Liverpool Bay, Vickers Armstrong, Department of Design, *Report No. V3031/HYDRO/04.*

Owen, M. W., 1970, Properties of a consolidating mud, Hydraul. Res. Stn., Wallingford, *Report INT 83*, 51 pp.

Parker, W. R., 1971, Aspects of the marine environment at Formby Point, Lancashire, Ph.D. Thesis, Liverpool University, 2 vols., unpublished.

Partheniades, E., 1964, A summary of the present knowledge of the behaviour of fine sediments in estuaries, Hydrodynamics Lab. M.I.T. *Technical Note No. 8*, 47 pp.

Partheniades, E., and J. F. Kennedy, 1966. Depositional behaviour of fine sediment in a turbulent fluid motion, *Proc. 10th Conf. cst. Engng., Tokyo*, Vol. 1, pp. 707–729.

Pierce, T. J., R. T. Jarman and C. M. de Turville, 1970, An experimental study of silt scouring, *Proc. Inst. civ. Engrs.*, **45**, 231–243.

Posey, C. J., 1969, Erosion prevention experiments, *Proc. 13th Cong. Int. Assoc. Hydraul. Res.*, Kyoto, Japan, Vol. 2, pp. 211–219.

Posey, C. J., 1971a, Protection of offshore structure against underscour, *J. Hydraul. Div. Amer. Soc. Civ. Engrs.*, **97**, 1011–1016.

Posey, C. J., 1971b, Discussion of paper by Muller, Gyr and Dracos, *J. hydraul. Res.*, **9**, 373–411.

Postma, H., 1967, Sediment transport and sedimentation in the estuarine environment, in G. H. Lauff (Ed.), *Estuaries*, pp. 158–179, Am. Assn. Adv. Sci., Publication No. 83.

Price, W. A., and M. P. Kendrick, 1963, Field and model investigations into the reasons for siltation in the Mersey Estuary, *Proc. Instn. civ. Engrs.*, **24**, 473–518.

Ramster, J. W., and H. W. Hill, 1969, Current system in the northern Irish Sea. *Nature, Lond.*, **224**, 59–61.

Reineck, H. E., 1960, Uber den transport des riffsandes, *Jahrb. 1959 Forsch. Stelle Norderney*, **11**, 21–38.

Reineck, H. E., 1961, Uber sandverlagerungen im bereich des nassen strandes, *Jahrb. 1960, Forch. Stelle Norderney*, **12**, 13–26.

Reineck, H. E., 1963, Sedimentgefüge im bereich der südlichen Nordsee, *Abh. sen. kenb. naturforsch. Ges.*, **505**, 1–138.

Seibold, E., 1963, Geological investigation of nearshore sand transport, in M. Sears (Ed.), *Progress in Oceanography*, Vol. 1, Macmillan, New York, pp. 1–70.

Sharaf El Din, S. H., 1970, Some oceanographic studies on the west coast of England, *Deep Sea Res.*, **17**, 647–654.

Simons, D. B., and E. V. Richardson, 1961, Forms of bed roughness in alluvial channels, *Am. Soc. Civ. Engrs., J. Hydraul. Div.*, **87**, 81–105.

Sleath, J. F. A., 1970, Wave-induced pressures in beds of sand, *J. Hydraul. Div., Am. Soc.* Civ. Engrs., **96**, 367–378.

Sly, P. G., 1966, Marine geological studies in Liverpool Bay and adjacent areas, Ph.D. Thesis, University of Liverpool, unpublished.

Sonu, C. J., and R. J. Russell, 1966, Topographic changes in the surf zone profile, *Proc. 10th Conf. cst. Engng., Tokyo*, Vol. 1, pp. 502–524.

Sonu, C. J., J. M. McCloy and D. S. McArthur, 1966, Longshore currents and nearshore topographies, *Proc. 10th Conf. cst. Engng., Am. Soc. Civ. Engrs., Tokyo*, Vol. 1, pp. 525–549.

Straaten, L. M. J. U. van, 1953, Megaripples in the Dutch Wadden Sea and in the basin of Arcachon (France), *Geologie Mijnb.*, **15**, 1–11.

Terwindt, J. H. J., 1967, Mud transport in the Dutch delta area and along the adjacent coastline, *Neth. J. sea Res.*, **3**, 505–531.

Terwindt, J. H. J., H. N. C. Breusers and J. N. Svasek, 1968, Experimental investigation on the erosion-sensitivity of a sand-clay lamination, *Sedimentology*, **11**, 105–114.

Tucker, M. J., 1963, Analysis of records of sea waves, *Proc. Instn. civ. Engrs.*, **26**, 304–316.

Watters, G. Z., and M. V. P. Rao, 1971, Hydrodynamic effects of seepage on bed particles, *J. Hydraul. Div., Am. Soc. Civ. Engrs.*, **97**, 421–439.

Work, L. T. de, and A. S. Kohler, 1940, Sedimentation of suspension, *Ind. Chem. Engng.*, **32**, 1329–1334.

Wright, J. E., J. H. Hull, R. McQuillin and S. E. Arnold, 1971, *Irish Sea Investigations 1969–70, Report 71/19*, Inst. Geol. Sci., H.M.S.O., 55 pp.

Discussion

Dr. P. H. T. Beckett, Department of Soil Science, University of Oxford. It is disturbing to observe that aeolian sand does not accrete on the seaward faces of dunes with nearly vertical faces, since this would have been the mechanism one would have wished to exploit to reclaim eroded dunes. The dunes presumably must have developed originally by aeolian accretion on sloping surfaces. Have we got to bulldoze the eroded face to a 30° slope or something like that, in order to encourage accretion?

The dune system at the north end of the area is, in part, artificial in that it was formed by the planting of brushwood breaks, in about 1921 when the coastline in that area was generally accreting. I don't know how the dunes in the central area were formed because it is very difficult to determine their age. They are younger than the 'horizon' of ironmongery that is reputed to have been placed there during the First World War but it is difficult to visualize what the seaward face of the dunes would have been like originally. To the south, and in the north where the coastline is accreting normally, you do get small linear ridges formed. As these gradually build up another one forms to seaward.

F. J. T. Kestner, Hydraulics Research Station, Wallingford. How are the 'wet areas' to which you referred at the beginning of your lecture, formed? Along the Holderness coast, north of the Humber, there are similar wet areas, which are called 'orbs'. When they occur cliff erosion is particularly severe. If the mechanism were known it might be possible to eliminate them and so lessen erosion.

I don't know what the prevention mechanism would be. All I can say is that very often the water which flows from this beach near the wet patches is of extremely low salinity. It is likely to be coming from the slack areas inshore behind the dunes.

The wet areas, themselves, are a response to the underlying Holocene sediment immediately seaward of them.

Dr. J. M. Thomas, Department of Geology, University of Exeter. How do the mud layers, which you describe in the runnels, accumulate when dune bedforms occur during the flood tide?

The dunes are usually formed when there is sufficient wave energy to develop currents in the runnels and under those circumstances mud does not accumulate. It tends to collect quite rapidly in the runnels during quiet weather conditions, and very often on neap tides. There is speculation about whether the mud is locally derived from the erosion of the underlying Holocene material or whether it comes in from the dredgings which are dumped to seaward. In the areas where the runnels are persistently muddy there are no dunes as such but there are layers of sand and mud. I can only presume that the sand is carried into the runnel in suspension by breakers etc. and then it adheres to the mud surface and further mud is added to it at times when conditions allow. You often get some mud settling out at low water in pools in the dunes because water is retained there; even the dune bedforms themselves may become clothed in mud. Then, perhaps, the whole lot may be sliced out by the currents which drain them at low water.

Dr. J. R. Hardy, Department of Geography, University of Reading. Were your current observations designed to detect, or give any indication of, ebb and flood currents moving in separate channels, of the kind described by Dr. Robinson in recent publications? It has been suggested that ebb and flood channels close inshore may account for some rapid shoreline changes, and it seems possible that your area might contain such a system.

The questions you pose should be considered under two headings, foreshore currents and currents in Formby Channel. If we take the foreshore first, the runnels are subject to two sources of currents. While tidal currents affect the runnels during the earliest part of the flood tide as the water fills the runnels, and the last part of the ebb tide as the water drains away, by far the most important type are the surf zone currents generated by oblique breakers in the way described by Per Bruun in his 1962 and 1963 papers and by Chiu and Bruun in their 1964 University of Florida publication. These surf zone currents flow in the same direction on the ebb and flood at Formby due to the general nature of breaker and wave approach directions. The tidal currents are not very significant.

In contrast, the currents in Formby Channel show a much more interesting relationship. If one examines the dune bedforms in the Channel, three distinct situations may be discerned. On the margins of the Channel there are some sites where dunes persistently have a flood tide orientation and other sites with a persistently ebb orientation suggesting classical ebb/flood avoidance within the Channel. Thirdly, the dunes on the floor of the Channel show ebb/flood reversal but the temporal asymmetry of the facing directions would suggest flood transport of sand. At low water these dunes face ebb. They reverse rapidly on the flood and maintain their flood asymmetry until the very end of the ebb when they reverse direction again. However, computed volume discharges for a large number of stations in this Channel show that water flow is either completely ebb-dominated at all levels or, in the case of one part of the Channel, there is almost no residual transport of water. As sediment transport, especially sand transport, relates to the budget of work rather than the budget of water volumes it suggests that relating net flow of water to net flow of sediment by simple mean transports or volume calculations may prove misleading. Although, as Robinson has suggested, some channels are dominated by ebb tide flow and others by flood tide flow, it is dangerous to assume that mass transport of sediment within these Channels has this same division. I agree that circulation within channels close inshore may account for some rapid shoreline changes (and this indeed is the case in this part of the southwest Lancashire coastline), but the correspondence between ebb and flood in terms of water flow and ebb and flood channels in terms of net sediment transport has yet, in my opinion, to be established.

CHAPTER EIGHT

Cyclical Changes in Shoreline Development at the Entrance to Teignmouth Harbour, Devon, England

A. H. W. ROBINSON

Abstract

The cyclical pattern of change at the mouth of the Teign Estuary in Devon has been recognized for over a century. Previous studies have been of limited duration so it is not surprising that estimates of the cycle period have varied between wide limits, with a maximum of about 7 years. There are also conflicting views as to the relative efficacy of tidal currents and waves in bringing about the observed changes of bottom configuration.

Since 1964 the morphology of the approaches to the estuary has been studied by mapping bank changes, tracing sediment movement and undertaking some direct measurements of the operative processes. For the most part the approach has been geomorphological and deductive. In spite of some inherent limitations in the method adopted, knowledge gained from 10 years of observations makes it clear that the cyclical pattern is both complex and irregular. Even in the small area of the estuary approach there are distinct current- and wave-dominated environments.

Introduction

The possibility that changes which occur in a marine environment follow a regular pattern has been noted from a number of areas and forms an attractive hypothesis of shoreline genesis. Thus, for example, Kestner (1970) invokes the cycle concept in his analysis of the changing bottom configuration of Morecambe Bay and, similarly, de Boer (1964) has adopted the same approach in his discussion of the evolution of Spurn Spit at the mouth of the Humber. In both these cases the cycle, as envisaged, has a comparatively long period, over 100 years in the case of Morecambe Bay and about 240 years in the Spurn cycle. This long time sequence means that, for Spurn, de Boer has had to rely on less accurate data for his earlier periods and while this does not invalidate the cycle hypothesis, the general conclusions obviously carry less conviction. For Kestner's analysis, the available hydrographic data began in 1845 and as only one complete cycle has taken place since that date, there is no guarantee that the same sequence of change he mentions will repeat itself in approximately the same period.

At the entrance to Teignmouth Harbour the suggested cycle is of much shorter duration, approximately 3 to 5 years, a fact recognized as long ago as 1856 by Spratt who made the first hydrographic study of the area.

The site (Figure 8.1) has the advantage of being comparatively small and hence changes along the shoreline and in the adjacent offshore zone can be readily mapped making a synoptic record possible. The area appears to form a largely self-contained unit with an almost constant sediment supply. Little is added either by cliff erosion of the Ness or from the coast to the north. The latter is protected by the railway and by a sea wall. The offshore zone also probably contributes little for the sediment grain size, even less than 1 km from the shore, is very different from that found on the inshore banks where medium to coarse sand and fine shingle are characteristic. Similarly, the river is unlikely to be a major source of sediment because the sand flats in the lower estuary appear to be formed of marine deposits carried upstream by a dominant flood residual which occurs along the Shaldon shore. The area can thus be regarded as in a state of dynamic equilibrium where a repetitive cycle of events can evolve,

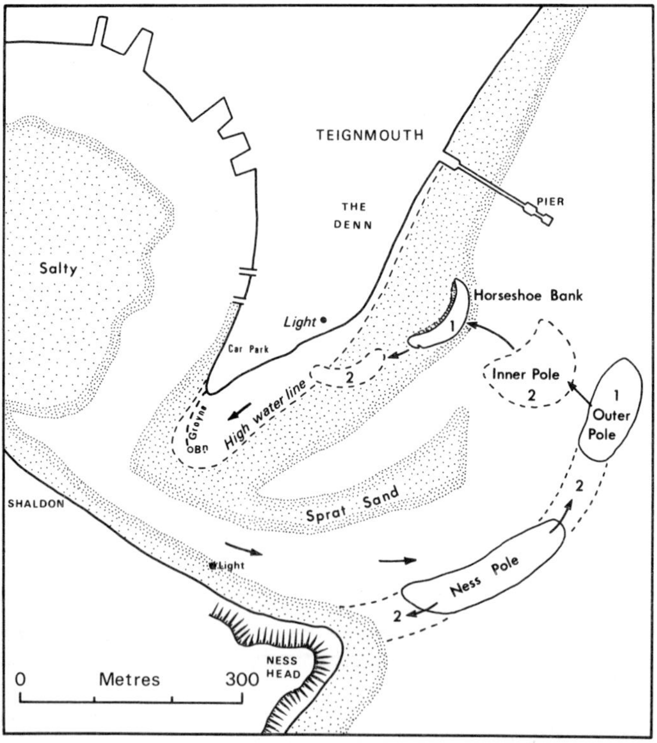

Figure 8.1. The general setting of the entrance of Teignmouth Harbour, Devon, England, with a diagrammatic representation of the principal changes in bank positions.

largely independent of those external forces which usually complicate and modify the picture in other areas. For these reasons the entrance to Teignmouth Harbour is an ideal field laboratory where studies of sediment movement, the interplay of marine processes caused by waves and tidal currents, and cyclical patterns of evolution can be examined.

Teignmouth estuary is comparable to others in East Devon in that it is a drowned river valley filled with sediment. The rock floor, composed of Permian sandstones and breccias, lies at a depth of up to 25 m below the present sand surface. The main channel winds across the sand flats between Newton Abbot and the mouth of the estuary at Teignmouth. At its exit the channel narrows considerably because of the southerly growth of the Denn Spit. The restricted width has enabled the current to erode a deep 'hole' in the beds of Permian breccia. Along the southern, or Shaldon, shore the breccias form a pronounced cliff culminating in the Ness which rises to 52 m and completely dominates the harbour entrance. Between the foot of the Ness and Teignmouth Pier the seabed is fashioned into a number of offshore banks through which the main estuary channel enters the deeper waters of the bay. It is the movement of these banks and the associated changes along the Denn beach which have given rise to the suggestion that a cyclical sequence exists in the approaches to Teignmouth Harbour.

Methods of study

The main approach to the problem by the present author has been a qualitative one and has involved the preparation of morphological maps of the estuary entrance over a period of 10 years from May 1964 to March 1974. The main stages of the continuously evolving pattern are recorded in Figures 8.2 to 8.7. The configuration has been determined by oblique photographic analyses with the individual photographs taken from the top of Ness Head. Linear measurements were used to provide a primary control in relation to fixed positions such as the Lucette and Denn lighthouses (Figure 8.2). On occasions a fuller plane table survey of the whole of the drying banks was made and the opportunity taken to run levelling profiles to determine the three-dimensional form.

In addition to the morphological mapping carried out at 2- to 3-monthly intervals throughout the ten-year period, observations on sediment movement were undertaken using both marked pebbles and fluorescent tracers. Those in 1969 were conducted jointly with a former colleague, Dr. S. J. Craig-Smith (1970); the experiments in 1973 by the present author.

Tidal currents were recorded at various points within the estuary and nearshore areas, and at differing stages within a tidal cycle.

Main diagnostic stages

Figures 8.2 to 8.7 provide a good illustration of the various stages in the cycle of development.

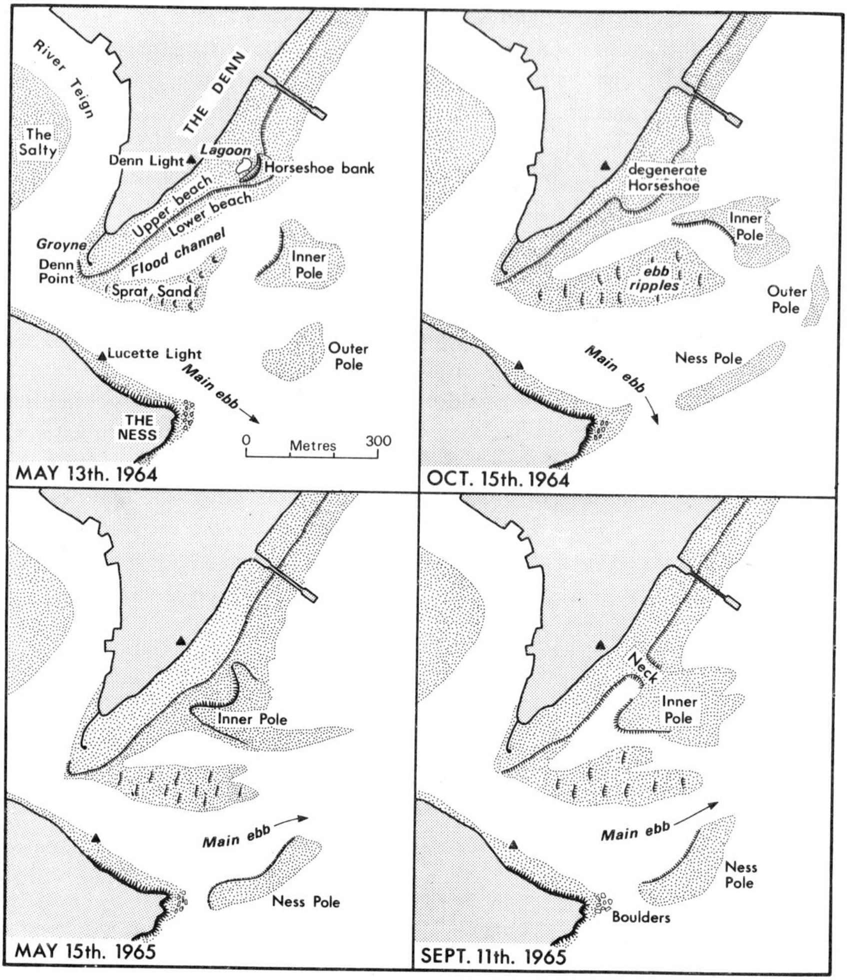

Figure 8.2. The changing pattern of banks during the first two years of observations.

Both the Outer and Ness Poles have similar characteristics. These are:

(1) That changes are fairly rapid, especially during the winter months, with the likelihood of completely different morphological characteristics following storm conditions.
(2) Growth is associated with elongation. The resulting instability ultimately leads to fragmentation mainly as a consequence of wave attack.

The Outer Pole can develop either as a result of the disintegration of the Ness Pole or quite independently from a supply of sediment brought down the ebb channel from the harbour mouth. As the Outer Pole becomes driven onshore it forms the Inner Pole. Its relief then increases from about 0·5 m above chart

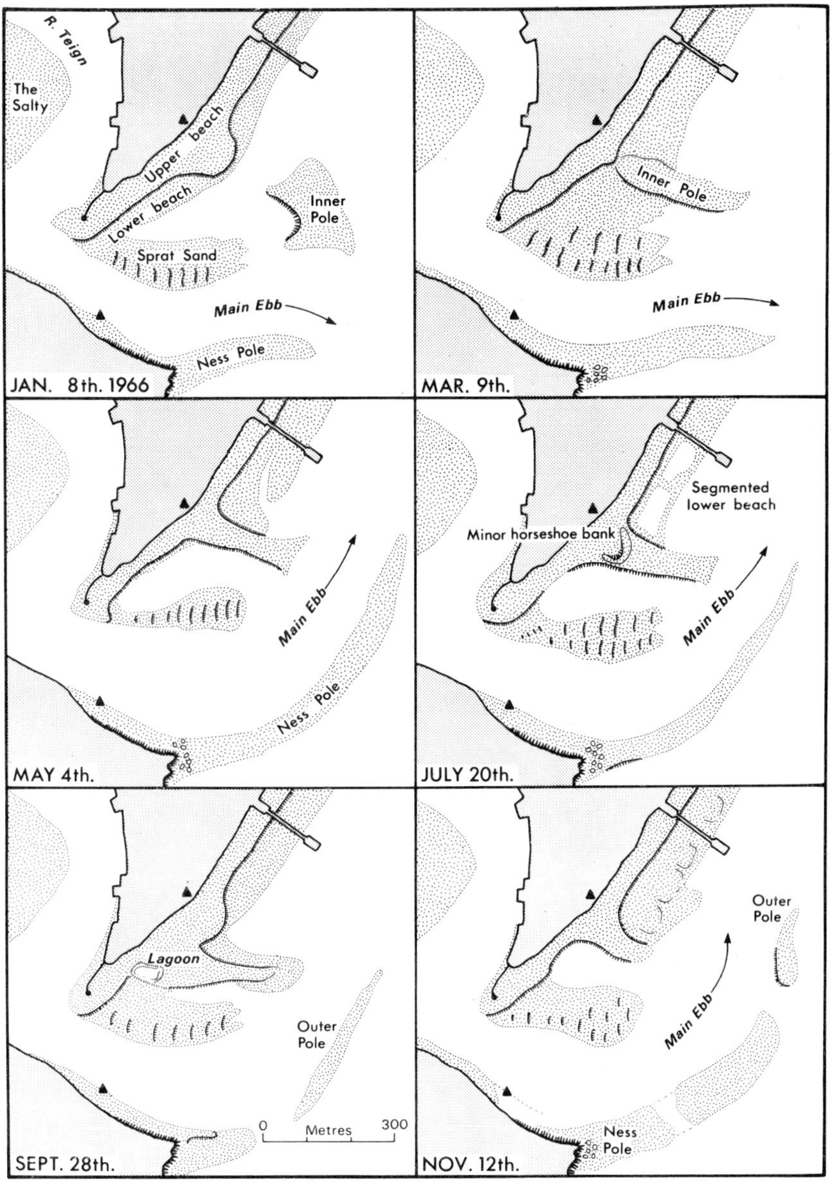

Figure 8.3. Changes in the bottom configuration throughout 1966. (May 4) and (July 20) show a prominent Ness Pole and depleted Denn Beach.

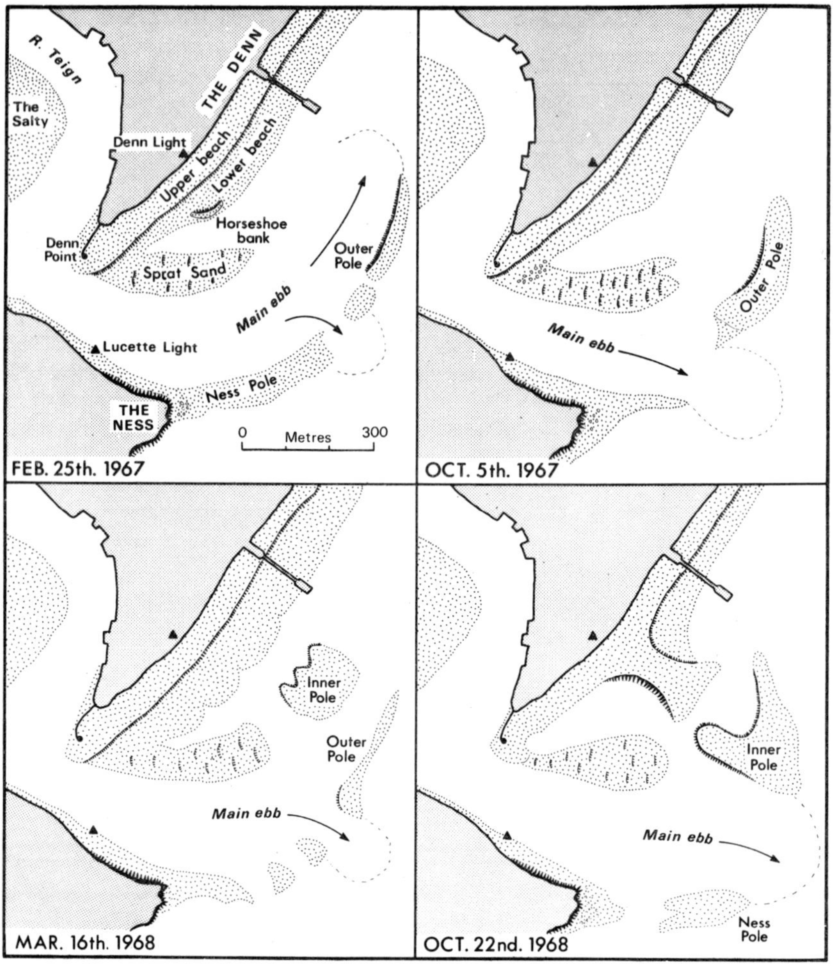

Figure 8.4. The pattern of banks and channels in the estuary approaches during 1967 and 1968. (Oct. 5) and (Mar. 16) show the transformation of the Outer Pole to the horn-shaped Inner Pole.

datum (approximately extreme low water mark) to some 2·5–3 m near its crest, while its former linear shape becomes barchan-like, triangular or, more rarely, cipher-shaped. An increase in height of up to 0·7 m has been observed between successive tides. Similarly, both Outer and Inner Poles have also been known to move laterally by some 5 m over a corresponding period. Clearly the overall amount of sediment in motion may be considerable.

The Inner Pole ultimately joins Denn Beach to form the arcuate Horseshoe Bank, a unique response to a combination of processes.

From Denn Point the low sandbank of Sprat Sand extends seaward at an angle to the spit. Throughout the 10 years of observations it is the only feature which

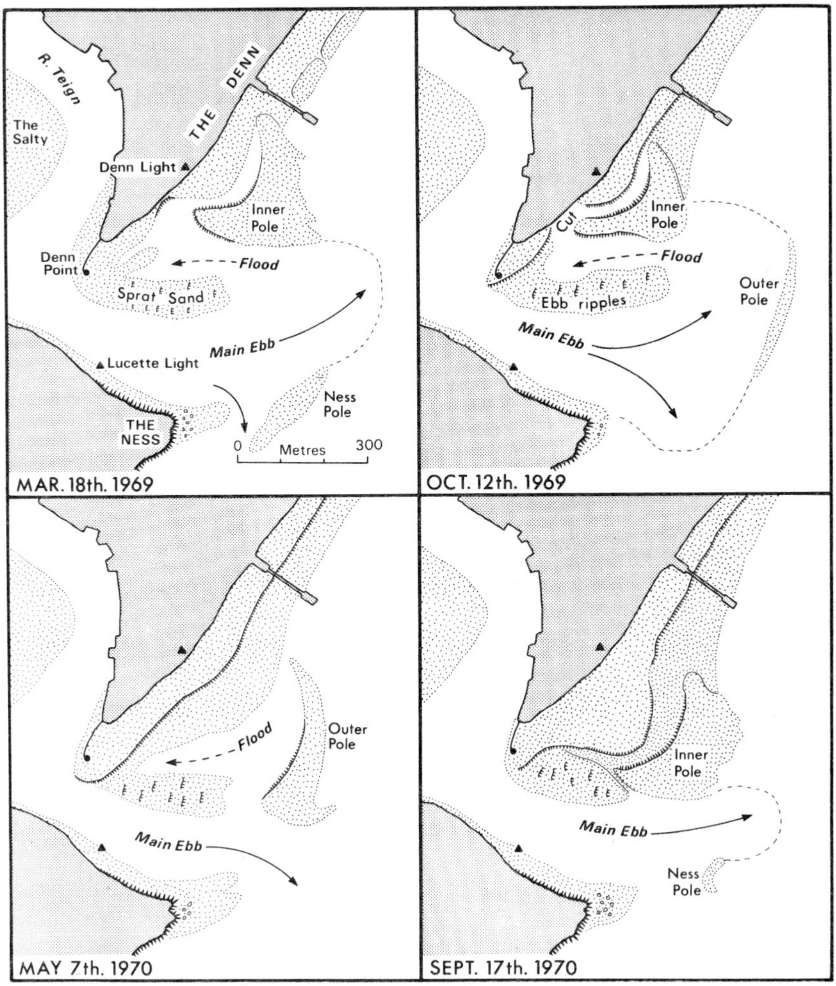

Figure 8.5. The main disposition of banks and channels throughout 1969 and 1970.

has shown any degree of stability though it, too, can change in length and breadth according to the vagaries of sediment supply. It is low throughout and even at low water spring tides it seldom rises more than 1 m above datum. The surface of coarse sand is usually fashioned into a series of prominent ebb ripples with an amplitude of 0·5 m and a wavelength of approximately 2 m though the ripple forms tend to become smaller towards the distal end of the bank. On the inner side there is usually a well marked channel or gut developed which comes in from seaward towards Denn Point. Like Sprat Sand it is an oft-repeated feature of the hydrography and though its depth may vary to the extent that the marsh clay base is sometimes exposed, it tends to persist while considerable changes of

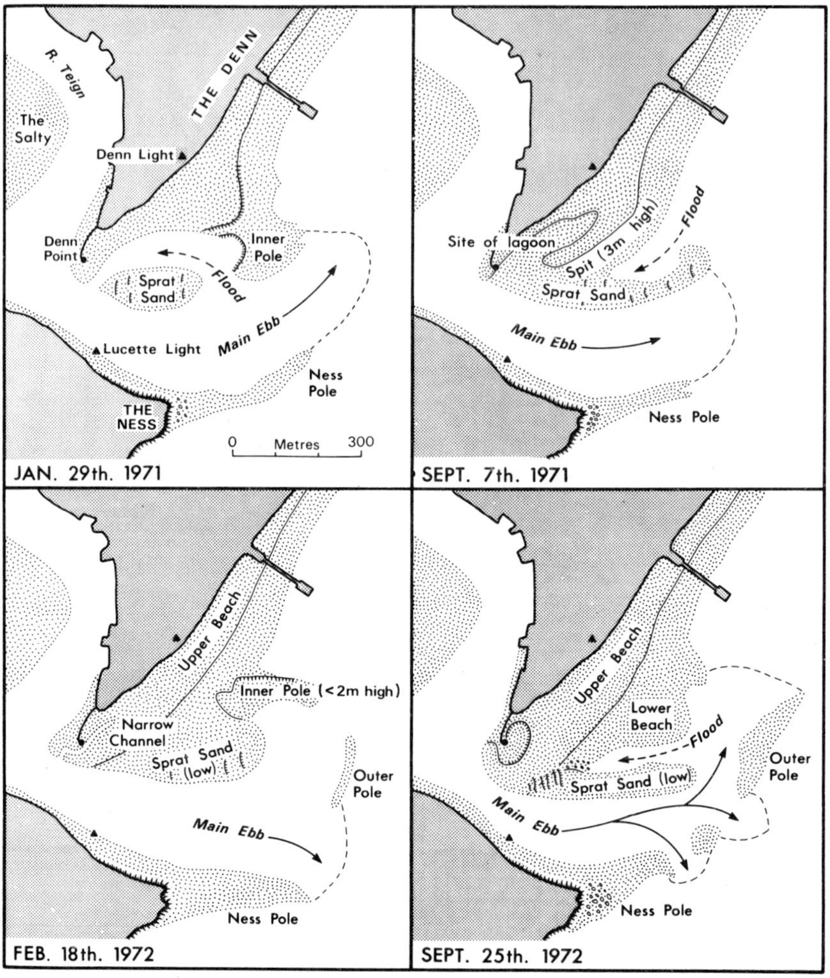

Figure 8.6. The changing configuration of the seabed in 1971 and 1972.

configuration are taking place all around. After low water it forms the earliest route for the rising flood tide which enters the channel while the ebb is still flowing out from the harbour entrance.

The remaining shoreline involved in the pattern of sediment movement at the entrance to Teignmouth Harbour is that which forms the southern margin to the estuary from Shaldon to Ness Head. Throughout the period of investigation there has been a noticeable deficiency in the amount of sand along this shore particularly at the seaward end between the Lucette Light and the Ness Head. In this section bare rock, with boulders, is often exposed with only a superficial covering of sand. Closer to Shaldon ferry the amount of sand increases so that opposite Denn Point the beach is broad and does not appear to change significantly from year to year.

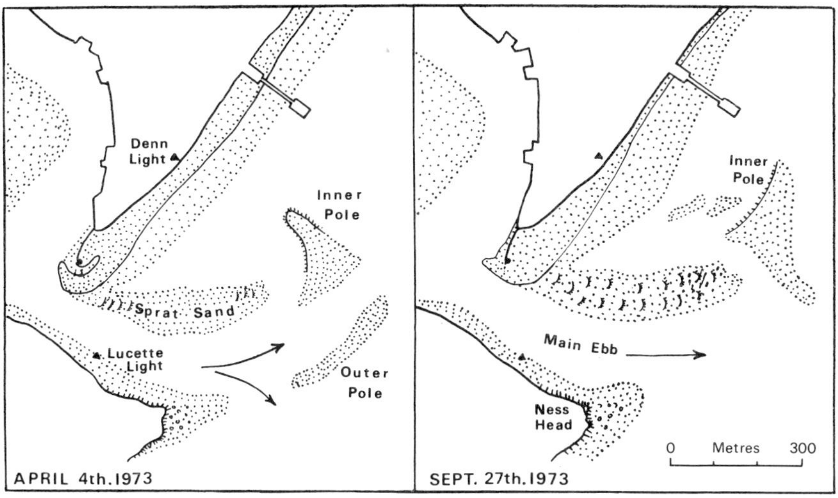

Figure 8.7. Two recent morphological maps of the approaches to Teignmouth Harbour.

In an estuarine environment such as this the sediment circulation is extremely complex and not surprisingly, therefore, the resultant bank and channel configuration can show unexpected changes over a relatively short period of time. It was for this reason that a long-term study, lasting a decade, was considered necessary in order to unravel the cycle of change as well as attempt to determine the reasons for minor perturbations within the general cyclic pattern.

The study of the changing hydrographic pattern confirms Spratt's view that a cycle of change exists with three main diagnostic stages. Stage 1 occurs when the available sediment is mainly in the vicinity of Ness Head, either as a projecting spit or more usually as the bar known as the Ness Pole lying athwart the main estuary channel which, in consequence, is forced to take a more north-easterly route seaward (Figure 8.3 (July 20)). Stage 2 is characterized by the development of the offshore banks of the Inner and Outer Poles lying seaward of the Denn Lighthouse (Figure 8.4 (Mar. 16)). The final Stage 3 occurs when these banks move onshore and a build-up of sediment takes place on Denn Beach, either in the form of the Horseshoe Bank (Figure 8.2 (May 13)), or as a triangular apron of sediment in front of the lighthouse. This sediment ultimately travels southwestwards towards Denn Point where it contributes to building up the beach by the car park and in the vicinity of the concrete groyne. Spratt was clearly puzzled by the variable length of the cycle which he estimated to be '3 to 5 years or even longer'. The recent study over the past decade throws some light on the reasons behind the variable length of the cycle. On average it seems to last about 40 months but it can be considerably shorter with a minor cycle occurring within the major period. Where the Inner Pole Bank comes ashore well ahead of the Outer Pole it will contribute its own quota of sediment to the Denn Beach and if this is then quickly dispersed because of winter storms, the sediment circulation passes immediately through Ness Pole Stage 1 and leads to the building up of the offshore banks again, all within a period of 6 months. Another minor cycle can occur when the accumulation of sediment forming the banks of the Inner and Outer Poles takes place close inshore. When this happens the anticlockwise circulation of sediment has a much smaller radius and takes less time to complete. This happened between December 1969 and December 1970 when, following a short phase of accumulation at the foot of Ness Head lasting only 3 months, the growth of the Outer Pole took place with great rapidity (Figure 8.5). During the summer of 1970 the bank quickly moved onto the beach. By the end of 1970 the Denn Beach was rapidly losing sediment at its distal end and this was quickly followed by the appearance of the Ness Pole, thus heralding the beginning of a new cycle. The minor cycle thus barely lasted 13 months (see Figure 8.8). It seems likely that Spratt's failure to recognize the existence of minor cycles superimposed on the major circulation patterns largely accounts for his estimate of a variable length of between 3 and 5 years.

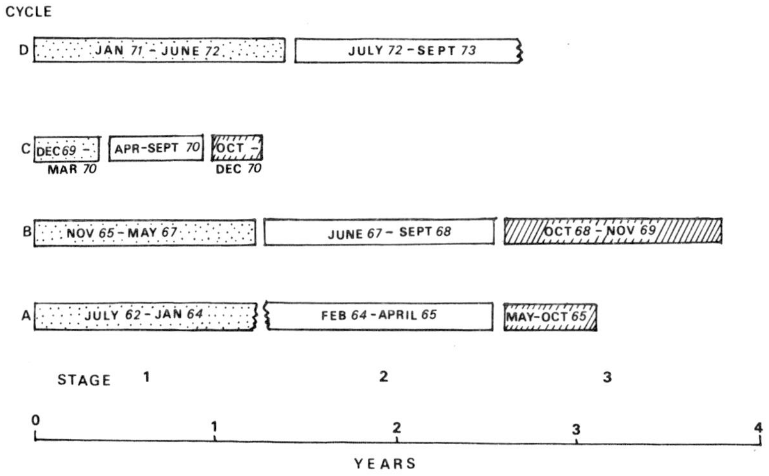

Figure 8.8. Proposed cycles in the pattern of banks and channels with their definitive stages throughout the period of the observations from May 1964 to September 1973.

Operative processes

Because of a dearth of both qualitative and quantitative field studies no general agreement exists about the main processes which affect the seabed configuration close inshore; the hydrography of the entrance to Teignmouth Harbour is no exception. When the engineer John Rennie reported on the condition of the harbour in 1838 he attributed the effects to wave action, but he was largely unaware of the sequential changes (Rennie, 1838). That wind direction is important is undisputed, for it obviously influences local wave generation. At

Teignmouth the coastal configuration has a considerable bearing on the problem. Because of the existence of Ness Head, that part of the Denn Beach between the pier and the distal point of the spit, together with the Shaldon shore in the vicinity of the Lucette light, is only affected by winds which approach from a direction within the overall sector between 025° and 190°, i.e. from north-northeast to slightly west of south. For individual sections of the shoreline, the wind approach sectors are even more limited. While changes do occur during periods of easterly gales it would be misleading to consider these conditions, as Rennie did, as the critical and overriding factor in the estuary and shoreline sedimentation sequence. The record of changing bank configuration over the past ten years suggests that the disturbance brought about by winter storm conditions is probably instrumental in causing a speeding up in the onshore migration of a bank like the Inner Pole when it is close inshore. Similarly the movement of sediment on the Denn Beach, often involving the destruction of the feature like the Horseshoe Bank and the translation of its sediment in a southwesterly direction towards the Denn Point, can take place very rapidly under winter storm conditions. On the other hand an easterly gale might have only a minimal effect on occasions.

In view of this, the role of tidal currents needs examination to see how far this process, either by itself or in conjunction with wave action, can account for the cycle of change which is associated with the sediment movement. It is undisputed that tidal currents, which reach velocities of almost 6 knots in the estuary approaches under certain conditions, can move even the coarsest grades of sediment. As in so many estuary environments the tidal characteristics at Teignmouth are slightly modified from the normal regular sinusoidal curve. Measurements made by the Hydraulics Research Station show that the period from low

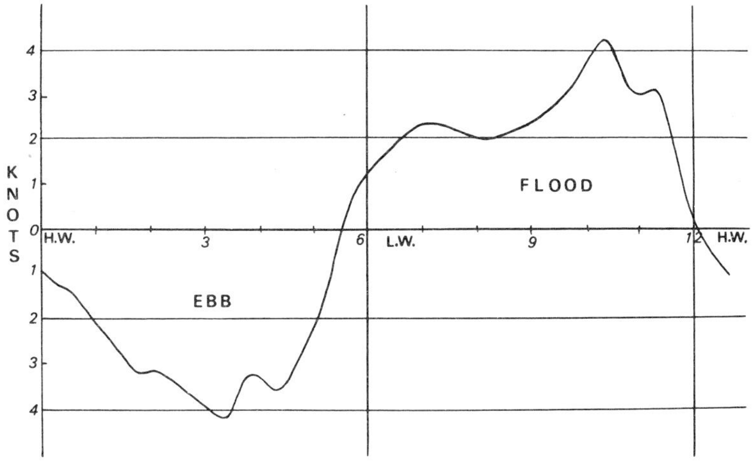

Figure 8.9. Tidal current curve for a station close to Denn Point, based in part on Hydraulics Research Station data. The measurements relate to a depth approximately 0·3 m above the seabed.

to high water is approximately 6½ hours compared with 5¾ hours for the following high to low water stage for a station in the harbour entrance (Figure 8.9) (Hydraulics Research Station, 1970). This means that the flood tide enters the estuary and runs approximately three-quarters of an hour longer than the ebb. If other factors are excluded, such as the relative strengths of the ebb and flood streams, this would result in a net sediment movement towards the estuary rather than in the reverse direction. There is some evidence that this is the case for marked pebbles deposited on the Denn Beach ultimately reached the inner part of the estuary both on the Salty Bank and the adjacent Shaldon shore. Much of the sediment which fills the inner estuary and forms banks like the Salty has probably been derived from seaward, being carried to its present situation by the dominant flood current along the Shaldon shore. This implies that a flood residual flow occurs very close inshore even though, in the main axial channel running seaward from the harbour mouth, flow is clearly in the opposite direction.

Only a small proportion of the sediment entering the inner estuary is allowed to remain there permanently, for the ebb stream, though of shorter duration, is quite capable to returning the bulk of it to the outer banks and beaches. Figure 8.10 shows that the ebb stream flows with velocities exceeding 3 knots for a much longer period than the corresponding flood.

The pattern of tidal flow is also of great importance in relation to bottom sediment movement in different parts of the estuary approaches (Figure 8.10). The velocities given refer to maximum values associated with spring tide conditions; neap tide velocities are approximately half these values. The measurements also relate to near surface flow and, while there is a decrease with depth, bottom velocities still amount to approximately 90 per cent of those of the surface streams. Drogue tests show that at the beginning of the ebb the stream flows out past Denn Point and then swings northeastward to join the anti-clockwise circulation in this western part of Lyme Bay. This direction continues for the second quarter but there is a steady increase in velocity so that speeds of 3·4 knots are reached off the Denn Spit after two hours from the beginning of the cycle. This increases still further so that, by the time of mid-tide, velocities are approaching 5 knots in the restricted exit between the Point and the Shaldon shore. Further seaward there is a reduction to about 4 knots. During the third quarter the stream turns more to the east to hug the Shaldon shore and flows out directly under the Ness Head. Speeds of the current are now beginning to decrease but they still exceed 3 knots. The final quarter, leading to the time of low water, is marked by a turning of the tidal stream to the south once it passes Ness Head. Because of the exposure of the banks near the time of low water the stream is now more and more restricted to the main estuary channel but even towards the end of the ebb, it still attains velocities of 2·8 knots. The change in direction of the stream out of the estuary during the third and final quarters is in response to the complete reversal of the main circulation of Lyme Bay which takes place about 2 hours before local low water. The same change takes place on the flood when the reversal of the streams in Lyme Bay is out of step by about

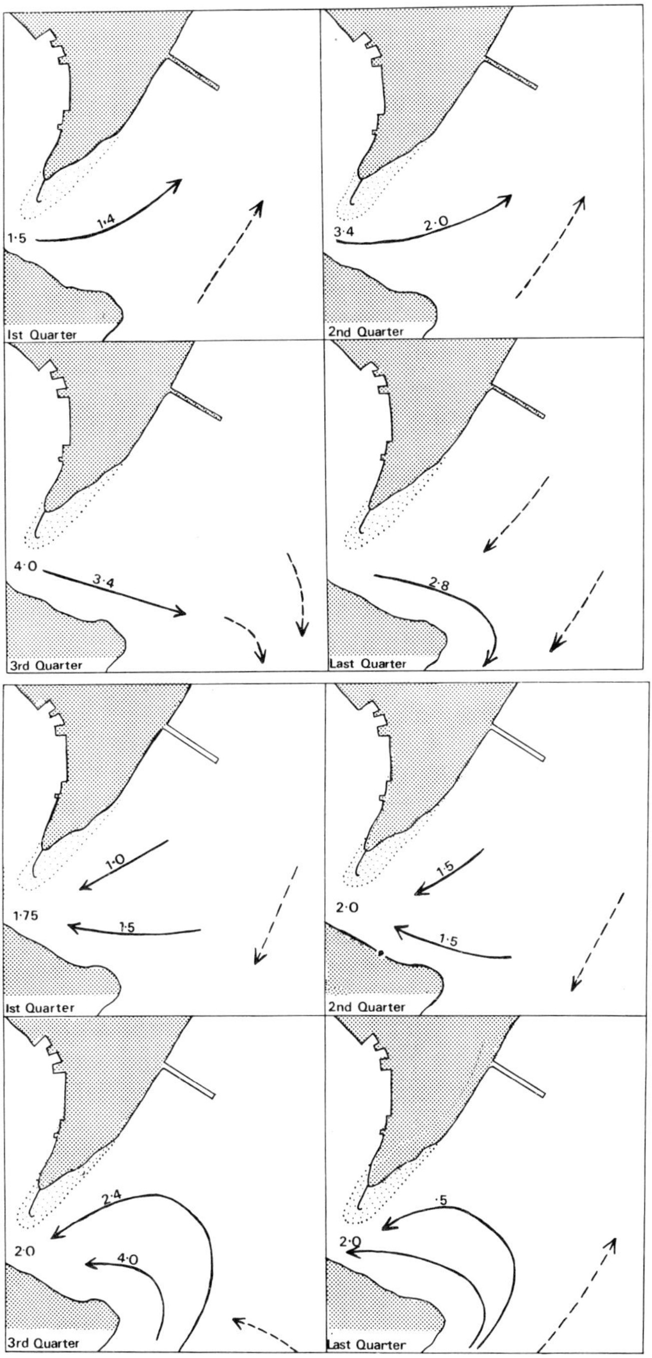

Figure 8.10. The pattern of tidal currents during the ebb (above) and during the flood (below). The velocities are in knots and relate to surface conditions.

2 hours with the changeover in the estuary mouth. This time lag leads to the stream which enters the estuary in the third and fourth quarters from the southward being forced to take a circuitous route. In general velocities are slightly less on the flood compared with the ebb but even so maximum values of 4·4 knots are reached off Denn Point at mid-tide.

It is evident from this pattern of water circulation that the northern part of the estuary is affected longer by a southwesterly flowing stream since it begins during the last quarter of the ebb and lasts for the whole of the succeeding flood. The result is that there is a definite water residual in the direction of the flood stream and this has resulted in the creation and maintenance of the flood channel which lies between the Denn Beach and Sprat Sand, one of the few constant elements in the hydrography of the estuary. The fact that the flood takes this more circuitous course during the third and fourth quarters means that its influence is less in the main channel of the harbour where an ebb residual water transfer exists. The effect of a flood residual trending southwest in the vicinity of the Denn Beach and an ebb residual nearer Ness Head shore is to create an anticlockwise water circulation in the estuary approaches and this must, of necessity, influence the movement of sediment.

It might be invoked that the pattern of the changing seabed configuration over the last decade and data gleaned from other sources afford some indication of the probable mode of sediment circulation. More specific studies using tracer techniques have been made both by the Hydraulics Research Station (H.R.S., 1966) and by the present author. After 24 hours, fluorescent tracer placed on both sides of Sprat Sand was found to have spread in a linear pattern parallel to the edge of the bank. That along the edge of the main channel showed a greater movement seaward while that on the inner side tended to move up the blind gut towards Denn Point (Figure 8.11). After a further 24 hours the tracer showed a similar tendency of dispersal though becoming more difficult to locate through burial. Sediment movement was of the order of 100 m in the 48-hour period and is an indication of the highly dynamic conditions which exist near the harbour entrance. Quantities of fluorescent sand were also placed at two points on the adjacent Denn Beach. The resulting envelopes of dispersion are shown in Figure 8.11. On the lower part of the beach near the flood channel the movement was basically linear and therefore similar to that on the Sprat Sand. On the upper part of the coarse sand and fine shingle beach the spread was much greater and this would seem to imply more effective wave action but with a longshore component added, hence the general translation towards Denn Point. It is possible that the sediment movement represents the effects of saltation where wave agitation lifts the sediment off the beach and then tidal currents carry it forward before it finally settles on the bed. On the lower part of the beach, wave action seems to have had only a minimal effect on dispersing the sediment. Here the observed movement more clearly reflects the influence of residual tidal currents which carry the sand along a narrow front towards the harbour mouth. During the period of the experiment the winds were offshore so that there was no evidence of beach drifting as a result of obliquely approaching waves.

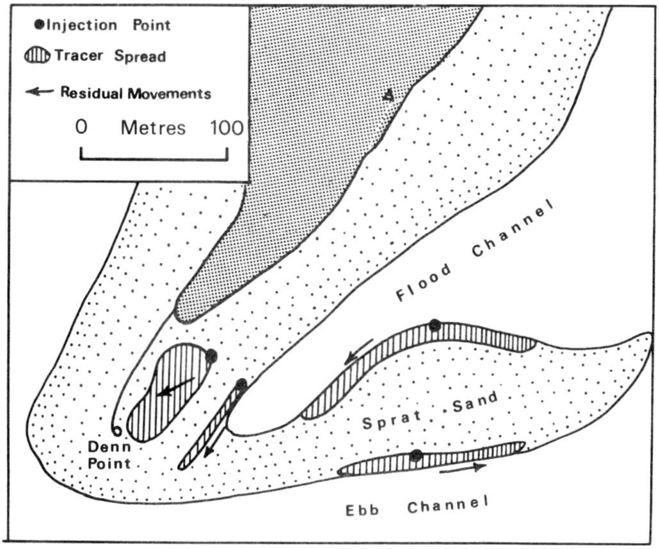

Figure 8.11. The results of various tracer experiments carried out simultaneously on Sprat Sand and in the vicinity of Denn Point.

It is apparent that in this particular estuarial environment both wave action and tidal streams operate, to varying degrees, either independently or sometimes in harmony, by inducing sediment movement (Figure 8.12). When waves break on the beach obliquely from an easterly or northeasterly quarter there is considerable beach drifting towards Denn Point. The flood residual water movement which, as has been noted above, operates immediately off the Denn Beach, will also give rise to sediment movement in this same direction toward the harbour mouth and is still effective when wave action is negligible as, for example, during periods of offshore winds. The result is that under most conditions there is a steady, consistent, movement of material on Denn Beach from north of the pier towards Denn Point at the harbour mouth. This consistent direction of sediment movement forms an important part of the cyclical pattern of change in the estuary.

The development of the Horseshoe Bank, resulting from the onshore migration of the Inner Pole, also illustrates the way in which wave and tidal currents can shape the seabed topography. While still well offshore the Inner Pole exists mainly as a result of wave action, for in the open sea, within a short distance of the estuary mouth, tidal stream velocities are relatively low, less than 0·5 knots, and therefore of little effect as regards sediment transport. Coming closer inshore, the bank assumes a crescentic shape with prominent projecting horns turned inwards towards the shore. This phase of development can be seen to be the result of waves and tidal currents working in harmony through saltation. Observations made on the Inner Pole during a rising tide show that along its

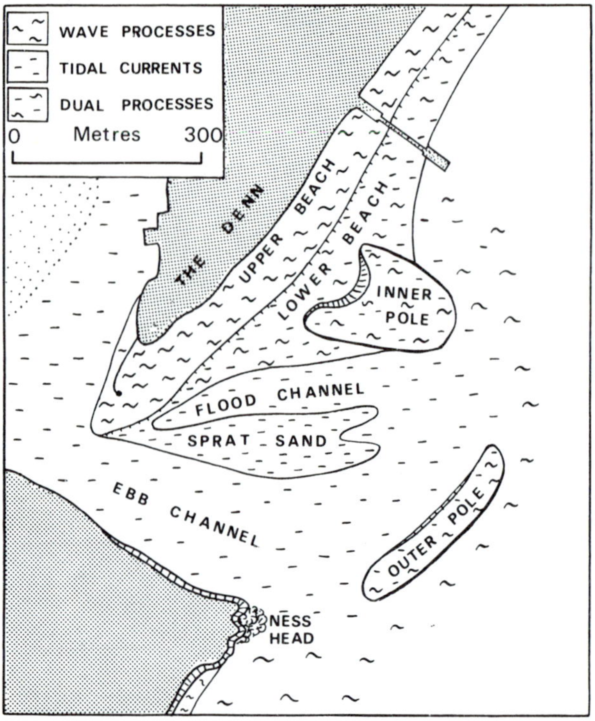

Figure 8.12. A diagrammatic assessment of the areas of influence of tidal current and wave processes.

margins wave agitation causes a considerable disturbance of the sediment which is then carried forward, i.e. shorewards, by the tidal currents which sweep around the bank. There is also considerable turbulence to add to the favourable conditions for mass sediment transport. As the bank is carried onto the upper beach the influence of tidal currents is reduced to zero and the more irregular wave processes ultimately bring about a complete destruction of the horseshoe shape leaving only a broad ness projection in front of the Denn Lighthouse.

The development of the Ness Pole, which extends from the foot of Ness Head, can also be explained in terms of the dual effects of wave and tidal current action. Spratt and, later, the Hydraulics Research Station, who studied the changing bank configuration of the Teignmouth approaches, considered that the Ness Pole grew as a spit from the base of the cliff of Ness Head. It has been suggested that it represents the effect of longshore drift carrying sediment northward from the Ness Beach which lies to the south of the headland. This is contrary to the findings of the author. A visual examination of Ness Beach—a typical bay-head beach structure—shows that it peters out against the rocky and boulder-strewn foot of Ness Head. There is no evidence that sand from Ness Beach is able to move across this rocky apron and then reach the harbour mouth. The grain size of the sediment in the two areas is different, that on the Ness

Beach being mainly medium sand of a median grain size of 270 microns which contrasts with the coarse sand and fine shingle of the sediment on the Ness Pole. A tracer test carried out in late 1973 on Ness Beach showed that the sediment movement here is entirely local with dispersion taking place up and down the beach but with little positive translation towards Ness Head, at least during the period of the 3 days of the experiment.

Other facts also make it unlikely that the Ness Beach acts as a supplier of sediment to the harbour circulation. The Ness Pole often begins to appear not as an ever-growing spit from the base of Ness Head but rather as a separate bank lying athwart the main channel running out of the harbour entrance, as was the case in 1965. The position of this bank varies and, if it is well to the northeast of the Ness Head, it will ultimately form the Outer Pole. If, on the other hand, its initial position is close to the Head though not actually joined to it, it can make a tenuous join later as more sediment arrives from the harbour mouth. It then takes on the appearance of a spit, the situation which existed in the early months of 1965. Once the link has been forged the growth of the Ness Pole can be quite rapid as sediment is carried along its margins. The bank also acts as a trap for any sediment moving out of the estuary, so that when there is a plentiful supply it develops into an elongated, tied bank almost 450 m in length. If there is a decrease in sediment supply from the estuary the neck of the bank becomes much lower and ultimately will be breached either by waves or by a strong ebb outflow from the estuary. The whole history of development of the Ness Pole is therefore intimately connected with the varying effect of waves and tidal currents.

The fact that both tidal currents and waves are instrumental in changes of bank configuration does not necessarily mean that both are of equal importance in bringing about the pattern of cyclic evolution. While the transfer of sediment along the Denn Beach from the pier to the point is a result of the combined effects of both processes, the return seaward along the main estuary channel is solely the work of the ebb tidal current which flows with great strength through the narrows and thence close to the Ness shore. Measurements taken by the Hydraulics Research Station, prior to building a model, show that by the time the current reaches Ness Head there has been a significant drop in its velocity and therefore deposition of sediment begins, a process aided by the opposing wave action (H.R.S., 1970). The subsequent movement northward of the sediment in the offshore zone is largely the work of tidal currents; and the fact that the incoming flood takes the circuitous path into the estuary means that sediment can follow a similar route. Wave action, particularly in bringing about the shoreward movement of banks like the Inner Pole during periods of easterly and northeasterly winds, aids this circulatory process.

Conclusions

The present study has demonstrated that the cycle which Spratt so carefully worked out in the middle of the last century, has existed in the form which he envisaged for the past 125 years. Although he failed to recognize minor perturbations and did not fully appreciate the exact role of tidal currents and wave forces in bringing about the changes he mapped, his work must stand as one of the earliest recognitions of the complexities of estuary dynamics. The fact that the cycle has changed so little during the past century is perhaps indicative of reasonably uniform conditions existing in this estuary approach and gives strength to the argument that a closed cell of sediment movement exists which is largely insulated from the changes taking place in the coastal and offshore areas around.

References

Craig-Smith, S. J., 1970, A hydrographic analysis of the approaches to Teignmouth, unpublished M.Sc. thesis, University of Leicester.

de Boer, G., 1964, Spurn Head, its history and evolution, *Trans. Inst. Brit. Geogr*, **34**, 71–89.

Hydraulics Research Station, 1958, Report on the situation in the harbour of Teignmouth and the stabilisation of the entrance channels, unpublished report.

Hydraulics Research Station, 1966, Teignmouth Harbour—fluorescent tracer experiment; unpublished report.

Hydraulics Research Station, 1970, Teignmouth Harbour model study to investigate improvements to the harbour entrance and their influence on the neighbouring beaches, unpublished report.

Kestner, F. T., 1970, Cyclic changes in Morcambe Bay, *Geogr. J.*, **136**, 85–97.

Rennie, J., 1838, Report on proposals for maintaining and improving the port of Teignmouth, unpublished.

Spratt, T., 1856, An investigation of the movements of Teignmouth Bar, John Weale, London.

Discussion

Dr. W. R. Parker, Institute of Oceanographic Sciences, Taunton, Somerset. Dr. Robinson's excellently illustrated and documented history of the morphological changes of the Teignmouth approaches contains a number of important points concerning the relevance of morphological changes to sediment movement. His conclusion that the changes so carefully monitored reflect the movements of the same population of sediment and the apparently small influence of offshore sediment movements or longshore supply are surprising in the context of usual models of nearshore sediment movement.

I would like to ask Dr. Robinson questions concerned with these points:

(1) How did he examine the relationship between the sediment volume he was studying and sediments farther offshore?

(2) Did he examine the thickness of the sediments comprising the banks? Some slides appeared to show the sandbanks on a boulder pavement.

(3) How did he calculate the total volume of sediment involved in this closed system?

(4) Did he examine the offshore sediment pattern of movement, and how did he calculate the residual movement of sediment in both the offshore and nearshore areas?

The method adopted for plotting the bank change in the approaches to the estuary, while giving an accurate picture of the changing configuration in a two-dimensional sense, could not provide data which would allow a quantitative assessment of the sediment involved in the circulation pattern. Only on rare occasions has there been an opportunity to run a series of level profiles and even then the offshore banks were not included because of difficulty of access. It must also be remembered that the banks are only exposed for about one hour on either side of low water spring tides and therefore the time available for mapping is very limited. For this reason I adopted the same approach as the Hydraulics Research Station employed during a preliminary investigation in the 1950s, namely to base the maps on the analysis of oblique photographs taken from a fixed position on Ness Head at the time of predicted low water. This photographic record was interpreted by means of a grid and supplemented by actual measurements on the beach itself to fix precisely such critical positions as the distal ends of banks, the size and shape of morphological features like the Horseshoe Bank and so forth.

I do not find that the small influence which the offshore zone seems to have is really surprising. The situation is by no means unique and one can point to many stretches of shoreline where the influence of the adjacent seabed area, and often the contiguous stretch of shoreline, is minimal. In Start Bay, for example, the beach at Hallsands has never been naturally replenished since it was artificially lowered at the beginning of the century and this in spite of the fact that a large bank of sand and shell, the Skerries, is located within 2 km of the shore, in direct line of waves and with a long fetch from the easterly quarter. The sediment of the Start Bay beaches is, in general, very different from that occurring some distance offshore. The same is true of the Teignmouth sediments. Those within the foreshore zone consist largely of coarse sand and small shingle (grain size 400 microns for the sand and between 0·5 cm and 5 cm for the shingle). This is in marked contrast to the sediment found only half a kilometre offshore which consists almost entirely of fine grain sand (grain size 200 microns). This strongly suggests that at Teignmouth the influence of the offshore zone is not of great significance in relation to the sediment budget. Tidal current rates are exceedingly low offshore. Measurements made so far show that they are broadly parallel to the shore, although as indicated, in the paper, the turn of the tidal streams does not coincide with the times of high and low water at the Teignmouth Harbour entrance. The amount of sediment movement was not measured although the indications are that with current rates seldom exceeding half a knot it is not likely to be very great. Clearly, if a more quantitative approach was considered necessary, direct measurements of this type, along with similar calculations of sediment volumes closer inshore, should be undertaken on the lines indicated by Dr. Parker. On the basis of the evidence presented here it has not been considered necessary to attempt a volumetric analysis. Further work is desirable, but it would need considerably more resources than I was able to command in what was only intended initially to be an exercise to determine significant parameters which might be worthy of more detailed investigation.

H. J. Geldof, Department of Civil Engineering, Delft University of Technology, The Netherlands. If I understand you correctly, you suggest that there is a closed sediment transport system. I am wondering if the system is really 'closed'; is there no longshore sediment flux into the system from the north?

Are there any data available on the total sediment volume in the system as a function of time?

As far as I am aware there is no sediment flux from the north at the present time. Any contribution which the eroding cliff-line might have made in the past was interrupted when the railway was built in the 1840s. There also appears to be a drift parting in the vicinity of the Parson and Clerk stack. North of this point the drift is apparently towards Exmouth Harbour entrance, although how significant this may be, in terms of sediment movement, is not really known.

There appears to be little contribution to the sediment in the harbour entrance by longshore drifting from the south. One of the fundamental errors made by Spratt in his classic study in the middle of the nineteenth century was the assumption that the Ness Pole was really a spit supplied by beach drifting from the south. Unfortunately the Hydraulics Research Station accepted this view which is demonstrably incorrect for reasons given in the paper. The mapping of the changing hydrography over the past 10 years shows that the Ness Pole often originates as a separate 'bar bank' athwart the exit of the main harbour channel and then later establishes a tenuous join with the foot of the Ness headland to give it the appearance of a spit as the paper states.

As indicated in my answer to Dr. Parker's questions, the mapping procedure does not allow total sediment volume to be calculated. In any case, a 10-year period would probably be inadequate to give a meaningful result and earlier measurements by Spratt and others are not helpful in this respect. The fact that the pattern of movement, as measured over the past 10 years, seems broadly similar to that seen by Spratt in the last century also seem to indicate that the sediment volume has not changed significantly during the intervening period.

F. J. T. Kestner, Hydraulics Research Station, Wallingford, Berkshire. Dr. Robinson has described to the meeting some changes in the configuration of banks and channels in the approaches to Teignmouth Harbour. I hope he will not mind if I proceed, by analogy, to call him a medical doctor and say that he has just described the symptoms of a patient who is suffering from a fairly well-known illness. I am a colleague, who has listened to his description of the symptoms and, although I have not examined the patient myself, I do not agree with his diagnosis. This disagreement happens not only in medicine, but is highly characteristic of the way in which the science of loose boundary hydraulics is still compelled to advance. It is compelled to deal with the pattern of the changes, because it is not yet able to deal with their causes. However, this is quite respectable. We have to study the regularities in the phenomena before we can advance to their causes. This is what Kepler did in astronomy before Newton could come along to advance the subject so greatly. He could not have done so without the observations of Kepler. Another example is the present state of meteorology, where long-term weather forecasts depend on the study of recurring patterns, not of causes. Loose boundary hydraulics, too, though a predictive science like meteorology, is still at this pre-Newtonian stage, and Dr. Robinson's paper illustrates this.

I agree that tremendous difficulties of interpretation often arise when loose boundary hydraulics are involved in any shoreline problem. Only detailed examination over a long period of time, not just using the morphological approach adopted by the author in order to identify areas of research worthy of more prolonged investigation along specific lines, but also using precise measurements in order to reach a realistic assessment of the sediment budget, are clearly necessary to fully understand what is happening. In my paper, I have drawn attention to the complexity of the changes taking place and stressed the role of both tidal currents and waves in shaping the bottom configuration, sometimes working in harmony, but often in opposition. Until this basic question of operative processes is fully studied and understood there will inevitably be divergence of opinion. All I can claim is that after 10 years of observation I am much better informed to come to certain preliminary conclusions than I would otherwise have been.

CHAPTER NINE

Dynamics and Sedimentation: The Tay in Comparison with other Estuaries

A. T. BULLER, C. D. GREEN and J. MCMANUS

Abstract

This paper begins with a brief section on estuarial hydraulics which provides a basis for discussion of estuarine sediment sources and distribution; erosion, transportation and deposition of fine-grained sediment; the formational mechanisms of zones of high suspended sediment concentrations; the threshold of motion of coarse-grained sediment; the genesis and migration of bedforms; general rates of transportation and sediment dispersion by waves. Examples from the Tay estuary, Scotland, are compared and contrasted with, and supplemented by, others from many parts of the world. This review provides a basis which can be used to isolate research topics which require special attention. Criticisms range widely, but two of the most important are (a) that expressions describing the erosion, transportation and deposition of estuarine sediments must accommodate unsteady state non-uniform flow conditions and (b) that knowledge of the mechanics of wave action and sediment entrainment by waves within estuaries is almost non-existent.

Introduction

The authors were encouraged to frame their topic within the style of an essay-cum-review of estuarine sedimentology rather than to produce an isolated contribution on the Tay estuary, Scotland. The result is a selective work within which examples from the Tay are complemented, supplemented and contrasted with others chosen from many parts of the world. In addition, it was suggested that some critical and discursive elements be incorporated which, although beyond the immediate scope of the Tay studies, might lead to an assessment of research problems which require further attention. The scope of such a wide brief is immense and the authors have had to severely ration themselves both in the topics described and in the numbers of examples used. The danger that such a work will neither be a comprehensive review of the Tay, nor a comprehensive review of estuarine dynamics and sedimentation is, paradoxically, the authors' aim—an aim to produce a selective essay centred on examples from the Tay which may complement other more conventional reviews (e.g. Emery and Stevenson, 1957; Postma, 1967; Schubel, 1971; Dyer, 1972).

Physical aspects

Introduction

Most sedimentary studies of shallow marine and littoral environments are concerned with textural and morphological changes in response to wind-induced waves and, to a lesser extent, tidal currents. Closer to the shore, in embayments and large inlets, the roles of waves and currents change in relative importance, although waves may still dominate. In estuaries the relative role of waves and currents is not only reversed, but the dynamics are further controlled and complicated by the progression and recession of waves corresponding to the frequency of the tides, the interactions between tidal and fluvial waters which induce various forms of gravitational circulation, and the development of chemical 'environments' conventionally described by degrees of fresh water dilution of sea salt. These physical and chemical characteristics are well summarized by Cameron and Pritchard (1963) who suggest that 'an estuary is a semi-enclosed coastal body of water which has free connection with the open sea and within which sea water is measurably diluted with fresh water derived from land drainage'.

Not all so-called estuaries, however, fulfil these requirements. Those formed by, or including, migratory bars and barriers at their mouths, may, according to season, rainfall and associated fresh water discharges, transgress from 'true' estuaries to lagoons as the requirement of 'free connection' is temporarily halted (e.g. some Australian estuaries; Jennings and Bird, 1967). Others almost empty out at low water, but freedoms of flows and mixing are usually ensured by well-defined channels (e.g. Barnstaple Bay, Mass. (Schubel, 1971), Bassin d'Arcachon, France, and the Scottish Eden estuary and Montrose Basin).

In this review only the dynamics and sedimentation of those estuaries which fulfil the requirements of the physical definition are discussed and examples from many parts of the world are compared and contrasted with the Tay estuary, Scotland.

Fluvial water

The quantities of fresh water entering estuaries depend on the climate, physiography, geology and vegetation cover of the hinterlands. Strong relief induces orographic rainfall; the porosity and lithological variability of the rocks affects the quantity, freedom of flow, and chemistry of groundwater; the density of vegetation cover determines the rate at which run-off is released into tributary streams. Estuaries with large and varied catchment areas may receive smoothed discharges with seasonal peaks, whereas small barren drainage basins tend to exhibit eccentric discharges controlled by extremes of dry periods interrupted by storms of varied intensities. Climatic controls on discharge are particularly strong for the Zaire River which is influenced by continuously heavy rainfall, and the Vellar which reflects monsoonal periods coincident with precipitation peaks (Dyer and Ramamoorthy, 1969).

The Tay estuary (Figure 9.1) receives most of its fresh water from a 6500 km^2 catchment area of varied rock types and vegetative cover, which is partly agricultural, partly forested and partly open moorland. The river flows are regulated as the waters pass through a series of lakes and reservoirs which suppress flood peaks and maintain enhanced dry weather flows (McManus, 1968). Even so, at the lowest gauging station peak flows of 1500 mean flows of 180, and minimum flows of 15 cumecs have been recorded. Thus the volume of fresh water entering the estuary can vary one hundredfold. Greatest discharges occur during the winter months and spring melts, although 'flashy' storms occur periodically during the summer.

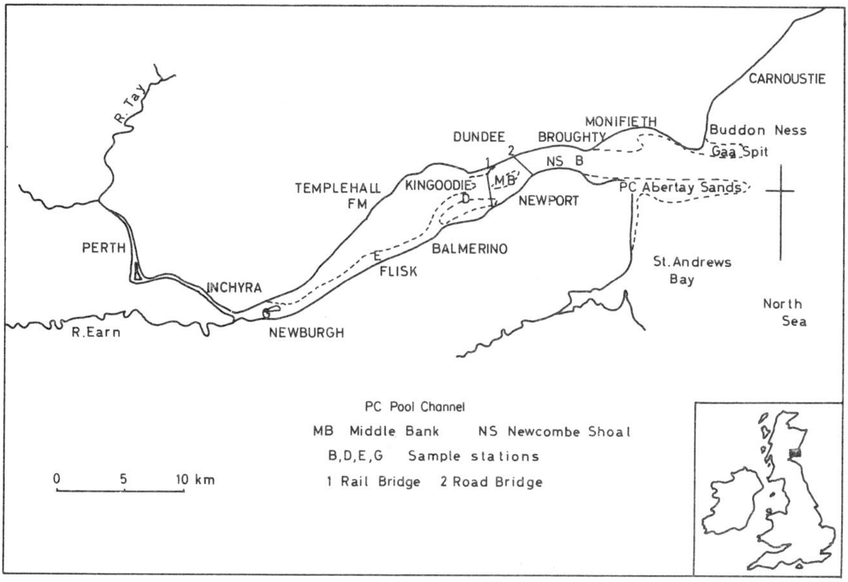

Figure 9.1. Location map of the Tay estuary, Scotland.

Marine water

In contrast to the considerable fluctuations and sometimes unpredictable nature of flows entering estuaries from rivers, the incursion of tidal waters is largely regular and predictable, altering according to the procession of spring–neap cycles occasionally modified by storms and sustained variations in barometric pressure. Normally two tides are experienced per day, but in some areas, for example the Limbang and Trusan estuaries, Borneo, only one occurs. Elsewhere, specific phase and amplitude relationships may permit the development of double maxima or minima of water level on each tide (e.g. Southampton Water, England). The tidal phases, mean amplitudes, and ranges vary according to local and global geographic constraints and the proximity of coasts to amphidromic points. The latter is important because tidal ranges along the same coastal enclosure under the influence of a single amphidrome can vary considerably. For example, under the influence of the southern North Sea amphidrome, the Rhine delta experiences semi-diurnal tides with ranges of approximately 2 m, whereas, on the complementary English coast, the ranges increase southwards from the mouth of the Deben (2–3 m), to the Stour and Orwell (3–4 m), the Blackwater (4 m) and into the Thames (5 m). In contrast some of the most intensively studied estuaries along the United States Atlantic Coastal Plain experience ranges of only 0·5–1 m (e.g. Chesapeake and Delaware Bays), which by British standards are almost 'tideless'. For any particular estuary the importance of ranges cannot be over-emphasized because they determine the volumes of water available, the tidally-induced flows within it and, in combination with fresh water discharges, the potential mixing characteristics.

The tidal range for the Tay is 4–6 m from neap to spring conditions.

Tidal progression

In an idealized state the speed of propagation of a tidal wave is roughly proportional to the square root of water depth and decreases as it moves headwards. During its advance, high water occurs first at the estuary mouth and is progressively delayed upstream. As the wave recedes the times of low water are subject to even greater delays (Table 9.1), and this temporal imbalance between high and low water results in the duration of the flood tide decreasing headwards, with a corresponding lengthening of ebb conditions (Francis-Boeuf, 1947). Other temporal imbalances may occur in extremely wide estuaries where tides recorded on opposite banks are slightly out of phase because of geostrophic forces (Coriolis), but in most estuaries these effects are negligible.

The maximum tidal delay in the Tay is about 2 hours between the times of high water at the mouth and high water near the landward limit of tidal influences at Perth—some 50 km from the North Sea (Figure 9.1). Delays are 4 hours in the Hooghly, 12 hours in the Plate and several days in the Amazon.

Table 9.1. Tidal delays and durations at five locations as the tidal wave enters and leaves the Tay estuary, Scotland; Buddon (mouth), Perth (near headward limit of tidal influence)

	Buddon	Broughty Ferry	Dundee	Balmerino	Newburgh	Perth
Low tide time	0820 hours	0910	0920	0930	1120	1300
Delay	0	50 min	1 h 0 min	1 h 10 min	3 h 0 min	4 h 40 min
High tide time	1420 hours	1430	1500	1520	1550	1610
Delay	0	10 min	40 min	1 h	1 h 30 min	1 h 50 min
Duration of rising tide	6 h 0 min	5 h 20 min	5 h 40 min	5 h 50 min	4 h 30 min	3 h 10 min

The tidal wave's motion and distortion depend mainly upon the buffering effects of the fresh water discharges, the morphology of the bed over which it passes and the roughness characteristics of the sedimentary deposits. These physical influences alter from one estuary to another, but the generalized form of progression and distortion may be described as follows. An initially symmetrical tidal wave at an estuary mouth becomes increasingly steeper as it enters the estuary. The steepening may increase to such a degree that the wave 'overshoots' and develops a bore (e.g. the Severn and the Hooghly (McDowell, 1970)). If the wave does not 'break' it may initially increase its range to a maximum within an estuary before it loses headward momentum and becomes increasingly asymmetric in profile (Figure 9.2). The progression of the tidal wave in the Tay has been simulated by a one-dimensional numerical model in which the frictional forces were reduced to a range of Manning roughness coefficients varying from 0·018 near the mouth to 0·024 towards the head (Routh, 1970).

Another, and more subtle, effect of an advancing tidal wave is that the mutual interaction between the wave and the river not only causes the fresh water outflow speeds to decrease to zero and then increase again as the flow is reversed upstream, but also that the 'ponding' effect results in the elevation of river water levels often many kilometres beyond the salt water/fresh water interface. This rise and fall commonly results in rivers being termed 'tidal', whereas *sensu stricto* the rises and falls are tidally-induced. The Amazon estuary is an extreme example. During periods of both low and high river discharge no salt water wedge enters the estuary (Gibbs, 1973) although tidally-induced effects can be measured well inland.

As mentioned in the previous sub-section, changes in atmospheric conditions can affect sea levels and therefore upset the normal daily or lunar tidal cycles, which in turn, alter the characteristics of 'normal' tidal wave incursions. In 1953, storms in the North Sea caused sea levels to rise to such an extent that much of the coast was flooded. The effect of estuaries is to enhance these levels further as greater heads of water attempt to translate extreme volumes through 'restricted' inlets. Balay (1961) reports tidal range enhancement of 4 m near Buenos Aires, during storms travelling along specific paths, and less dramatic effects are reported by S. L. Davies (1973) who shows that tidal ranges in the lower Tay may alter by 0·5 m during modest barometric changes.

Intermixing of fresh and salt water

The gross circulation patterns, flow intensities and formation of density- or gravitationally-induced flows within estuaries, depend on the relative quantities of fresh and salt water entering them. As previously described, the volumes of salt water available for entry depend on tidal ranges, and the volumes of fresh water are controlled by seasonal rains. For example, the tidal volume entering the Tay varies from 650 (spring tide) to 170 m^3 (neap)—a ratio of 1 to 4. A fresh water spate coincident with neap tides produces a fresh to salt water ratio of 1 to 2, whereas on spring tides this ratio is 1 to 10. In the summer, when fresh water

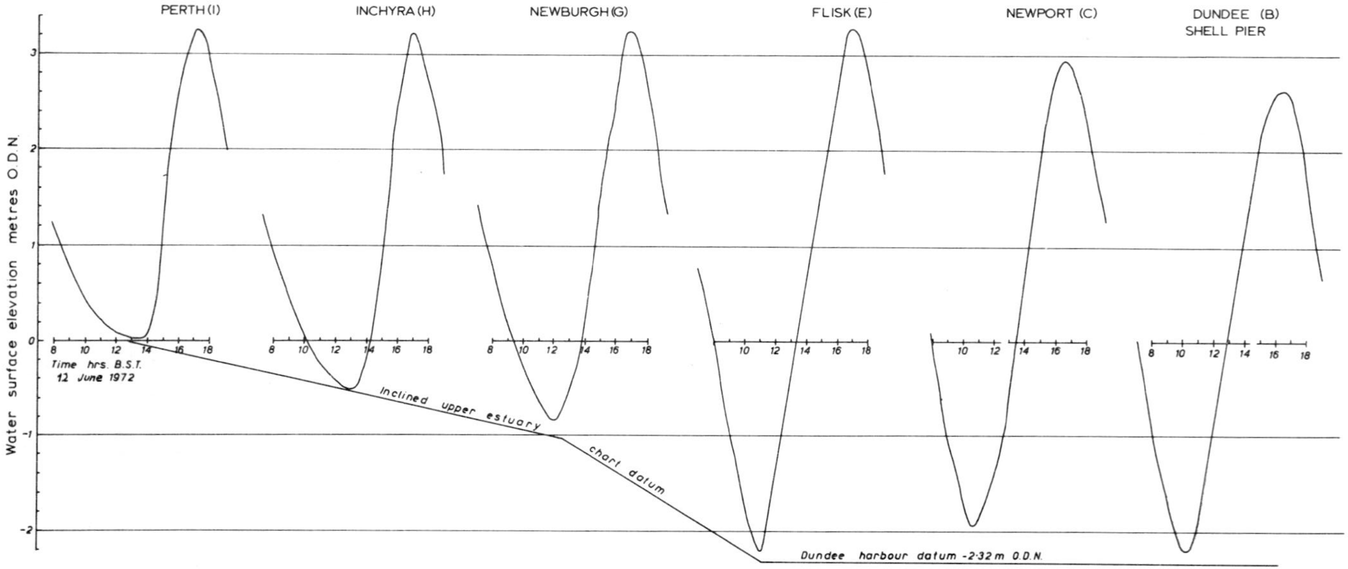

Figure 9.2. Progression and recession of a spring tidal wave in the Tay estuary, Scotland. Ordance Datum, Newlyn (O.D.N.) is approximately mean sea level.

flows are low the ratio can increase to 1 to 50, and exceptionally, 1 to 200 (after Cunningham, 1896). Other estuaries with low tidal ranges and high river discharges will be fresh water dominated; those with high tidal ranges with negligible fresh water inputs will be dominated by tidal waters—but variations will occur for each variety.

Tidal flow intensities not only depend on the volumes of sea salt available, but are also affected by estuary configuration. If an estuary narrows, the water passing through the reach is accelerated in a fashion analogous to jet flow with an expansion and slowing of flow speeds as the estuary widens again. These effects are most pronounced towards the mouths of estuaries, because farther upstream the flows are increasingly influenced by fluvial discharges. If an estuary has the classic 'trumpet-like shape', flows will tend to decrease progressively in intensity towards the head.

Flow intensities at a point fluctuate from zero at stillstands to maximum values as the tides become fully established. In a single channel system the outgoing flows are stronger because they are backed, rather than buffered, by the river flows. The resistance, or enhancement, to flows is caused by the fresh water which controls the vertical distributions of flow speeds. On flooding tides the greatest flows are commonly recorded below the water surface, whereas on ebbing tides the surface flows are greatest because the less dense fresh water tends, at least initially, to remain in the upper portions of the water columns.

In 'braided' estuaries these simple relationships may be modified. Depending upon the disposition of banks and channel networks, individual channels may be dominated by either flood or ebb flows. Accordingly, such preferences may be reflected by textural differences in the sediments. In the Tay, the Queens Road channel is lined with medium to fine sand whereas the main channel to the south of the intervening Middle Bank is lined with coarse sand. The former is weakly flood dominant and the latter, ebb dominant. More obvious sedimentary responses to ebb and flood dominant systems are the formation of bedforms whose orientations indicate disparate flow preferences. Nevertheless, if any estuary is viewed as a whole it is ebb dominant because there is a net seaward mass transfer of a volume of water equal to the volume of fresh water entering from the river.

The characterization of estuaries, as stratified, partly stratified or homogeneous types, varies according to the generation of density or gravity induced circulations. Stratified estuaries are produced when the denser sea water penetrates headwards along the bed while the lighter fresh water flows seawards over the top of it. The dense water forms a 'wedge', bounded by an upper surface which slopes headwards and oscillates with the tides, across which little mixing of salt and fresh water occurs. During the initial period of generation 'unbalanced downstream pressure forces near the fresh-water end overcome the frictional resistance at the interface and contribute to an increase at the velocity head; at the salt-water end an unbalanced pressure force drives salt water upstream near the bottom' (Partheniades, 1973, p. 29).

This simplest state can develop when river flows constantly predominate over tidal influences (e.g. some Mississippi distributaries (Scruton, 1956)) or when rivers are in spate and tidal amplitudes are small (Dyer and Ramamoorthy, 1969).

When fluvial discharges decrease and tidal amplitudes increase, mixing (by turbulent eddy diffusion, gravitational convection and longitudinal dispersion), in combination with oscillatory upstream and downstream motions of salt water, results over a time-averaged period in the creation of net non-tidal landward flows of salt water which terminate headwards at the fresh/salt water boundary. If the tidal range is very large and river discharges sufficiently small, then little variation of salinity occurs between the tops and bottoms of the water columns and an estuary is termed well-mixed (Pritchard, 1955). With the homogenization of vertical salinity changes, the major detectable changes are longitudinal. Any lateral variations may be caused by inflows from small streams, run-off from marsh drains or, in very wide estuaries, by geostrophic forces.

Williams and West (in press) have shown that the waters of the Tay estuary are partially mixed. Examples of other estuaries which display partially mixed characteristics are the James, the Mersey and Southampton Water, whereas the wider reaches of the Delaware and Raritan are well mixed.

Residual motion

During the foregoing discussion several types of residual motions have been described (e.g. net non-tidal flows and ebb or flood dominant flows). These motions are of such importance in estuarine sedimentation studies that they merit further amplification both below and in the following major sections.

Sediment transport paths are often determined indirectly from measurements of flow intensity profiles during entire tidal cycles. Differences in intensities between flood and ebb tides, or residual flows, are often used to indicate the directions of sediment motion and to calculate the quantities of material transported. The accuracy of directly measuring bed load transport using 'labelled' grains or bed load samplers, is still open to technical objections; and the measurement of suspended load, even if it is tagged with radioactive isotopes, is an expensive and difficult procedure. Nevertheless, suspended load can be effectively calculated if water samples are collected at the same time as flow intensities are being measured and if concentrations of suspended sediment are linked mathematically to the flow determinations. Using the method of Inglis and Allen (1957) flux determinations of this type can be integrated for half-tidal cycles and the difference between flood and ebb quantities give an accurate measure of the net motion.

Many studies using the flux system tend, however, to concentrate only on longitudinal measurements (e.g. Buller (in press)) but Halliwell and O'Connor (1966) showed for the Mersey that they should be combined with transverse monitoring because suspended sediment is rarely homogeneously distributed throughout estuarine cross-sections. Fischer (1972) uses data from the same estuary and suggests that there is some form of transverse net non-tidal circulation which may greatly affect the motion and distribution of fine grained sediment. Clearly this criticism is applicable both to sedimentologists and hydraulicians.

Waves

Of all the physical parameters measured within estuaries, wind-induced waves have received least attention although they are important erosive agents (especially over intertidal flats, drying sandbanks and channel margins) and may in extremely shallow estuaries modify or destroy circulation patterns. Only for estuarine mouths and approaches are wave studies reasonably well-documented because the regimes are nearest those of open coasts.

The importance of waves at estuary mouths cannot be over stressed. They transport sediment and 'mould' it into subtidal and intertidal bars and spits. The bars and spits function as natural 'training walls', necessary to maintain a 'free connection' for channels between the estuary waters and the open sea, and protect the inner estuary by reducing the waves' power and erosive capacity. For example, a wide spectrum of waves generated both in deep oceanic areas and in shallow waters of shelf or restricted seas, may reach the outer limits of estuaries. The effects of bars and spits are twofold: they prevent wave penetration by 'filtering' and may induce waves to break.

Sediment: sources and distribution

Introduction

In contrast to other environments the study of estuarine sedimentation is extremely complicated. This complexity is not governed by some special properties of the sediment, rather, it is imposed upon it by the interaction and transition between marine and fluviatile regimes which create the characteristic estuarine conditions which control erosion, transportation and deposition. Within this framework freshly derived sediment from the sea, rivers and supratidal cliffs and slopes, mingles with resident estuarine sediment which itself may be periodically remobilized. Once within the estuarine system sediment has two terminal 'choices': it may remain in the estuary until finally deposited, or it may 'escape' out to the sea. Between these two end-points there are often intricate and cyclical histories of erosion, dispersion and temporary deposition. In this section the derivation and distribution of sediment in the Tay are described and related to findings in other estuaries.

River input

The rates of erosion of terrestrial materials, and therefore the yields of sediment entrained by rivers, are controlled by such general multivariate parameters as relief, lithology, climate and human interference. Of the total quantity denuded from the landscape only about one quarter of it is sluiced through the river networks, mainly in suspension (Miller and Piest, 1970). In the Rivers Niger and Benue only 5–6% of the total load is carried along the bed (NEDECO, 1959), although as much as 30% has been recorded entering the Columbia River estuary (Hauschild and coworkers, 1966).

More precisely quantities of suspended sediment, and probably bed load, depend on two major factors which are largely independent when one river, or river system, is compared with another (Buller and McManus, 1974). These are fresh water discharge and availability of sediment supply. Meade (1969) demonstrated that along the Eastern Atlantic Coastal Plain rivers to the north of Cape Lookout have relatively higher fresh water discharges than those to the south, yet carry lesser quantities of sediment. Similar relationships are evident when drainage basin areas, average annual suspended sediment loads and fresh water discharges for the major rivers in the world are compared (Holeman, 1968).

In contrast there is close correlation between the two factors for individual rivers. For example the hydrographs and suspended sediment curves for the Colville River, Alaska, and for the Brahmaputra, India, describe similar shapes (Figure 9.3(b) and (c)). Interdependence of this form allows the construction of rating curves by plotting the logarithm of water discharge against the logarithm of sediment discharge. Usually these curves are linearly related for both bed (NEDECO, 1959) and suspended sediment loads (Miller and Piest, 1970). For the Alpenrhein and other rivers in the same geographical area Müller and Förstner (1968) have shown that the suspended sediment concentration C (mg l^{-1}) is related to fresh water discharge Q ($m^3\ s^{-1}$) by the equation $C = aQ^b$ where a ranges between 0·004 and 80,000 and the exponent b ranges from zero to 2·5. But occasionally these relationships do not hold, especially in areas which experience a very dry season abruptly terminated by monsoonal rains or flash floods. In such circumstances the sediment versus discharge curves take the form of a hysteresis loop. Seasonal flood hysteresis curves have been plotted for the Niger and Benue Rivers (NEDECO, 1959, p. 480) and storm-induced hysteresis curves are given by Miller and Piest for the Pigeon Roost Creek Watershed near Holly Springs, Mississippi (1970, p. 1314).

From worldwide considerations of erosion rates (Fournier, 1960), climate, physiography and geology, the small quantities of river sediment entering the Tay estuary are unexceptional, but from local examination they are very small especially as the River Tay is Britain's foremost river in terms of average discharge (180 $m^3\ s^{-1}$). For example, concentrations of suspended sediment during 1973 summer conditions of above average discharge were only within the range 12 to 22 mg l^{-1} (Buller and coworkers, 1972). This dearth of sediment highlights the independence of discharge and availability of sediment supply. The Tay's 6500 km^2 catchment area is not conducive to supplying vast quantities of sediment for transport. A cool temperate climate (Berry, 1968) is superimposed on a rugged hinterland of resistant rocks covered by mantles of glacial and post-glacial unconsolidated debris (Ramsay, 1968). In addition the Tay's catchment contains over 70 fresh-water lochs through which more than a quarter of the

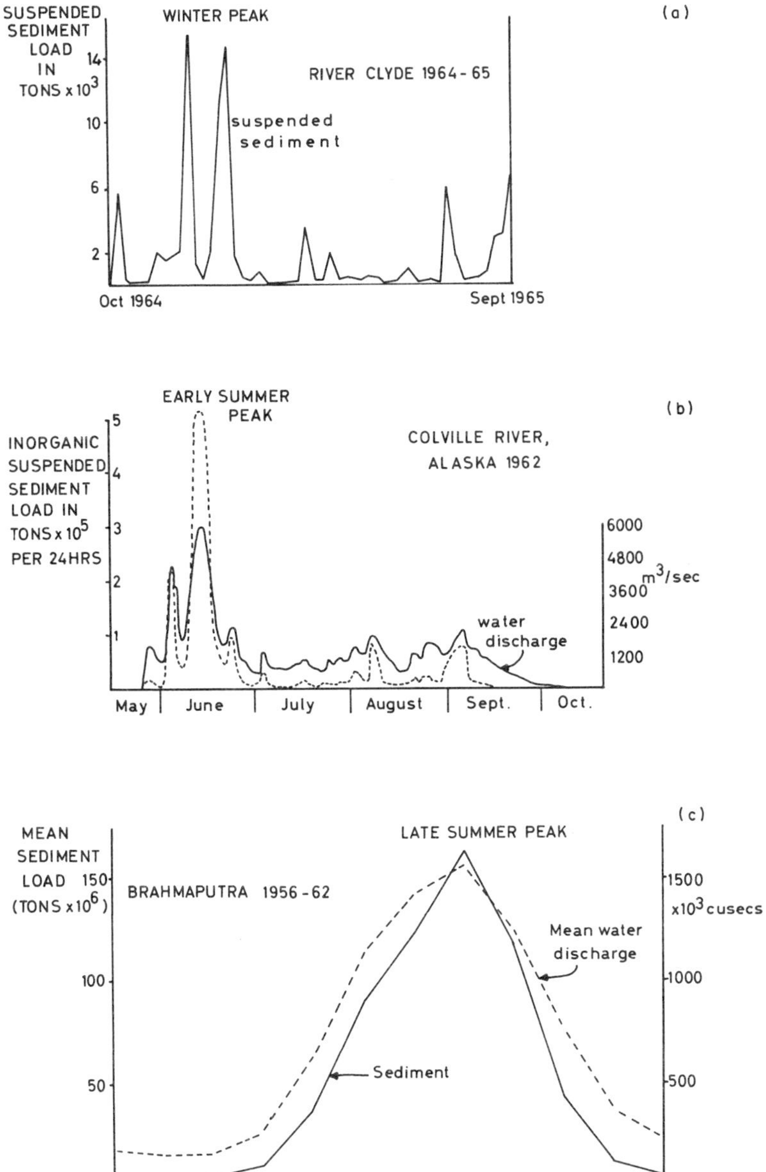

Figure 9.3. Suspended sediment load and water discharge curves for: (a) the River Clyde at Daldownie, Scotland, for 1964–1965 (based on Fleming, 1970, Figure 3, p. 2224); (b) the Colville River, Alaska, 1962 (based on Arnborg, Walker and Peippo, 1967, Figure 3, p. 133); (c) the Brahmaputra River at Bahadurabad, India, mean water discharge 1956–1962, mean sediment discharge 1958–1962 (based on Coleman, 1969, Table II, p. 156).

total water passes. These and other man-made reservoirs probably serve as sediment 'traps', especially for the coarser materials. Underwater examination of the river bed entering the head of the estuary shows that it contains residual pebbles, boulders and scoured de-watered post-glacial clays. In contrast the River Earn (area 307 km^2; average discharge 31 $m^3 s^{-1}$), which also rises in rugged terrain but passes in its lower course through a wide plain of unconsolidated post-glacial sediments and alluvium, is floored by coarse sands and granules. It appears, therefore, that the two rivers play distinctive and complementary roles: the River Tay probably supplies most of the suspended sediment load to the estuary because of its superior discharge, whereas the Earn supplies most of the bed load because this grade of material is more readily available (Buller and coworkers, 1971; Buller and McManus, in press).

Systematic seasonal data on annual quantities of suspended and bed load sediment inputs into the Tay estuary will be available shortly, but a body of occasional data suggests that variations in loads from each of the rivers are related to their individual hydrographs. Rainfall, run-off and discharge are closely correlated within the region which experiences most of its cyclonic rainfall during the winter months, supplemented by short-duration storms during July and August (McManus, 1968; Buller and coworkers, 1971). On the west coast of Scotland the River Clyde area experiences broadly similar climatic conditions which are reflected by the monthly variations of suspended sediment discharge shown in Figure 9.3(a). The Clyde's winter peaks contrast with those for other rivers subject to very different climatic regimes. The early summer maximum water and sediment discharge peak for the Colville River occurred in a 3-week period just preceding, accompanying, and just after the break-up of the ice-locked river and hinterland (Figure 9.3(b)). The Brahmaputra late summer peak reflects the influence of monsoonal floods (Figure 9.3(c)). Summer storm-induced hysteresis relationships may occur for the Tay and Earn, but they are thought to be unlikely as the region, apart from agricultural land, is well vegetated.

Marine input

All estuaries probably receive some sediment from the sea, but this source is not easily proven. Two of the most convincing European examples for suspended sediment are the Scheldt (Crommelin, 1949) and the Ems (Crommelin, 1940; Van Straaten, 1960). For the United States, Meade (1969) produces a strong argument for the landward motion of both suspended and bed sediments of the estuarine Atlantic Coastal Plain. His evidence for suspended sediment is based on flux criteria linked with net non-tidal circulation, and for bed load he suggests that the landward movement of continental waters must be accompanied by the carriage of bottom sediments. Beach sand moves towards and into the mouths of some estuaries at a rate of several thousand cubic metres per annum and mineral assemblages in the lower reaches suggest offshore derivation. Kulm and Byrne (1967) also use evidence of littoral drift, mineralogy and sediment texture to show that marine sands penetrate 10 km into the Yaquina Bay estuary, Oregon.

For the Tay the present-day lack of sediment entering from the rivers is believed to be counterbalanced by small quantities of sand entering from the sea and accumulating in the central reaches. Qualitative evidence of residual flood oriented dunes in this region, together with fresh evidence of sandflat and sand-bank heavy mineral assemblages more typical of sands found near the estuary mouth (as described by Mishra, 1969), strongly suggest some form of marine incursion (Buller and McManus, in press). The exact processes of introducing marine sand into the estuary are not fully known, but available evidence suggests that the headward net motion may be related to disparate sediment transport in flood and ebb dominated channel systems and through flood and ebb dominant sections of a single channel.

On the northern side of the estuary mouth marine sand is transported by a combination of littoral drift and flood currents across the Gaa Spit into Monifieth Bay. From here it is temporarily caught in a local anticlockwise tidal 'gyre' which shifts it towards the lower middle estuary. Further headward motion into the central reaches takes place via a flood dominated portion of the main channel (West, 1973 (from flow data); Buller, in press (from suspended sediment flux calculations, station B)) which links with a more complicated system of flood and ebb dominant channels well within the estuary.

On the southern side of the estuary mouth different short-term mechanisms resulting in no net headward motion of sand are described by Green (1974). Based on results of fluorescent tracer experiments linked to calculations of residual flows, he suggests that on the flood tide marine sand is transported from St. Andrews Bay (over the Abertay recurved spit) into the main estuarine channel where it is flushed seawards again during the ensuing ebb tide.

Suspended sediment entering the Tay from the North Sea has not been studied, but some derivation is inevitable, especially during easterly storms.

Marginal input

Schubel (1971) states that one of the most underestimated sources of fresh sediment is from the erosion of supratidal cliffs and slopes. In upper Chesapeake Bay 13% of suspended matter is derived from dry shore erosion, and 52% from the middle Bay (Biggs, 1970). Other examples of marginal estuarine supply include slumping and rain-washing of periglacial 'head' into Breton estuaries and the sliding of loosely consolidated masses of siltstone and shale in the marshlands of Newport Bay, California (Guilcher, 1967; Stevenson and Emery, 1958).

Marginal sources are unimportant in the Tay estuary (Buller, in press). Some minor patches of erodible post-glacial material occur along its southern shore, but their area and volume are negligible in comparison with the almost unbroken range of andesitic cliffs. The northern shore is lithologically different. Founded mainly upon softer sediments of Upper and Lower Old Red Sandstone age most of them have been covered by post-glacial sediment capped by modern intertidal flats, intertidal marshes and supratidal areas of reclaimed ground. Only washings from the reclaimed agricultural land, occasionally adding fine-grained sediment to the estuary, constitute a marginal input. Seaward of the railway bridge the waterfront is largely artificial or rocky and input of material comes from industrial and municipal wastes.

Distribution of sediment

The temporary or final distribution of sediment-types in an aquatic environment results from two main interdependent factors—'hydraulic sorting' and 'deposition'. 'Hydraulic sorting consists of the grouping together by fluid flow of particles that respond to the flow in a similar manner and, at the same time, the separation of such particles from those that respond differently to the flow' (Blatt and coworkers, 1972, p. 102). Those particles that respond differently to the flow may remain as lag deposits, or they may drop out and be deposited because the flow can no longer maintain their motion, or they may be transported laterally or vertically into a more compatible flow field. This selectivity continues until the fluid ceases all motion, and if the decay in flow intensity and turbulence is gradual the deposited sediment should reflect its decline by showing a gradation from coarse to fine, from poorly sorted to well sorted.

In an estuary such regular conditions cannot exist. The flow intensity and associated turbulence alters from maximum flows to stillstands with the tides; maximum flows vary according to the lunar tidal cycle; the progression and distortion of the tidal wave can be modified by variation in fresh water discharges; and the entire flow characteristics can be altered by severe flooding and the superimposition of wind-induced waves (see 'Physical aspects'). In addition to these flow variables and their 'normal' effects on sediment dynamics, there are other complex sedimentological variables which depend more on (a) the size, shape and density of particles for coarser sediment, (b) the chemistry of the water and number concentration of fine particles for fine sediment and (c) the relative concentrations of organic to inorganic materials, rather than on flow intensity and turbulence alone. These include particle to particle interactions which affect the motion of highly concentrated dispersions of coarse sediment near, or on, the bed, the probabilities of fine grain collision which may result in flocculation or bioflocculation, and the aggregation of mixed organic and inorganic materials (Schubel, 1971). The combined action of these physical, chemical and biological variables produces sedimentary deposits whose textures may be viewed as representing their integration over a period of time sufficient for the deposit to become 'characteristic' of its particular hydraulic and hydrodynamic setting.

Buller and McManus (in press) use this premise in their sub-environmental, textural and inferential hydrodynamic analysis of sediment distribution in the upper and upper middle reaches of the Tay estuary. Their technique is based on the *QDa–Md* system (Buller and McManus, 1972) in which the arithmetic quartile deviation values (QDa in mm) are plotted against the median diameter (Md in mm) on double-log paper. The position of estuarine values plotted on such a graph are interpreted as being representative of sediment having been subjected to a series of compounded processes which form and characterize sediments from other environments, e.g. rivers, beaches, 'quiet-water' seas and lakes. This unconventional approach allows estuarine sediments to be classified indirectly according to gross process responses. The data are then transposed onto a facies-like distribution map linking sediment-types to the compounded processes which formed them. Examples of the relationships between sediment-types, gross processes and sub-environments are given in Table 9.2. Allen (1971) uses a similar technique for the Gironde estuary to delineate areas dominated by bed load and graded suspension transport based on the C–M textural system of Passega (1957 and 1964). By this method the first coarsest percentile (C) is plotted against the median (M) on double-log graph paper.

Table 9.2. Relationships between sub-environments and hydraulic and sedimentary processes in the Tay estuary, Scotland

Tay estuary subenvironments	Simple qualitative hydraulic and sedimentary processes for each sub-environment
Marsh-edge	Very low flow speeds; slight surface wave and surge action; dominantly deposition of silt and clay from suspension
Upper intertidal flats	Low flow speeds; slight surface wave action; deposition of fine grains from suspension and coarser grains from graded suspension
Lower intertidal flats	Moderate flow speeds; moderate wave action; deposition from suspension and bed load
Scoured channel	Very strong flows; extreme turbulence; scouring; present-day coarse sediment mixed with older consolidated 'muds' and 'muddy' sands
Channel and sandbanks	Moderate to strong flows; turbulent; moderate to strong wave action; deposition from bed load and coarse grains from 'jumping' trajectories
Minor channels; shoals; major sandbanks; edge of some intertidal flats	Moderate flows; strong wave action; extremely turbulent; deposition from bed load and coarse grains from graded suspension

Figure 9.4 is an example of a QDa–Md representation for samples from estuaries and marshes along the French coasts (data from Verger, 1968). This graph shows that few of the QDa–Md values for samples from the conventional estuarine sub-environments, e.g. slikke, schorre, infratidal channels etc., plot in well-defined groups—an observation consistent with the Tay; but many of the QDa–Md values for samples of slikke and schorre areas occupy a position within, or close to, the trend envelope for 'quiet-water'. For the Tay only one or

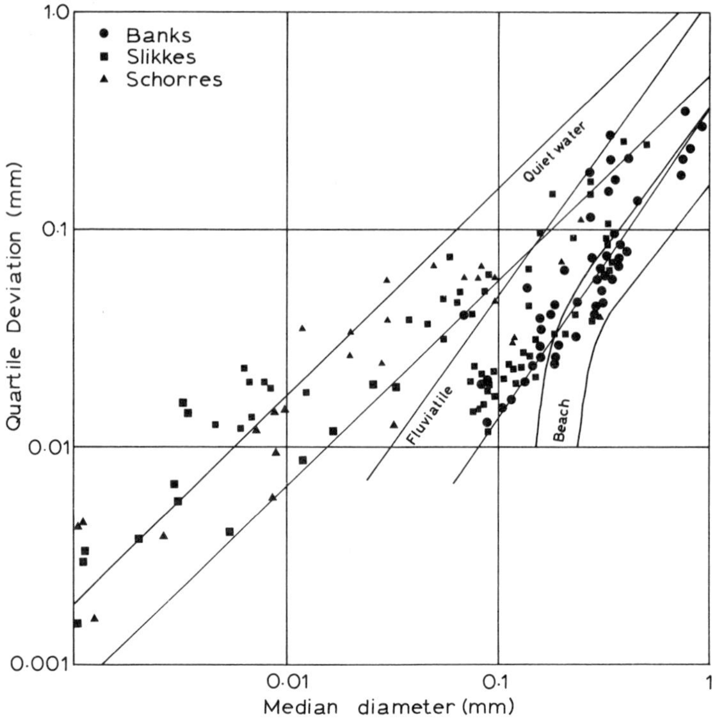

Figure 9.4. A quartile deviation (*QDa*)–median diameter (*Md*) representation of samples taken from various estuarine and marsh sub-environments along the French coasts (data from Verger, 1968). The 'quiet-water', 'fluviatile' and 'beach' envelopes are taken from Buller and McManus (1972).

two samples from the upper 'mud' flats close to the northern shore marsh are fine enough for their *QDa–Md* values to enter this field.

From the marsh-edge in the Tay the intertidal sediments become progressively coarser towards the main channel, but this progression is textural rather than morphological. From infrared black and white air photographs (see Figure 9.5) a tonal division separates the 'darker grey' upper flats characterized by featureless fine sandy and silty sediments, from the 'lighter grey' rippled sands of the lower flats which are traversed by numerous drainage channels and runnels. The subtidal portion of the upper estuary consists almost entirely of the main channel which runs close to the southern shore. Channel bottom sediment-types vary considerably as parts of the channel have laid bare coarse relict materials, whereas other parts are covered by mobile rippled sands.

The upper middle reach, just west of the railway bridge, is characterized by a thick cover of mobile rippled sands, whose sizes decrease shorewards. This is the area in which sand derived from the sea is believed to be accumulating.

From the railway bridge to Broughty Ferry (the lower middle reach) the

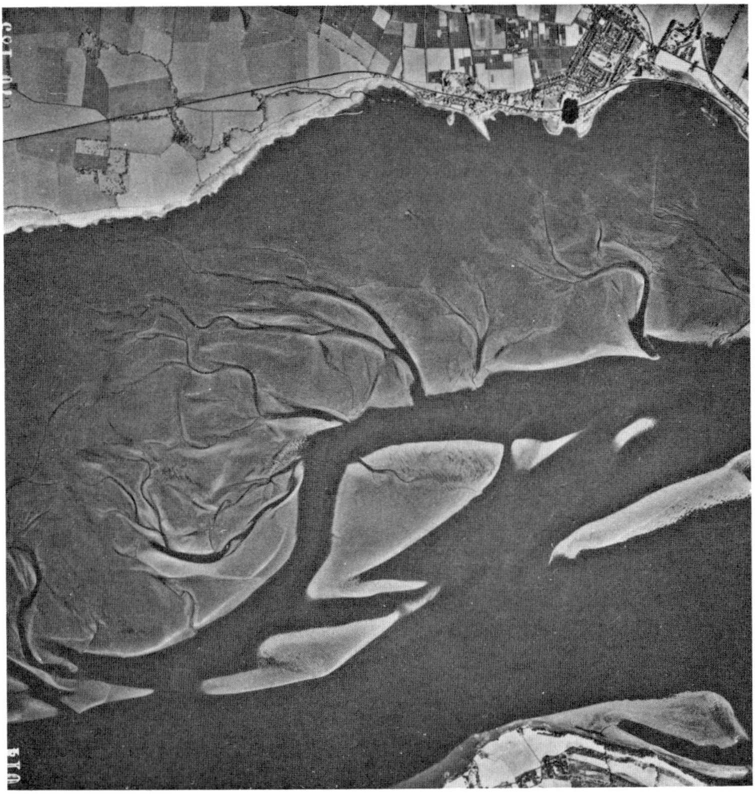

Figure 9.5. An infrared black and white air photograph of the Tay estuary between Temple Hall Farm (top left) and the railway bridge (just out of frame on right). Note the tonal divisions between the upper intertidal flats (dark grey), the lower intertidal flats (intermediate grey) and the sandy areas and banks (light grey)—some of which show ripples. The 'streaks' in the water are plumes and streamers of suspended sediment being transported by weak ebb tidal flows just before low water stillstand. (Reproduced by kind permission of the Controller General of the Ordnance Survey.)

channel again exhibits erosional and depositional contrasts, but in its lower course it is dominated by rippled coarse sand and broken shells. On the north bank the sediment rapidly decreases in grain-size as the water shallows and flow intensities decrease, but the distribution towards the south is more complicated. Irregular patches of coarse sands and gravels, related to 'gravel mounds' and 'ledges', have been identified from transit sonar records (Figure 9.6). They are interrupted by a triangular spread of coarse materials covering the Newcombe shoal and elsewhere by largely featureless deposits of intertidal and subtidal fine sands and silts. Seawards of Broughty Ferry the lower estuary is dominated by a stable channel which, in part, has cut into late-glacial tills and stiff clays. The lateral margins of the channel are gradually covered by rippled sand which increases in area and thickness towards the entrance beaches, spits and bar.

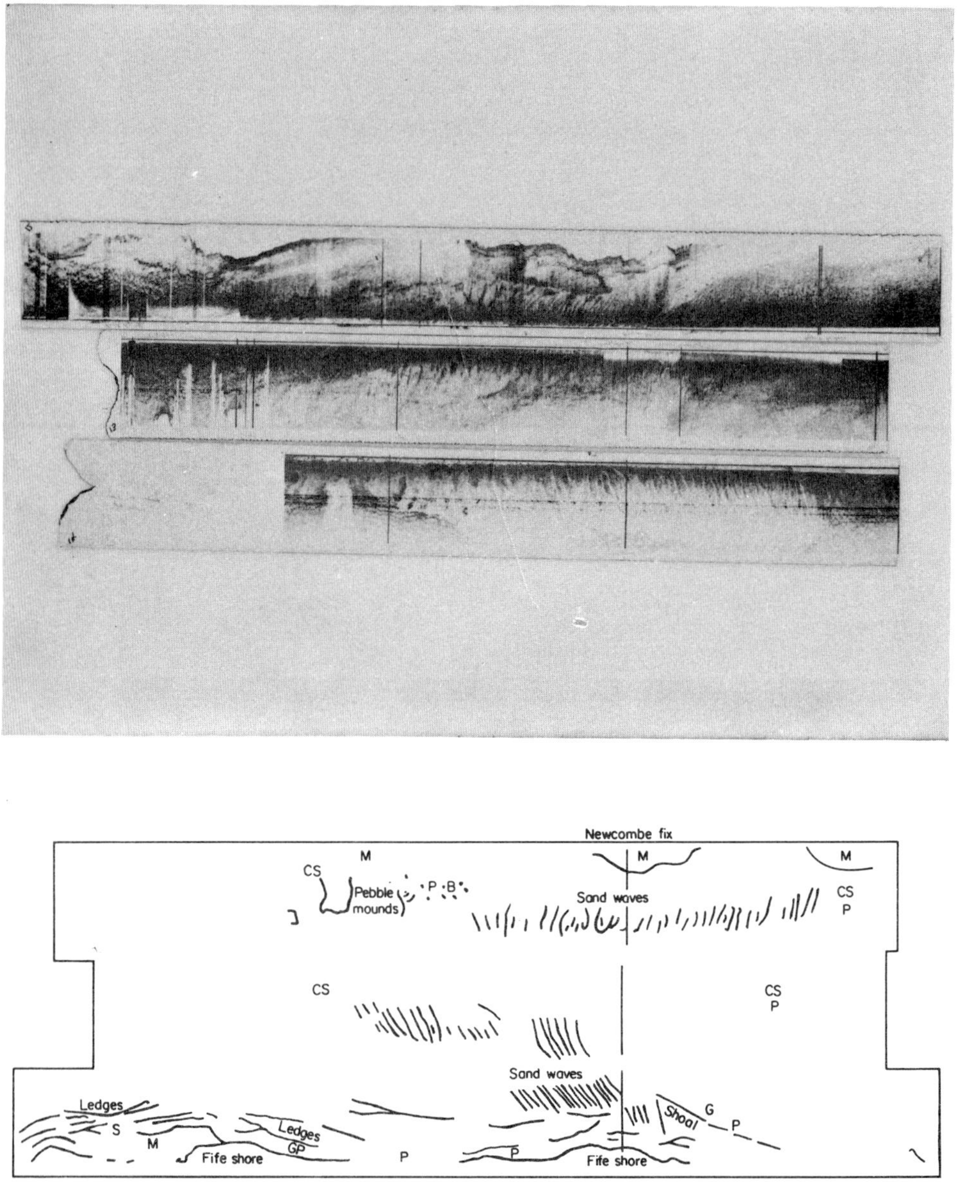

Figure 9.6. Sonograph and legend of the area between the road bridge and Broughty Ferry, Tay estuary, Scotland (see Figure 9.1). *P* = pebbles; *G* = gravel; CS = coarse sand; M = 'mud'. (Data from Green (1974).)

This brief description typifies a 'high energy' estuary set in a temperate glaciated area which is receiving little sediment input.

Two additional processes which affect the distribution of estuarine sediments outside cool temperate areas are ice action and the trapping of fine-grained material by mangrove swamps. Ice can form a seasonal intertidal 'protective' cover thus inhibiting all-the-year-round sedimentation (Anderson, 1970) and the wasting and floating of ice during break-up can add coarse material to the fine-grained sediment as well as eroding large masses of sediment and marsh and redistributing them out of sedimentological 'context' (e.g. the St. Lawrence (Dionne, 1969a and 1969b). Mangroves act as 'baffles' which effectively reduce current flows and damp turbulence to such a degree that fine-grained sediments quickly build up between the roots where they are 'trapped' and bound by additional organic matter. In tropical estuaries of the Freetown Peninsula, Sierra Leone, the mangroves are so effective in trapping sediment that they tend to build platforms up to 30 cm above the surrounding intertidal flats (Tucker, 1973). The nearest temperate equivalent to mangroves are marshes. Salt marshes at or above spring high water in the Tay cause the accretion of fine-grained, darkly coloured sediment around plant stems and roots thereby raising the level of the marshes above the bordering tidal flats (McManus, 1968).

Fine sediment (silt and clay)

Erosion

Fundamental differences controlling the behaviour of beds of coarse and fine particles are largely attributable to the presence of electric charges which are strong enough to 'bind' together the fine sediments, but are less effective as particle size increases. Resistance to erosion is also dependent on the presence of water between the particles and the way in which it is 'trapped'. For fine sediments water may act as an electrolyte which enhances the strength of bonding between elementary particles, but if the water content during initial phases of shear is high then the bed is more easily mobilized (Postma, 1967).

Problems of erosion have been investigated in two, very different, ways. Some researchers (e.g. Dunn, 1959) working with field data, or with the complexities of field relationships in mind, have produced empirical relationships relating the onset and rate of erosion to a variety of 'mass properties' such as bulk density and shear strength. Others, working mainly in the laboratory (e.g. Partheniades, 1962), produce simulated low density beds by allowing particles to settle out before subjecting them to carefully controlled increases in flow. A principal difference between the two approaches is the acceptance, or rejection, that 'mass properties' markedly affect the onset and rate of erosion. For example, Partheniades (1962), and later Odd and Owen (1972), show from experimental work that the rate of erosion varies linearly with the shear applied by the flowing water for a low density bed according to the empirical expression

$$\left(\frac{\mathrm{d}m}{\mathrm{d}t}\right)_{\mathrm{crit}} = M\left(\frac{\tau}{\tau_{\mathrm{crit}}} - 1\right) \tag{9.1}$$

where m is the mass of fine sediment eroded, t is time, τ is the shear stress acting on the bed, τ_{crit} is the critical shear stress which must be exceeded before erosion can occur, and M is a constant equal to the rate of erosion at $\tau = 2\tau_{crit}$. (See also notation at end of paper.)

For the onset of erosion Dunn (1959) shows that the critical shear stress (τ_{crit} in lbs/ft^2) is empirically related to the vane shear strength of the material (τ_v in lbs/ft^2) by

$$\tau_{crit} = \frac{k_1 \tau_v}{1000} + 0{\cdot}180k_1 \qquad (9.2)$$

where k_1 is a dimensionless coefficient. For materials of silt and clay sizes the value of k_1 approximates to unity, but increases with increasing mean size of the sediment. Migniot (1968) also suggests that the yield strength (τ_y) of a variety of muddy sediments is related to the critical shear velocity U_{*c}(cm s^{-1}) by

$$U_{*c} = 0{\cdot}5\,\tau_y^{1/2} \text{ for } \tau_y > 15 \text{ dynes cm}^{-2} \qquad (9.3)$$

and

$$U_{*c} = \tau_y^{1/4} \qquad \text{for } \tau_y < 15 \text{ dynes cm}^{-2} \qquad (9.4)$$

and Allen (1971) includes a term E_{FS}, the ratio of a suitable force intensity to the bulk density of the bed, called the specific transfer-energy in an expression for the rate of fluid-stressing in deference to mass physical properties. Yet Partheniades (1972) maintains his original (1965) claim and emphatically chooses the net attractive forces between particles as the critical variable impeding erosion, rather than a 'mass property'. He states that 'although in the average there might be a slight increase in the resistance to erosion with increasing soil density and shear strength for medium to high strength clays, critical eroding shear stresses for clay soils of nearly similar strength may differ by orders of magnitude; soils with relatively low shear strength may resist shear stresses appreciably higher than other clayey soils of much higher shear strength. Moreover, the eroding shear stresses are smaller by several orders of magnitude than the macroscopic cohesive shear strength of the clays as measured by a conventional shear strength test (pp. 20-5–20-6).

Motion

'The ability of a mathematical model to simulate a complex tidal process such as "suspended sediment" (mud) transport, depends on the degree to which the various physical processes can be described in mathematical terms and on a knowledge of the laws involved' (Odd and Owen, 1972, p. 177). The expressions which describe the motion of suspended sediment are generally included within two partial differential equations: (a) that for the conservation of mass and (b) the dynamic equation of motion. These expressions are flexible enough to offer solutions for a variety of boundary conditions. In unidirectional shear flows, which may be established in rivers, expressions describing the motion of suspended sediment largely depend on the fulfilment of the following assumptions: (1) that concentrations of suspended sediment decrease exponentially towards the water surface; (2) that the velocity increases exponentially toward the water surface; (3) that the turbulence intensity and thus diffusivity varies with distance from the bed (Graf, 1972). These define the boundary conditions which may occur in some estuaries (e.g. the Thames); they are not, however, universally applicable. In the Tay the first two assumptions are usually invalid and the third is inapplicable for stratified forms.

During motion the gravitational settling of grains is resisted by an exchange of momentum between the turbulence and particles. The balance between the downward gravitationally induced motion and the upward 'buoyant' effect of turbulence can be expressed for low concentrations as

$$0 = wC + \varepsilon_s \frac{\partial C}{\partial z} \tag{9.5}$$

where w is the fall velocity, C is the time-averaged suspended sediment concentration, ε_s is a turbulent diffusion coefficient, and z is a height above the bed. Furthermore a concentration C_a at a height z *relative* to a height a can be predicted by

$$\frac{C}{C_a} = \left(\frac{h-z}{z} \cdot \frac{a}{h-a}\right)^y \tag{9.6}$$

where d is the total depth and y is a shortened form of the relation

$$y = \frac{w}{kU_*} \tag{9.7}$$

where k is the Von Karman constant and U_* is the shear velocity. In open-channel flow the Karmán value is 0·4 but in the presence of sediment there is a tendency for the value to decrease. Field data from Einstein and Chien (1955) show a k value range of 0·38 to 0·19 as sediment load increases.

Equation (9.5), however, arises from an *a priori* assumption that the rate of erosion from a bed is constant at all times between $t = 0$ and $t = \infty$ (i.e. $\partial C/\partial t = 0$) in an appropriately time-averaged convective dispersion equation

$$\frac{\partial C}{\partial t} = 0 = \frac{\partial}{\partial x_1}\left[\varepsilon_1 \frac{\partial C}{\partial x_1} + wC\right] \tag{9.8}$$

whence

$$\varepsilon_1 \frac{\partial C}{\partial x^1}\bigg|_{\text{bed}} = -wC\bigg|_{\text{bed}} \tag{9.9}$$

where ε_1 is a convective dispersion coefficient and x_1 is distance along the bed. These conditions are valid only for unidirectional, steady state, flow, and as mentioned above are not always applicable to estuaries. Furthermore, from a mathematical viewpoint, Equation (9.6) can be criticized because where $z = 0$

(at the bed) the concentration $C = \infty$, a condition which is impossible. It is reasonable that suspension does not exist close to the bed because particles are not *sensu stricto* within the water (Graf, 1972).

Experimental, theoretical, and field applications of these basic relationships have been refined by Einstein and Chien (1955) and Odd and Owen (1972) who show a preference for two-layer models when near-bed concentrations are *high*. In an investigation of suspended sediment transport in the Thames estuary Odd and Owen use a lower layer of constant thickness within which the concentrations of suspended sediment decrease from an extremely high value to an 'average' value at the upper limit of the layer. Einstein and Chien's rationale for a two-layer system is that the high concentrations in the lower 'heavy-fluid' zone affect the density of the sediment–fluid mix, whereas the density of the upper 'light-fluid' zone is unaffected because concentrations are lower. Graf (1972) suggests that the 'heavy-fluid' zone is an important physical concept because the zone acts like a filter reducing the turbulence level.

Deposition

Depending on the intensity of tidal flows there is a period centred about peak flows when the shearing rates at the boundary layer near the bed are too high to allow deposition. As the flow intensities decrease towards high and low water stillstands the shear stresses reduce and deposition takes place. Expressions for deposition have been given by many authors, notably Sundborg (1956 and 1958), Einstein and Krone (1962) and Odd and Owen (1972), and each is similar. That used by Odd and Owen is based on Krone (1962) who postulated that the probability, p, of 'flocs' adhering to the bed increases from zero to unity as the shear falls below a limiting value (τ_d) where

$$p = \left\{1 - \frac{\tau}{\tau_d}\right\} \tag{9.10}$$

Using this probability expression Odd and Owen express the net rate of deposition of suspended sediment as

$$\left(\frac{dm}{dt}\right)_d = C_1 w\left(1 - \frac{\tau}{\tau_d}\right) \tag{9.11}$$

where C_1 is a depth-mean concentration. This relationship also depends on concentration values because, as with the heavy-zone mentioned in two-layer models, if concentrations are exceedingly high, individual or flocculated particles not only damp turbulence and increase the sediment–fluid mix density, but they may ultimately present a coherent mass of 'fluid mud' which behaves as a layered unit. Examples of fluid muds are recorded for the Thames (Inglis and Allen, 1957) the Gironde (Allen, 1971(b)), the Chao Phya estuary (Allersma and coworkers, 1966), and the Bristol Channel (Kirby and Parker, personal communications).

Reports of measured deposition rates in the field are difficult to find, but

Skempton (1970) gives the following rates for two British estuarine reaches: 2·5 m per 1000 years for Tilbury, and 2·0 m per 1000 years for Avonmouth. On a local and global scale these values will alter according to climate, season, suspended sediment input, estuarine load, and the hydraulic-hydrodynamic conditions prevailing in each estuary.

Zones of high suspended sediment concentrations

Many studies of the distribution and motion of suspended sediment in estuaries are descriptive, illustrative and deductive, with occasional simple arithmetic representations such as flux diagrams and computations of residual motions (e.g. Allen and Castaing, 1973). The reluctance of sedimentologists to adopt mathematical models is easily defended: the assumptions fundamental to mathematical representations cannot always be justified from prototype data; the basic equations have to be modified for use in the field; and the realization that many subtle field changes have to be averaged out to fit the equations engenders fears not only that something may be lost in the translation but also that the real environment will be reduced to an abstraction. For example, Buller (in press) who gives a description of suspended sediment transport in the Tay estuary during conditions of low concentrations does not attempt to model the system, even as one 'layer', because vertical distributions of concentrations do not fulfil the assumption that concentrations decrease exponentially towards the water surface; and independent ongoing studies of turbulence structure in the Tay suggest that turbulence intensity does not vary systematically throughout the water column. Furthermore, measurements were made during the summer months and the results may be peculiar for that time of the year.

During such periods of moderate river discharge and relatively quiet marine conditions the quantities of fluviatile suspended sediment entering the estuarine circulation system of the Tay are extremely low. That material which is in suspension is largely derived from the estuary margins. At high tide the intertidal flats are covered by a shallow body of water. As the tide recedes fine-grained surface sediment from the upper flats is resuspended by weak tidal flows, a process which is greatly enhanced by the superimposition of wind-induced, short wavelength, low amplitude waves. As the ebb becomes fully established the swift fall in tidal height (Figure 9.2) leaves the upper flats dry and water derived from them passes over the lower flats. During late-stage exposure this water increases its sediment load, not by resuspension of lower flat surface sediments (which are sandy), but by erosion of the drainage runnels whose banks of sand and bioturbated silts slump into the channelled flows where the silt 'blocks' disintegrate releasing dense clouds of fine grains into suspension (Figures 9.7 and 9.8).

Throughout this period the falling-stage water, highly charged with suspended sediment, drains laterally into the surface and mid-depth waters of the main channels. This process acting along the entire length of the channels flanking the intertidal flats intensifies toward low water until the lower flats are dry. The

Figure 9.7. Low water exposure of estuarine intertidal drainage channel margin (Kingoodie Bay) showing the erosive effects of late-stage drainage by weak tidal flows and wind-induced waves. Note the erosion face and the widespread distribution of 'silty' cuboidal blocks which decrease in size towards the water edge.

remaining ebbing water is then contained by the main channels and moves a short distance downstream until it is ponded by the tide flooding from the sea while the upper estuarine water is still weakly ebbing. The lateral input of high concentrations of suspended sediment into a decreasing volume of water results in the cumulative formation of low-tide zones of high suspended sediment concentrations. As the flood tidal wave progresses headwards the zones are initially swept upstream following the passage of the channels, but as the volumes of tidal water increase the zones are progressively diluted and suspended sediment is dispersed over the entire cross-sectional area. At high water the zones have disappeared.

The number of zones, their definition and the concentrations of suspended sediment which form them, vary according to (a) the lunar tidal cycle which partly controls the flow intensities and the area of intertidal flats covered at high water, (b) the efficiency of erosion which is enhanced by wind-induced waves and (c) the network patterns of drainage runnels (or 'catchments') which determine the locations of late-stage input into the channels. From aerial photographs two catchments are recognized. The first extends on the north bank from Newburgh to Flisk and the second extends from Temple Hall Farm almost to the railway bridge (Figure 9.5).

Figure 9.8. Detail of erosion face showing horizontal laminations intersected by vertical burrows which result in the eventual production of 'silty' cuboidal blocks (see Figure 9.7). One such block has fallen leaving the evident cavern.

The formation of a zone of high suspended sediment concentrations fed by water draining from the first catchment is illustrated in Figure 9.9. Comparison of the ebb tide flux profiles and concentrations shows the downstream transportation and accumulation of suspended sediment through stations G and E, reaching a maximum at station D about two hours before low water when the entire tidal flats are dry and the lateral input of suspended sediment has ceased. The concentration diagrams illustrate the heterogeneity of sediment profiles and the bulging shapes of the flux profiles emphasize that most of the sediment is transported in the surface and mid-depth sections of the water columns. The symmetry of the flux and concentration values about low water at station D shows the return passage of the zone, as the tide floods, and its complete dilution and disappearance towards high water.

The salinity distributors demonstrate that this section of the Tay estuary is very well mixed, with only slight stratification towards high water, and that the location of the zone is situated seawards of the low tide limit of measurable sea salt (Figure 9.9).

Contoured distribution diagrams of this type are sufficient to gain a broad understanding of suspended sediment transport styles different from that described for the Tay, especially if their interpretation is supplemented by knowledge of the distribution of bottom sediment-types and the class of circulation. For

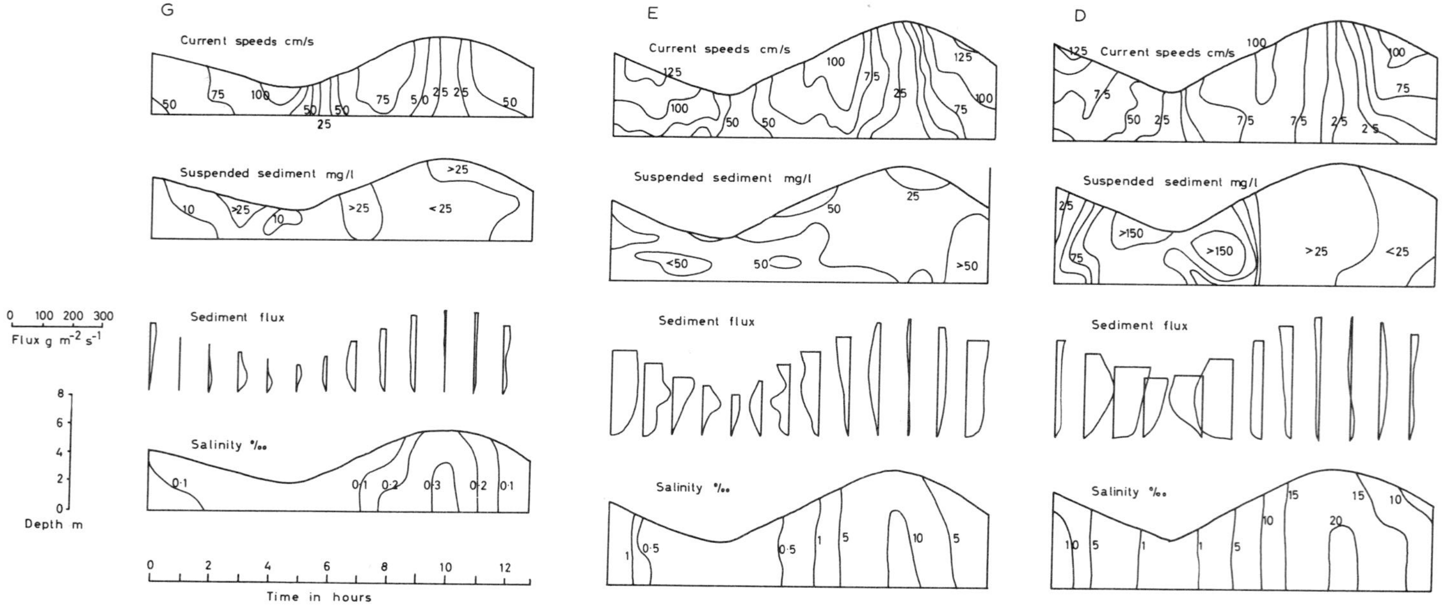

Figure 9.9. Relationships between water speed, suspended sediment concentrations, suspended sediment flux and salinity, spring tidal cycle at stations, G, E and D (Figure 9.1), in upper Tay estuary, Scotland (data from Buller, Charlton and McManus, 1972; Buller (in press)).

moderately stratified or vertically homogeneous estuaries Meade (1969 and 1972) maintains that 'reservoirs' of fine-grained bottom sediments accumulate near the headward limits of sea salt, areas which also correspond with the headward limits of the net non-tidal flows which have transported the sediments. In these 'muddy' reaches the sediments are resuspended during incoming and outgoing tides with concentrations increasing and decreasing with flow intensity. Such periodic resuspensions near the landward limit of sea salt result in the formation of zones of high suspended sediment concentrations (the classic 'turbidity maxima') which decay both seawards and headwards. Examples include the moderately stratified Savannah River estuary and Sacramento–San Joaquin River delta area, and the vertically homogeneous Thames estuary (Inglis and Allen, 1957). In a highly-stratified or salt wedge estuary such as the Southwest Pass, Mississippi River, a mid-depth zone develops along the upper part of the halocline. It is best defined during high water, or on the rising tide, when the landward flow of salt water has been active for several hours; but during the ebb tide when the basal flow is weak the zone becomes diffuse (Meade, 1973, p. 113).

Two other types of zone are common and largely independent of estuarine circulation systems. The first of these form at river mouths during floods and freshets. Depending on the availability of fine-grained sediment from the hinterland and the discharge of individual rivers, vast quantities of suspended material can be transported into estuary headwaters. As the river water meets and mixes with tidal water concentrations become progressively diluted seawards with some deposition in the upper reaches (e.g. the Susquehanna entering upper Chesapeake Bay (Schubel, 1968)). If, however, the fresh water discharges are extremely powerful they can not only destroy any form of density-induced estuarine circulation by forcing water to flow predominantly seawards at all depths, but much of their suspended load may be flushed through the estuary and out to sea with little or no estuarine residence, trapping or deposition (Meade, 1969).

The second type of zone forms by the resuspension of fine-grained intertidal sediment during the incoming and outgoing tides. Figure 9.10 illustrates the formation of one of these short-lived lateral zones along the southern margins of the Tay. The weak tidal flows were superimposed by wind-induced waves whose heights gradually built up to a maximum of 1 m as the wind speeds increased. Concentrations of resuspended sediment increased linearly with flow speeds, wind speeds and wave heights, until the bed was covered by a depth of water too great for the waves to 'touch bottom'. At this critical stage the resuspended material began to settle out from the water column and the zone became increasingly diffuse until it disappeared before high water.

The relative dependence of estuarine circulation systems on the formation of zones of high suspended sediment concentrations is shown in the style of longitudinal graphical 'models' and examples (Figure 9.11). The dilution model (a) shows the theoretical case where river-borne suspended matter is simply diluted by less turbid water as it enters the estuary without deposition of individual particles or enhanced deposition caused by flocculation or bioflocculation

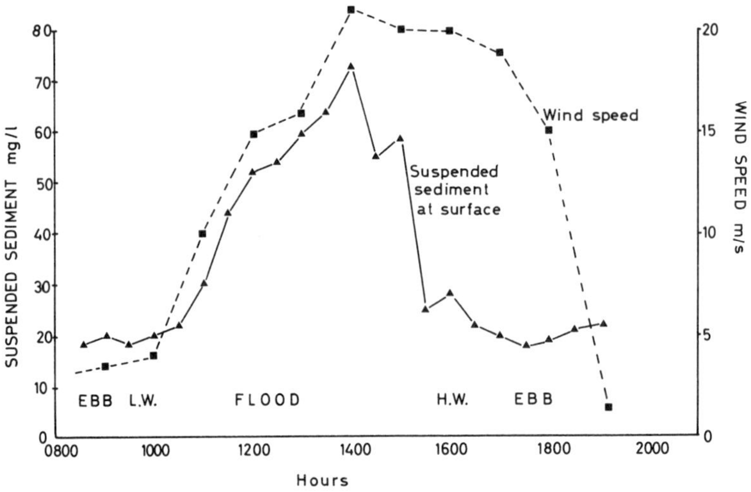

Figure 9.10. Relationships between surface water suspended sediment concentrations and wind speed. Formation of a short-lived lateral zone of high suspended sediment concentrations formed by wind-induced waves resuspending fine-grained intertidal material, Tay estuary, Scotland.

(based on Meade, 1972, p. 100). Model (b) shows the situation where river-borne material starts to be diluted and deposited before it reaches salty water and approximates to the case described above for the Susquehanna freshet. The third model (c) illustrates the 'turbidity maximum' centred about the chloride range of zero to three grams per kilogram. Field data from the York River, Virginia, illustrate this 'classic' situation which has been recorded in many American and European estuaries. The final model (d) shows a zone, seawards of the low chloride location of 'turbidity maxima', formed by the intertidal input of high suspended sediment loads into estuarine channels during ebbing tides.

Of the four models only (c) is wholly dependent on estuarine circulation, (a) and (b) are largely independent, and (d) is free from net non-tidal controls, although the final location of the zone is dependent on the strengths and directions of outgoing tidal flows. Any estuary which is subject to considerable annual variations in river discharge and river-borne suspended sediment may grade from one situation (model) to another.

The generality and simplicity of these synoptic diagrams largely depend on whether flocculation (and bioflocculation) can be expected in the field as judged by arguments and indirect extrapolations based on experimental laboratory results, whether flocculation can be proven in the field by observation and sampling, and whether the role of flocculation is important in comparison with the role and behaviour of elementary fine particles. Krone (1972), in a field-based study of the Savannah Harbour, argues strongly in favour of flocculation as an important factor because changes in concentrations during slack-water periods, and magnitudes of concentration profiles in relation to flow strengths, suggest

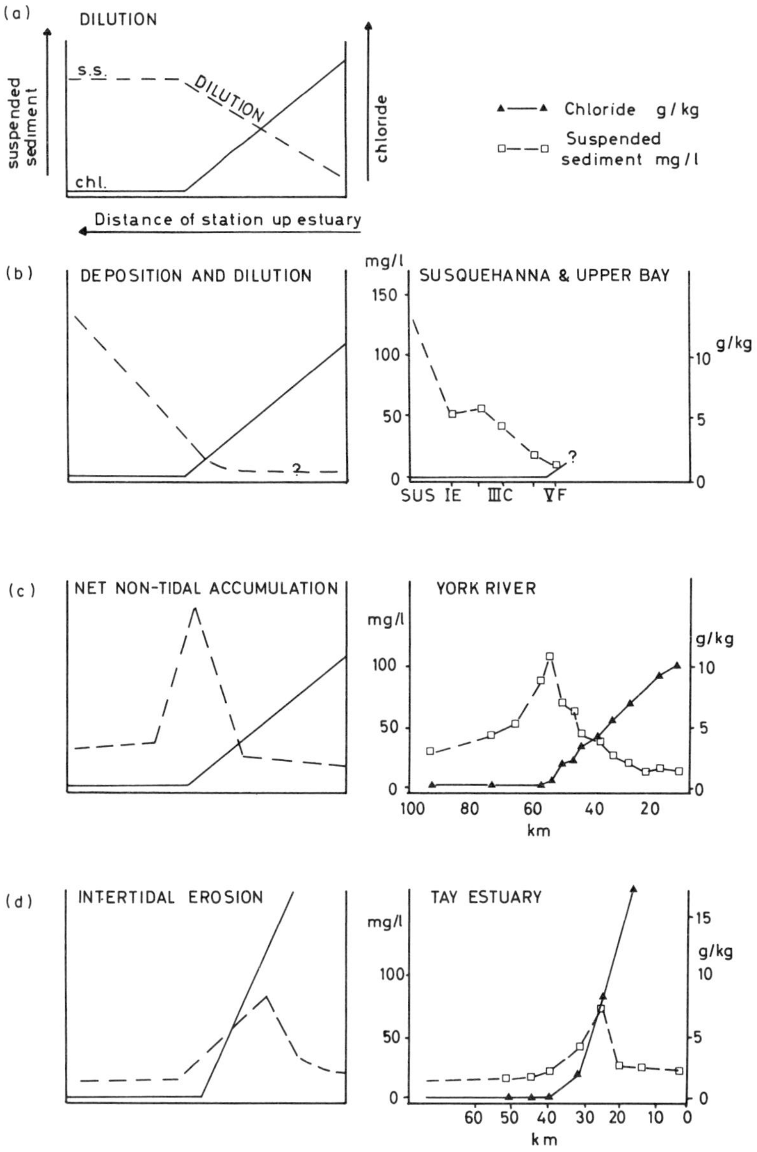

Figure 9.11. 'Models' and field examples of longitudinal variations between suspended sediment concentrations and salinity (chloride) in estuaries. (a) Dilution model (based on Meade, 1972, Figure 3, p. 100); (b) deposition and dilution model with field example from the Susquehanna and Upper Bay, Chesapeake, USA (based on Schubel, 1968, Figure 36, p. 73); (c) net non-tidal accumulation model with field examples of York River, Virginia, USA (based on Meade, 1972, after Nelson, 1960); (d) internal erosion and accumulation model with field examples of Tay estuary, Scotland (data from Buller, Charlton and McManus, 1972; Buller (in press)).

that settling velocities are greater than those of individual particles and must, therefore, be enhanced as a result of aggregation.

Of the three main mechanisms of interparticle collision the most common is due to Brownian motion where particles are agitated in response to thermal motions of the suspending medium. The second mechanism is caused by internal shearing where particles in differentially moving layers are brought into contact along the layers' interfaces, and the third results from interparticle collision by the geometric and physical inhomogeneity of particles settling at different rates.

Details of flocculation processes, and the arguments for and against the relative role of flocculation in the field, are topics too broad to be included in this section; but the validity of the descriptions above, which have been based openly, or by inference, on the *a priori* assumption that fine particles are neither flocculated in fresh water entering an estuary, nor increasingly flocculated after entry as they become part of a complex physico-chemical hydrodynamic 'environment', must be critically assessed by the reader who should be aware that the role of flocculation, at least in some estuaries, may be a predominant factor in the modes of transport and deposition of fine-grained materials.

Coarse-grained sediment (sand and gravel)

Introduction

In the two previous sections, sediment sources, distribution, and the motion of fine-grained sediments have been discussed. In this section the initiation of motion and transportation of coarse sediments are described.

The distinction of the total transported load into bed and suspended load is an idealization of actual conditions. There is in practice no distinct border separating the paths of particles that constitute the bed load from those constituting the suspended load. In general, however, grains transported close to the bed by rolling or jumping are known as bed load (Yalin, 1972, p. 16) while grains transported by the flow and the forces of turbulence in random paths within the main body of the flow are termed the suspended load.

Although gross physical processes (such as flow intensity and turbulence) directly affect the sand-sized material forming the bed load, rapid changes in salinity and diurnal variations in water temperatures also affect the hydraulic constraints associated with bed load transport, although to a lesser degree. It is these minor variables, however, that differentiate the transportation of bed materials in an estuary as compared with those in shallow marine or fluvial environments. Furthermore, rapid changes in the concentrations of suspended load also constantly change the estuarine hydraulic regime, but all these differences are small when compared with the more fundamental problems that exist in differentiating between transportation of bed load in an estuarine setting of reversing flow and variable bathymetry to the more classic studies made in flumes and alluvial channels. For example, the latter imply experimentation under uniform flow and steady state conditions. Although upstream sections of estuarine reaches may approximate to these conditions the study of bed load motion in estuaries involves working in non-uniform unsteady flow conditions. Even today published data are mainly related to flume studies and their direct application within the estuarine environment is questionable.

Threshold of motion

The threshold of motion of coarse particles can be considered in terms of a 'critical velocity' or a 'critical boundary shear stress'. Such measurements have been made, notably by Gilbert (1914), Hjulström (1935), Shields (1936) and Sundborg (1956), who have mainly experimented with conditions of uniform flow in flumes and alluvial channels. There is, however, little verification of the results for flow conditions associated with 'marine' environments.

A number of techniques have been used to estimate the boundary shear stress in marine and estuarine environments and all are associated with expressions for determining the boundary shear stress in two-dimensional flow in open channels.

Sternberg (1967, 1971 and 1972) derived his results from the quadratic stress law and velocity profile technique. In the former, boundary shear stress (assuming turbulent flow) is proportional to the fluid density and the square of the mean velocity or

$$\tau \propto \rho \bar{u}_z^2 \tag{9.12}$$

where ρ is the density of the fluid and $\bar{u}$ is the mean velocity at a distance z above the bed. As field research usually involves measuring flow velocity at 1 m above the bed and introducing a proportionality constant relating the velocity near the bed to the force exerted by the fluid on a unit area of the bed, we establish

$$\tau = C_{100} \rho \bar{u}_{100}{}^2 \tag{9.13}$$

where C_{100} is the drag coefficient at one metre from the bed; or in terms of a friction velocity

$$U_* = C_{100}{}^{1/2} \bar{u}_{100} \tag{9.14}$$

Sternberg produced threshold criteria for the initiation of grain movement as a function of grain diameter linking his studies to the competency curves of Sundborg's threshold mean velocity at $\bar{u}_{100}$ and Inman's threshold drag velocity (1963). His work in Puget Sound, Washington, was done on a rippled sandy bed and, although there was agreement with the competency curves mentioned above, his data points only cover a minimal size range.

Sternberg also calculated bottom shear stress using the Von Karman–Prandtl velocity profile equation which relates the mean velocity at a given distance from the bed to the boundary shear stress. Boundary shear stress may be obtained from the slope of the velocity profile as

$$\tau = \rho \left(\frac{u_{z_2} - u_{z_1}}{5{\cdot}75(\log z_2 - \log z_1)} \right)^2 \tag{9.15}$$

The Von Karman–Prandtl logarithmic distribution only holds for flow close to the bed. Schubauer and Tchen (1961) pointed out that the logarithmic distribution does not hold beyond 0·15δ where δ is the thickness of the boundary layer. This negates the usefulness of the method in the region termed as 'outer flow'. The distribution of velocity in outer flow has been studied in flumes and has led to the velocity defect law (Schlichting, 1968)

$$\frac{U - u}{U_*} = f\left(\frac{y}{\delta}\right) \tag{9.16}$$

where U is the velocity at the outer limit of the boundary layer. Hama (1954) provided an empirical velocity defect formula, later used by Ludwick (1973 and in press), which allows calculation of a boundary shear stress. Ludwick (1973) working in the entrance to Chesapeake Bay found that the critical value for the initiation of motion of fine sand approximated to 0·0035 lb/ft^2. More recently Green (in press) has provided data for a critical boundary shear stress for medium sand (mean size 0·2 mm) from experiments in a tidal channel at the mouth of the Tay estuary. He used the Von Karman–Prandtl formula.

$$\tau_0 = \rho(ky)^2\left(\frac{du}{dz}\right)^2 \tag{9.17}$$

to determine boundary shear stress 25 cm from the bed. A critical value for the initiation of motion for sand lining the channel is 300 g cm^{-1} s^{-2}.

There are unsatisfactory features to all these techniques because in tidally controlled waters flow is unsteady and it never becomes fully developed. However, the expressions do allow calculation of a bouddary shear stress, and the Von Karman–Prandtl formula allows the greatest precision in determining boundary shear stresses assuming that the flow can be monitored sufficiently close to the bed. Until more precise formulae and monitoring apparatus are available this facility may be considered sufficient to overcome unsatisfactory features of all the methods outlined.

Bedforms

At certain times during motion the surface of a mobile bed may be formed into irregularities or 'bedforms'. The quantitative study of bedform development and the prediction of bedform type has also been mainly examined in flumes and alluvial channels. Most field documentation of bedform mechanics only considers qualitative and descriptive aspects. However, general inferences can be made about the distribution of various bedform types associated with the transportation of bed load in fluvial, tidal and estuarine channel systems. Descriptive accounts of minor estuarine forms have been given by Van Straaten (1953), McManus and coworkers (1969) working in the Tay, Allen (1970) in the Gironde, and Farrell (1970) and Hartwell (1970) in estuarine inlets of the seaboard of Maine and Massachusetts, U.S.A.

More general studies of bed morphology and bedform type include those of

Terwindt (1970) and Green (1974). Both rely on echo sounding and transit sonar records, the former in tidal and estuarine inlets of the southwest coast of the Netherlands and the latter in the lower reaches of the Tay estuary. In both locations asymmetrical ripples of varying heights are identified as well as large featureless areas. The smaller-scale bedforms (30–100 cm in height) reverse in response to the bi-directional currents, but no reversal is evident for the larger bedforms. In the Tay the crestal orientations are perpendicular to the alignment of the channel and the sense of asymmetry is linked to residual flow patterns associated with individual flood and ebb dominant channels and sections of single channels. In the Haringvliet estuary, however, crestal orientations make an angle with the channel axis. A possible explanation for this is the presence of secondary currents causing 'helicoidal flow'. Alternatively the crestal orientation could be due to a response to the along channel fluid stress and the across channel slope. This could give a cross channel component on the sediment movement without necessarily a lateral water flow.

In the tidal channels of the Dutch coast, Terwindt (1970) found no significant textural differences, in terms of mean size, between rippled and non-ripple areas; but in the Tay estuary the occurrence of bedforms is closely related to textural controls. Rippled areas coincide with fine to coarse sand while a coarser bed-lining is associated with gravel mound and gravel ledge topography (Figure 9.6).

These largely descriptive reports of the interaction of moving water and sediment load can be compared with studies of the hydrodynamic interpretation of bedforms. These are again allied to experimental data from flume studies. Gilbert (1914) provided some of the earliest data on the relationships between bedform development and flow velocity. More recent studies by Simons and Richardson (1961) showed that a 'classic' sequence of bedforms and bedform stages can be directly related to various hydraulic parameters.

Predictive approaches attempting to define the limits of flow conditions associated with individual bedform stages have been made by Yalin (1964), Bogardi (1965), Znamenskaya (1965), Garde and Ranga (1966) and Simons and Richardson summarized in Guy and coworkers (1966). For example, Bogardi (1965) related bedform type to the mean diameter of the sediment and to the parameter

$$\frac{gd}{\tau\rho} \tag{9.18}$$

where g is the acceleration due to gravity. Garde and Ranga (1966) produced a graph showing the ordinate

$$S(Y_s - Y_f/Y_f) \tag{9.19}$$

where S is the water surface slope, Y_s is the specific weight of the sediment and Y_f the specific weight of the fluid. The abscissa gives values of the ratio r/d where r is the hydraulic radius of a channel. Plots of the data from Guy and coworkers (1966) by Allen (1968) show the relationship of the bedform to 'streampower' and median diameter of the bed material, where 'streampower' is the product of mean velocity and shear stress ($\tau\bar{u}$).

Green (in press) provides one of the few examples of a field-based study attempting to define ranges of flow conditions associated with bedform stages in a tidal estuarine channel. He proposes a fourfold basic division of bedform stages (Figure 9.12). Separation is possible using

$$\rho(kz)^2\left(\frac{du}{dz}\right)^2 \times \frac{\bar{u}}{\sqrt{gd}} \tag{9.20}$$

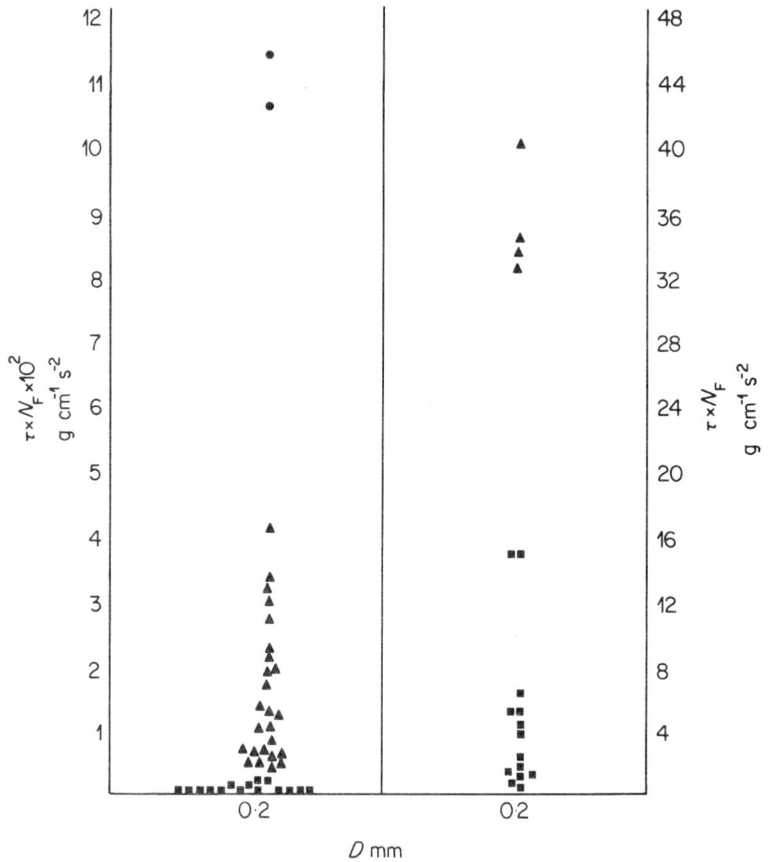

Figure 9.12. Bedform stages as functions of $\rho(kz)^2(du/dz)^2 \times \bar{u}/\sqrt{gd}$ (Equation 9.20) for the median diameter (0·2 mm) of bed material examined. The stages are: (1) no motion (squares), (2) motion with the generation and destruction of bedforms (triangles), (3) transition to (4) (circles). The fourth bedform stage of upper flow regime with plane beds and antidunes is associated with Froude number $N_F > 1$, and is not shown. For the left-hand diagram the scale is from 0–1200 g cm^{-1} s^{-2}. The right-hand diagram is an extended version for the range 0–50 g cm^{-1} s^{-2} (data from Green (in press)).

plotted against the median diameter of the bed material. This equation (9.20) has been applied in areas in which little density stratification exists. Such conditions do not hold throughout all estuarine reaches and in these instances a gradient Richardson number might be more appropriate.

Comparison of the field data from the Tay estuary with the separation criteria given above, all of which have been established from flume data, indicates the limitation of relying on traditional steady state derivations for the parameters characterizing the conditions associated with bedform development.

Bedform migration

The physical investigation of bedform migration rates is again related to flume rather than field studies (e.g. Kennedy, 1963 and 1969). Laboratory investigations have been necessitated by such unsolved problems as the initiation of motion and turbulent transport; and this has resulted in a series of essentially descriptive studies of bedform migration.

Bedform migration rates have been studied in estuaries and marine areas, using underwater photography, surveying and echosounding, and the results have only been associated with process in very general terms. Field investigations vary from the study of small-scale bedforms (2–5 cm in height and 10–40 cm in wavelength) to large sand waves in excess of 3 m in height and 270 m in wavelength (Ludwick, 1972). Kachel and Sternberg (1971), using underwater photography, measured the rate of migration of small-scale ripples in Puget Sound as $2{\cdot}5 \times 10^{-3}$ to $2{\cdot}2 \times 10^{-2}$ cm s^{-1}. Similarly Klein (1970) and Klein and Whaley (1972) identified the hydraulic parameters controlling bedform migration on intertidal sand bars in Minas Basin, Nova Scotia. The bedforms varied from 3 cm to 6 m in height and from 5 cm to 40 m in wavelength. They considered that bedform migration was controlled by the time–velocity asymmetry of bottom tidal currents and changes in mean sediment size. The distances of bedform migration increased with grain size and increasing differences between maximum flood and ebb velocity.

Ludwick (1972) working on the migration rates of sand waves at the mouth of Chesapeake Bay noted seasonal changes in their shapes. Summer conditions resulted in steepened forms and winter conditions in lower rounded forms. Seaward migration rates were estimated from repeated echo sounder observations and these varied from 35 to 150 m per year. The processes of migration were believed to be the construction of sediment 'steps' or secondary waves superimposed on the main sand wave. In contrast, Langhorne (1973) indicated that sand waves in the outer Thames estuary showed little progressive movement. Sand wave crests tended to 'flex' up to 25 m per year, but side-scan surveys indicated that the greatest morphological changes take place during the periods of northeasterly gales.

General transport

Studies of the movement of bed load are primarily of two types: (1) those using bed load equations to predict the amount of bed load moving in a natural channel and (2) investigations relying on the use of fluorescent and radioactive tracers to determine the directions and residual patterns of bed load movement.

The three main types of bed load equations consider transport rates from relationships of shear stresses, discharge and statistical considerations of lift forces. All equations (e.g. DuBoys (1879), Schocklitsch (1926) and Einstein (1950)) rely essentially on experimental data from flume studies to determine their various integral coefficients. Field data allowing re-evaluation of these coefficients are scarce and imprecise definitions make bed load hard to measure. All the bed load equations consider the maximum bed load that a stream or channel can carry; and this 'transporting capacity' may not necessarily coincide with the actual load of a channel at a particular instant of time. Problems are also created if the bed load transport fluctuates as it does in fluvial, tidal and estuarine channels.

One of the few published examples of applying a bed load equation to the marine environment is given by Kachel and Sternberg (1971). They used Bagnold's (1963) equation

$$\frac{\rho_s - \rho}{\rho_s} gj = K\omega \tag{9.21}$$

where ρ_s is the density of the sediment, j is the mass discharge of sediment (g cm^{-1} s^{-1}), K is a proportionality coefficient that expresses the ability of the flow to transport sediment, and ω is the measure of the power expended on the bed by the fluid. The fluid power is expressed in terms of a friction velocity

$$\omega = \rho U_*^3 \tag{9.22}$$

and the equation becomes

$$\frac{\rho_s - \rho}{\rho_s} gj = KU_*^3 \tag{9.23}$$

The authors apply the equation in their Puget Sound study by allowing K to vary as a function of the excess boundary shear stress ($\tau - \tau_{crit}/\tau_{crit}$) and sediment size. The final results conform with the flume data of Guy and coworkers (1966). However, the technique has limitations in terms of the range of sediment size, and the part of a tidal cycle in which it may be used. Similarly it may not be used in areas where there are continuous fluctuations in suspended sediment concentrations.

Investigations using tracer techniques to study bed load movement in estuaries include the Hydraulics Research Station (1958), Courtois and coworkers (1970) and Green (1974). The Hydraulics Research Station used radioactive tracers to determine the movement of fine sand in Liverpool Bay and the Mersey estuary. The results showed that during the study period sediment moved headward from Little Burbo Bank (Liverpool Bay) into the estuary. This net drift was estimated as $\simeq$150 m per day. The study of Courtois and coworkers (1970) also included radioactive tracers to determine sand movement on the bars at the confluence of the Dordogne and Garonne. Initial displacement of the tracer was

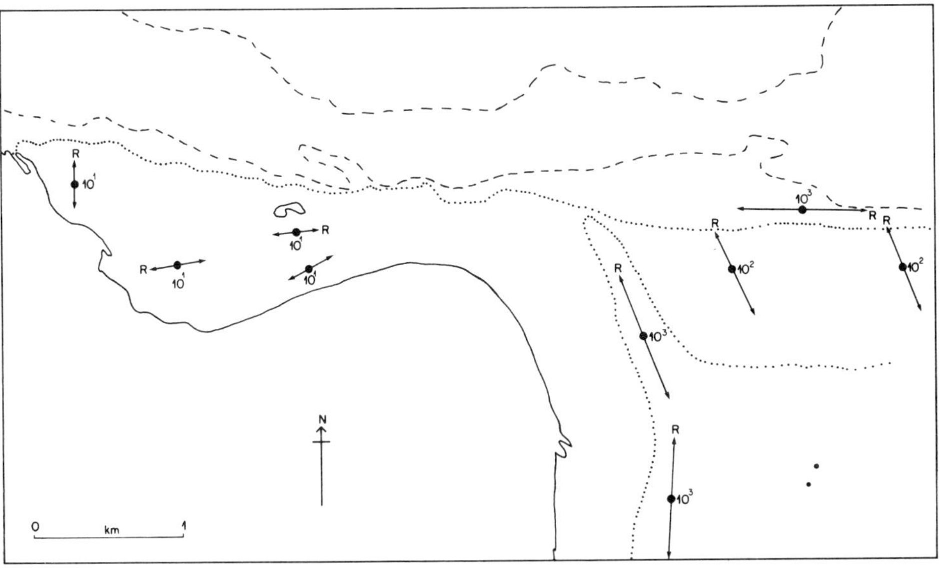

Figure 9.13(a). Sedimentary pathways along the southern shore of the Tay estuary, Scotland, showing: (1) locations of tracer injections (black circles), (2) flood and ebb tide sediment pathways (arrows), (3) diurnal rates of sediment migration (e.g. 10, 10^2, 10^3 m) and (4) the residual directions of sediment movement indicated by 'R'. For location see Figure 9.1. (All data from Green (1974).)

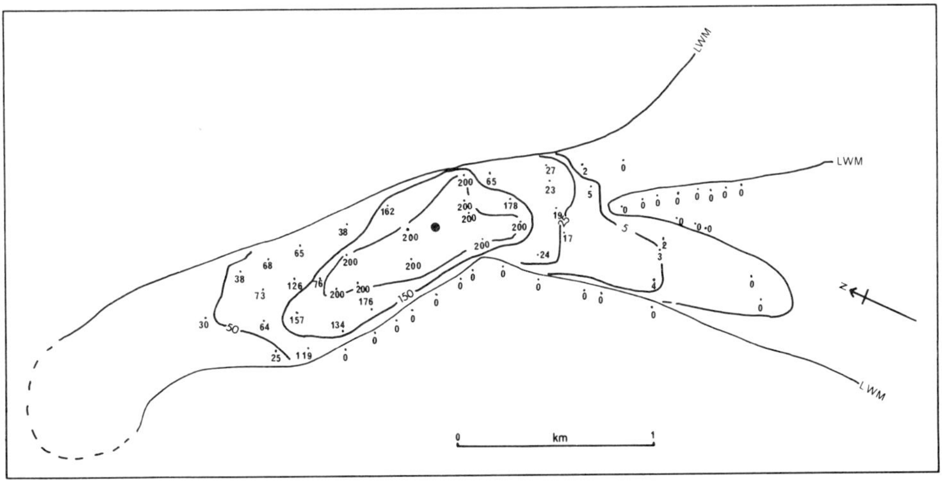

Figure 9.13(b). A tracer dispersion diagram showing residual sediment motion to the northwest—one flood and one ebb tide after injection; Pool Channel, station X. (All data from Green (1974).)

upstream as the tide was flooding, but final results after nine days showed that the *net* drift of sediment was upstream during periods of large tidal amplitude, but reversed during periods of decreasing amplitude proportionally with effects from fluvial discharges.

In contrast, Green (1974) employed fluorescent stained sand to trace the sedimentary pathways at the mouth of the Tay estuary (Figure 9.13). Tracer dispersion diagrams indicated a circulatory system of bed load movement; a net flood residual movement in St. Andrews Bay contrasts with a net ebb residual drift of sand in the main channel of the Tay. The overall rates of sand movement were in the order of 10^2 m/day, on the major banks and bars, and 10^3 m/day in the main tidal channels of St. Andrews Bay and the estuary proper.

Until there is greater understanding of the phenomenon of bed load transport, including the modification of bed load equations for field use, fluorescent and radioactive tracers remain the best methods for identifying the directions, rates and residual drifts of material in estuarine channels.

Sediment dispersion by waves

The influence of waves upon the erosion, transportation and deposition of sediment is one of the least understood aspects of modern sedimentary dynamics (see 'Physical aspects'). Wave activity can transport sediment by inducing longshore currents, translating material over beaches by breaking waves, and move sediment solely by the influence of oscillatory movement of water particles contained within the waves.

The standard equation which relates ideal wave velocity, wavelength and depth for oceanic wind waves is

$$c^2 = \frac{g\lambda}{2\pi} \tanh \frac{2\pi d}{\lambda} \tag{9.24}$$

where c is the celerity and λ the wavelength. The water particles follow orbital paths in which the circumferential velocity $\pi H/\lambda$, where H is the wave height, may rise to several metres per second in large waves. Ocean wind waves are modified as they move into shallow water and ultimately may become solitary waves such that as the depth decreases to half the wavelength ($d \rightarrow \lambda/2$), $\tanh 2\pi d/\lambda$ approaches $2\pi d/\lambda$ and

$$c^2 = gd \tag{9.25}$$

The study by Inman and Nasu (1956), measuring orbital velocities at Scripps Pier, La Jolla, indicated that particle velocities conformed to values such that $c^2 = g(H + d_1)$ where d_1 is the depth of water below the wave trough. Observed values of c were 1·83–2·13 m s^{-1} for waves 1·52 m high with sub-trough depths of 2·14–2·44 m.

As waves continue into shallow water the orbital paths of particles become progressively flattened. The horizontal velocity components are maintained, but the vertical components decrease until ultimately the water circulation paths cease to close and mass movement of water takes place in the direction of wave motion. The crestal waters move with the celerity of the wave, but the base of the wave is decelerating so that the crest spills forward and breaking occurs.

Galvin (1972) recognized four principal types of beach breaker: spilling, plunging, collapsing and surging. They are identified according to height, period and beach slope, and may form a gradational series.

The breaker zone is one of high energy dissipation; frictional energy loss increases with distance travelled and further losses are caused by percolation. The remnant energy is expended on the beach. Early works attempting to relate the energy flux in the breaker zone to sediment transport failed to produce realistic results (e.g. Putnam and Johnson, 1949). Longuet-Higgins (1952) estimated the mean wave energy (E) per unit area of water surface as

$$E = \tfrac{1}{8}\rho g H^2 \text{ r.m.s.} \tag{9.26}$$

The energy flux per unit length of crest is the product of E and the group velocity of the waves.

However, from tracer studies of sand longshore drift Inman and coworkers (1968) have shown that the calculated energy input for waves of a single narrow band of frequencies (using significant wave height) is too high by a factor of two.

Attempts to quantify turbulence in the breaker zone are virtually impossible because there are difficulties in assessing the vertical and horizontal components of motion. Also aeration and incorporation of sediment load induces changes in the nature of the fluid. Teleki (1972) suggested that examination of the water/sediment interface and boundary shear stresses is preferable to using equations of energy and frictional loss.

Although under conditions of oscillatory flow the temporal mean shear stress at any level will be zero, the local instantaneous shear stress may exceed the critical shear force (0·7 of the submerged weight of the particle (Müller and coworkers, 1971)) necessary for entrainment. If shear stresses are to be derived then it is important that instantaneous, rather than time-averaged, values of velocity should be determined.

The detailed structure of the boundary layer is of considerable importance to sediment motion under the influence of waves. Einstein (1971) stressed that both turbulence and load are founded in this layer. As the load increases the nature of flow changes away from the interface. The inner, heavily loaded and slowly moving layer of low turbulence is superseded by a more mobile but clearer outer layer. He also suggested that vorticity, which controls the lifting forces and the heights to which particles move, is of extreme importance in entrainment.

Ingle (1966) demonstrated that in the breaker zone particles of 0·14 mm diameter were most common in suspension, and these grains are not deposited in zones of strong wave activity; but in offshore areas such as Start Bay (England), where the wave characteristics differ from those of the breaker zone, grains of this size are found (McManus, 1975).

As waves cross estuary entrance bars and spits, shoaling causes increases in wave steepness. Upstream, waves continue to advance with decreased steepness but the energy losses at the bars lead to a decrease of wave period. The number of wave fronts per unit area increases in the 'over bar' position and Byrne (1969) determined that a doubling of numbers occurred where the bar depth to trough depth ratio was approximately 0·6. Chandler and Sorensen (1972) demonstrated that a non-linear transformation might allow other multiples to occur with the production of the 'solitons' of Galvin (1972).

The interaction of ebb tidal currents and waves shortens wavelength but radically increases wave height. For example, Hales and Herbich (1972) noted that on a flood tide the wavelength increased and height decreased with an accompanying increase of energy and competency to transport sediment.

Oertel (1972) examined the interaction of currents and waves on spits of Georgia Coast estuaries and recognized that at low water shoreward migration of sediment is only caused by wave activity. As water levels rise wave refraction around shoals produces an intermeshing of wave crests to form a diamond pattern (cf. Figure 9.14). Strong turbulence occurs below these intermeshing crests and the 'bores' so produced move water and sediment landward. Before and after high water, wave action planes shoal surfaces and any sediment migration is induced by currents.

Fluorescent tracer studies indicated that during a complete tidal cycle the combined processes moved the sediment along a series of short-term gyres. This

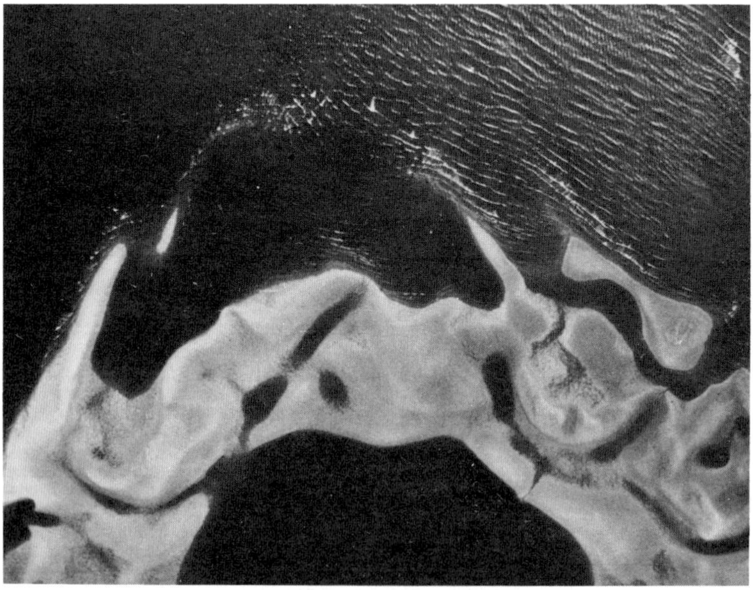

Figure 9.14. Low water air photograph showing intermeshing of wave crests forming distinctive diamond patterns on the outer bar near the mouth of the Tay estuary, Scotland.

produced a 'dynamic sediment trap' reflected in the production of a 'swash platform', that is, a relatively flat-topped feature. Longer-term tracer investigations (Blackley and coworkers, 1972; Green, 1974) indicated that the residence time of individual grains may be very brief but the geomorphological form remains as a permanent feature.

Siefert (1972), working in the entrance of the Elbe estuary, found that waves running onto tidal flats from deep water lost up to half their height, although their period remained constant or slightly increased. This led to a drastic reduction in the steepness of the waves breaking on the flats, and many failed to break at all. The maximum wave heights depended only upon the depth of water and not on the steepness of the wave profile.

Waves may be generated within estuaries, the spectrum depending on fetch, wind strength, duration and water depth. Bretschneider (1965) and the Coastal Engineering Research Center (1966) have produced theoretical curves for the prediction of waves generated in shallow water. However, few field examples from estuaries are known but J. T. Davies (1973) measured waves developed in the central reaches of the Tay estuary in response to southwesterly winds. With a maximum fetch of 20 km, known wind speeds and durations, he found that for rising spring tides the significant wave height and period were 1·3 m and about 3·9 seconds respectively

The mechanics of wave action are as yet only partly understood, equally it is evident from discussions of Bagnold (1963), Kalkanis (1964) and Einstein (1972) that our knowledge of the mechanics of load entrainment by waves is inadequate. Laboratory and field investigations have been mainly confined to examination inside the breaker zone. Field experiments of wave-induced sediment motion outside the breaker zone and within estuaries, have received little attention.

Concluding remarks

From this review some theoretical and experimental expressions for describing the erosion, transportation and deposition, of both fine- and coarse-grained sediments have been shown to be mainly applicable to flumes and alluvial channels, whereas others have been applied with some success in the field although further modifications are still required. The separatism between theoretical and field approaches does not, however, denigrate the intrinsic value of largely quantitative field studies because they are capable of describing multivariate processes which cannot be easily quantified. A mutual exchange of theoretical and practical material is the basis for more sophisticated research of both types.

Of the many aspects discussed, criticized or intimated, the following are isolated as research areas which require special attention:

(1) There is a lack of quantitative data on the relative amounts of coarse and fine sediment entering estuaries from rivers, seas and the erosion of supratidal

cliffs and slopes. Without additional information on sources, residence times and budgets, problems of siltation and the modern dangers of pollutant sequestration cannot be accurately assessed and geological problems of erosion and estuarine infilling cannot be satisfactorily modelled.

(2) Theoretical and empirical expressions describing the erosion of fine-grained sediment are not contradictory, in the sense that some include mass properties whereas others do not; but usually they are mutually exclusive depending on whether the investigator is concerned with the erosion of sediments which have suffered some degree of post-depositional change (e.g. compaction, de-watering, re-orientation of fabrics) or is more interested by the onset and rate of erosion of sediments which are freshly deposited, either in the field or laboratory flume. Some form of theoretical holism may eventually satisfy all requirements. Such expressions must, however, be sufficiently flexible to include a variety of unsteady non-uniform flow conditions typical of most estuaries or estuarine reaches. This requirement is also applicable to the motion and deposition of fine-grained sediment.

(3) Enhancement of the onset and rate of fine-grained sediment erosion by the superimposition of shallow-water waves on unidirectional flows has been largely neglected.

(4) The relative role of flocculation in river-borne suspended sediment, and sediment entering the physico-chemical complex of estuaries, is still contested. Unequivocal direct, rather than inferential, evidence of natural flocculated systems is required from carefully controlled, sophisticated, field experiments.

(5) Many problems exist in differentiating between bed load transportation in estuaries and transportation studies from research in flumes and alluvial channels. Because estuaries largely exhibit unsteady non-uniform flows (see (2) for fine sediments) the application of data from flume studies to the estuarine environment is questionable.

(6) Similarly, knowledge of the threshold of motion of coarse particles, of the genesis of bedforms and of predictive approaches to define the limits of flow conditions associated with individual bedform stages in estuaries is confined by reliance on traditional steady state derivations.

(7) Also bed load equations rely on experimental data from flume studies to determine their various integral coefficients. Field data allowing re-evaluation of these coefficients are scarce.

(8) Knowledge of the mechanics of wave action and sediment entrainment by waves within estuaries is totally inadequate.

Acknowledgements

We would like to thank Dr. J. R. Hails for inviting and suggesting the form of this paper, and Dr. D. J. A. Williams and Dr. K. R. Dyer for their constructive criticisms of the penultimate manuscript. The authors, however, remain responsible for the final production.

Notation

c = wave celerity
C = time-averaged concentration of suspended sediment
C_1 = depth-mean concentration of suspended sediment
C_a = concentration of suspended sediment at a height, a
C_{100} = drag coefficient related to $z = 100$
d = depth of flow
d_1 = water depth below wave trough
E = energy flux
E_{FS} = specific transfer energy
g = gravitational constant
H = wave height
j = mass discharge of sediment
k = Von Karman constant
k_1 = dimensionless coefficient
K = proportionality constant
m = mass of fine sediment
M = empirical constant of proportionality
r = hydraulic radius
S = water surface slope
t = time
$\bar{u}$ = mean velocity
u = velocity
U = velocity at outer limit of boundary layer
U_* = shear velocity
U_{*c} = critical shear velocity
w = fall velocity
Y_s = specific weight of sediment
Y_f = specific weight of fluid
z = distance from the bed
δ = thickness of the boundary layer
ε_s = diffusion coefficient
λ = wavelength
ρ = density of the fluid
ρ_s = density of the sediment
τ = shear stress
τ_0 = shear stress acting on the bed
τ_{crit} = critical shear stress
τ_d = limiting shear stress for deposition
τ_v = vane shear stress
τ_y = yield strength
ω = fluid power per unit area

References

Allen, G. P., 1970, Utilisation d'un écho-sondeur pour l'observation des facies et des structures sédimentaires dans l'estuaire de la Gironde, *Bull. Inst. géol. Bassin Aquitaine*, **9**, 193–200.

Allen, G. P., 1971a, Relationships between grain size parameter distribution and current patterns in the Gironde estuary (France), *J. sedim. Petrol.*, **41**, 74–88.

Allen, G. P., 1971b, Deplacements saisonniers de la lentille de 'crême de vase' dans l'estuaire de la Gironde, *C.R. Acad. Sc.*, Paris, **273**, 2429–2431.

Allen, G. P., and P. Castaing, 1973, Suspended sediment transport from the Gironde estuary (France) onto the adjacent continental shelf, *Mar. Geol.*, **14**, M47–M53.

Allen, J. R. L., 1968, *Current Ripples*, North Holland, Amsterdam, 433 pp.

Allen, J. R. L., 1971, Transverse erosional marks of mud and rock: their physical basis and geological significance, *Sedim. Geol.*, **5**, 167–385.

Allersma, E., A. J. Hoekstra and E. W. Bijker, 1966, Transport patterns in the Chao Phya estuary, *Proc. 10th Conf. cst. Engng., Tokyo, Japan*, Vol. 1, pp. 632–650.

Anderson, F. E., 1970, The periodic cycle of particulate matter in a shallow temperate estuary, *J. sedim. Petrol.*, **40**, 1128–1135.

Arnborg, L., H. J. Walker and J. Peippo, 1967, Suspended load in the Colville River, Alaska, 1962, *Geogr. Annlr.*, **49A**, 131–144.

Bagnold, R. A., 1946, Motion of waves in shallow water, interaction between waves and sand bottoms, *Proc. R. Soc. Lond.*, **A1187**, 1–18.

Bagnold, R. A., 1963, Mechanics of marine sedimentation, in M. N. Hill (Ed.), *The Sea: Ideas and Observations*, Interscience, New York.

Balay, M. A., 1961, El Rio de la Plata entre la atmosfera y el mar, *Publn. No. 621*, Servicio de Hidrografia Naval, Buenos Aires.

Berry, W. G., 1968, Climate, in S. J. Jones (Ed.), *Dundee and District, Advmt. Sci., Br. Ass.*, David Winter, Dundee, pp. 39–61.

Biggs, R. B., 1970, Sources and distribution of suspended sediment in northern Chesapeake Bay, *Mar. Geol.*, **9**, 87–201.

Blackley, M. W. L., A. P. Carr and R. Gleason, 1972, Tracer experiments in the Taw-Torridge estuary, *N.E.R.C. Unit of Coastal Sedimentation Report U.C.S., 1972/22*, 19 pp.

Blatt, H., G. Middleton and R. Murray, 1972, *Origin of Sedimentary Rocks*, Prentice-Hall, Englewood Cliffs, New Jersey, 634 pp.

Bogardi, J. L., 1965, European concepts of sediment transportation, *J. Hydraul. Div., Proc. Am. civil Engrs.*, **91**, 29–54.

Bretschneider, C. L., 1965, *The generation of waves by wind: state of the art*, Natn. Engr. Sci. Co., Off. Naval Res., SN 134–6, 96 pp.

Buller, A. T., (in press), Sediments of the Tay estuary. Part II. Formation of ephemeral zones of high suspended sediment concentrations, *Proc. R. Soc. Edinb.*, B.

Buller, A. T., and J. McManus, 1972, Simple metric sedimentary statistics used to recognise different environments, *Sedimentology*, **18**, 1–21.

Buller, A. T., and J. McManus, 1974, Factors influencing the formation of 'turbidity maxima' with examples from the Tay estuary, Scotland, Proceedings International Symposium on Interrelationships of Estuarine and Continental Shelf Sedimentation, *Mémoir Institut de Géologie du Bassin d'Aquitaine*, **7**, 34–44.

Buller, A. T., and J. McManus, (in press), Sediments of the Tay estuary. Part I. Bottom sediments of the upper and upper middle reaches, *Proc. R. Soc. Edinb.*, B.

Buller, A. T., J. McManus and D. J. A. Williams, 1971, Investigations in the estuarine environments of the Tay. No. 1. Physical aspects: an interim report, *Dundee Univ. Tay Estuary Research Centre, Res. Rept.*, *1*, 62 pp.

Buller, A. T., J. A. Charlton and J. McManus, 1972, Data from physical and chemical measurements in the Tay estuary for neap and spring tides, *Dundee Univ. Tay Estuary Research Centre, Res. Rept.*, *2*, 53 pp.

Byrne, R. J., 1969, Field occurrences of induced multiple gravity waves, *J. geophys. Res.*, **74**, 2590–2596.

Cameron, W. M., and D. W. Pritchard, 1963, 'Estuaries', in M. N. Hill (Ed.), *The Sea*, Vol. 2, Wiley, New York, 306–324.

Chandler, P. L., and R. M. Sorensen, 1972, Transformation of waves passing a submerged bar, *Proc. 13th Conf. cst. Engng., Vancouver*, Vol. 1, 385–404.

Coastal Engineering Research Center, 1966, Shore protection, planning and design, *Tech. Rept. 4*, 3rd edition.

Coleman, J. M., 1969, Brahmaputra River: Channel processes and sedimentation, *Sedim. Geol.*, **3**, 129–239.

Courtois, G., B. Jeanneau, G. Allen and A. Klingebiel, 1970, Étude de la stabilité d'un banc de sable sur le sit du bec d'amber par l'emploi de traceurs radioactifs, *Bull. Inst. géol. Bassin, Aquitaine*, **8**, 155–208.

Crommelin, R. D., 1940, De herkomst van het zand van de Waddensee, *Tijdschr. K. ned. Aardrijksk Genoot.*, **57**, 347–361.

Crommelin, R. D., 1949. Quelques aspects granulométriques et minéralogiques de la sedimentation le long de l'estuaire de l'Escaut, *Sedimentation et Quaternaire, La Rochelle*, pp. 63–71. Led. Sam, Bordeux.

Cunningham, D., 1896, The Estuary of the Tay. *Minut. Proc. Instn. civ. Engrs.*, **120**, 299–313.

Davies, J. T., 1973, Waves in the western part of the Tay Estuary, B.Sc. thesis, University of Dundee.

Davies, S. L., 1973, A preliminary investigation with the meteorological effect on the tide in the Tay Estuary, B.Sc. thesis, University of Dundee.

Dionne, J-C., 1969a, Tidal flat erosion by ice at La Potcatière, St. Lawrence estuary, *J. Sedim, Petrol.*, **39**, 1174–1181.

Dionne, J-C., 1969b, Erosion glacielle littorale, estuaire de Saint Laurent, *Revue Géogr. Montr.*, **23**, 5–20.

DuBoys, P. F. D., 1879, Le Rhône et le rivier a lit affouillable, *Annls. Ponts Chauss.*, **18**.

Dunn, L. S., 1959, Tractive resistance of cohesive channels, *J. Soil Mech. Fdns Div. Proc. Am. Soc. civ. Engrs.*, **85**, 1–24.

Dyer, K. R., 1972, Sedimentation in estuaries, in R. S. K. Barnes and J. Green (Eds.), *The Estuarine Environment*, Applied Science Publ., London, pp. 10–32.

Dyer, K. R., and K. Ramamoorthy, 1969, Salinity and water circulation in the Vellar estuary, *Limnol. Oceanogr.*, **14**, 4–15.

Einstein, H. A., 1950, The bed load function for sediment transportation in open channel flows, *U.S. Dep. Agric. Soil Conserv. Serv., T.B., No. 1026.*

Einstein, H. A., 1972, A basic description of sediment transport on beaches, in R. E. Meyer (Ed.), *Waves on Beaches*, Academic Press, New York, pp. 53–93.

Einstein, H. A., and N. Chien, 1955, Effects of heavy sediment concentration near the bed on velocity and sediment distribution, *Univ. Calif. Inst. Engr. Res., No. 8.*

Einstein, H. A., and R. B. Krone, 1962, Experiments to determine modes of cohesive sediment transport in salt water, *J. geophys. Res.*, **67**, 1451–1461.

Emery, K. O., and R. E. Stevenson, 1957, Estuaries and lagoons, *Mem. geol. Soc. Am.*, **67**, 673–750.

Farrell, S. C., 1970, *Sediment distribution and hydrodynamics Saco River and Scarboro estuaries, Maine*, Contrib. No. 6-CRG, Dep. of Geology, Univ. of Massachusetts, 129 pp.

Fischer, H. B., 1972, Mass transport mechanisms in partially stratified estuaries, *J. Fluid Mech.*, **53**, 671–687.

Fleming, G., 1970, Sediment balance of Clyde estuary, *J. Hydraul. Div., Proc. Am. Soc. civ. Engrs.*, **96**, 2219–2230.

Fournier, F., 1960, *Climat et érosion: la relation entre l'érosion du sol par l'eau et les précipitations atmosphériques*, Paris, 201 pp.

Francis-Boeuf, C., 1947, Recherches sur le milieu fluvio-marin et les dépôts d'estuaires, *Annls. Inst. océanogr., Monaco*, N.S., **23**, 150–344.
Galvin, C. J., 1972, Wave breaking in shallow water, in R. E. Meyer (Ed.), *Waves on Beaches*, Academic Press, New York, pp. 413–456.
Garde, R. J., and K. R. Ranga, 1966, Resistance relationships in alluvial channel flow, *J. Hydraul. Div., Proc. Am. Soc. civ. Engrs.*, **92**.
Gibbs, R. J., 1973, Amazon river estuarine system, in B. W. Nelson (Ed.), *Environmental framework of coastal plain estuaries*, Mem. geol. Soc. Am., **133**, 85–88.
Gilbert, G. K., 1914, The transportation of debris by running water, *Prof. Pap. U.S. geol. Surv.*, **86**, 1–263.
Graf, W. H., 1972, *Hydraulics of sediment transport*, McGraw-Hill, New York, 513 pp.
Green, C. D., 1974, Sedimentary and morphological dynamics between St. Andrews Bay and Tayport, Tay Estuary, Scotland. Ph.D. thesis (unpublished), University of Dundee.
Green, C. D., (in press), A study of hydraulics and bedforms at the mouth of the Tay estuary, Scotland, *Proc. 2nd Int. Estuarine Res. Conf.*, Myrtle Beach, Carolina.
Guilcher, A., 1967, Origin of sediments in estuaries, in G. H. Lauff (Ed.), *Estuaries*, Am. Assoc. Advmt. Sci., Washington, D.C., pp. 149–157.
Guy, H., D. B. Simons and E. V. Richardson, 1966, Summary of alluvial channel data from flume experiments 1956–1961, *Prof. Pap. U.S. geol. Surv.*, 462-J.
Hales, L. Z., and J. B. Herbich, 1972, Tidal inlet current—ocean wave interaction, *Proc. 13th Conf. cst. Engng., Vancouver*, Vol. 1, 669–688.
Halliwell, A. R., and B. A. O'Connor, 1966, Suspended sediment in a tidal estuary, *Proc. 10th Conf. cst. Engng., Tokyo*, Vol. 1, pp. 687–706.
Hama, F. R., 1954, Boundary layer characteristics for smooth and rough surfaces, *Trans. Soc. nav. Archit. mar. Engrs.*, **62**, 333–358.
Hanis, D. L., 1972, Characteristics of wave records in the coastal zone, in R. E. Meyer (Ed.), *Waves on Beaches*, Academic Press, New York, pp. 1–52.
Hartwell, A. D., 1970, Hydrography and Holocene sediment of the Merrimack River estuary, Massachusetts, *Contrib. No. 5-CRG*, Dept. of Geology, University of Massachusetts, 166 pp.
Hauschild, W. L., R. W. Perkins, H. H. Stevens Jr., G. R. Dempster, Jr. and J. L. Glenn, 1966, Progress report: Radionuclide transport in the Pasco to Vancouver, Washington reach of the Columbia River, July 1962 to September 1963, *U.S. geol. Surv. Open file Rept.*, 188 pp.
Hjulström, F., 1935, Studies on the morphological activity of rivers as illustrated by the river Fyris, *Bull. geol. Instn. Univ. Uppsala*, **25**, 221–527.
Holeman, J. N., 1968, The sediment yield of major rivers of the world, *Wat. Resour. Res.*, **4**, 737–747.
Hydraulics Research Station, 1958, Radioactive tracers for the study of sand movement. A report of an experiment carried out in Liverpool Bay in 1958, *Hydraul. Res. Stn.*, Wallingford.
Ingle, J. C. Jr., 1966, *The movement of beach sand. Developments in Sedimentology*, Vol. 5, 221 pp. Elsevier, Amsterdam.
Inglis, C. C., and F. H. Allen, 1957, The regimen of the Thames estuary as affected by currents, salinities and river flow, *Proc. Instn. civ. Engrs.*, **9**, 193–216.
Inman, D. L., 1963, Sediments: physical properties and mechanics of sedimentation, in F. P. Shepard (Ed.), *Submarine geology*, 2nd ed., Harper and Row, New York, pp. 101–151.
Inman, D. L., P. D. Komar, and A. J. Bowen, 1968, Longshore transport of sand. *Proc. 11th Conf. cst Engng., London*, Vol. 1, pp. 298–306.
Inman, D. L., and N. Nasu, 1956, Orbital velocity associated with wave action near the breaker zone, *Tech. Memo. Beach Eros. Bd. U.S., 79.*
Jennings, J. N., and E. F. C. Bird, 1967, Regional geomorphological characteristics

of some Australian estuaries, in G. H. Lauff (Ed.), *Estuaries*, Am. Assoc. Advn. Sci. Publ., **83**, 121–128.
Kachel, N. B., and R. W. Sternberg, 1971, Transport of bed load as ripples during an ebb current, *Mar. Geol.*, **19**, 229–244.
Kalkanis, G., 1964, Transportation of bed material due to wave action, *Tech. Mem. U.S. cst Engng. Res. Center, No. 2.*
Kennedy, J. F., 1963, The mechanics of dunes and anti-dunes in erodible bed channels, *J. Fluid Mech.*, **16**, 521–544.
Kennedy, J. F., 1969, The formation of sediment ripples dunes and anti-dunes, *Ann. Rev. Fluid Mechs.*, **1**, 147–168.
Klein, G. de V., 1970, Depositional and dispersal mechanics of inter-tidal sandbars, *J. sedim. Petrol.*, **40**, 1095–1127.
Klein, G. de V., and M. L. Whaley, 1972, Hydraulic parameters controlling bedform migration on an inter-tidal sand body, *Bull. geol. Soc., Am.* **83**, 3465–3470.
Krone, R. B., 1962, *Flume studies of the transport of sediment in estuarial shoaling processes*, Univ. of California, Berkeley Hyd. Eng. Lab. and Sanit. Eng. Res. Lab., 110 pp.
Krone, R. B., 1972, A field study of flocculation as a factor in estuarial shoaling processes, Committee on Tidal Hydraulics, Corps. of Engineers, *U.S. Army Tech. Bull. 19*, 62 pp.
Kulm, L. D., and J. V. Byrne, 1967, Sediments of Yaquina Bay, Oregon, in G. H. Lauff (Ed.), *Estuaries*, Am. Assoc. Advn. Sci. Publ. 83, 226–238.
Langhorne, D. N., 1973, A sandwave field in the Outer Thames Estuary, Great Britain, *Mar. Geol.*, **14**, 129–143.
Longuet-Higgins, M. S., 1952, On the statistical distribution of the heights of sea waves, *J. mar. Res.*, **11**, 245–246.
Ludwick, J. C., 1972, Migration of tidal and sand waves in Chesapeake Bay entrance, in D. J. P. Swift, D. B. Duane and O. H. Pilkey (Eds.), *Shelf Sediment Transport*, Dowden, Hutchinson and Ross Inc., Stroudsburg, Pa., 377–410.
Ludwick, J. C., 1973, *Tidal currents and zig-zag sand shoals in a wide estuary entrance*, Old Dominion University, *Tech. Rept.*, *No. 7*, 89 pp.
Ludwick, J. C., (in press), Tidal currents, sediment transport and sand banks in Chesapeake Bay entrance, Virginia, *Proc. 2nd Int. Estuarine Res. Conf.*, Myrtle Beach, S. Carolina.
McDowell, D. M., 1970–1971, Broad and Deep, *Mem. Proc. Manchr. lit. phil. Soc.*, **113**, 1–11.
McManus, J., 1968, The hydrology of the Tay basin, in S. J. Jones (Ed.), *Dundee and District*, Advmt Sci., Br. Ass., David Winter, Dundee, pp. 107–124.
McManus, J., 1975, Quartile deviation—median diameter analysis of surface and core sediments from Start Bay, *J. geol. Soc. Lond.*, **131**, 51–56.
McManus, J., J. West and D. J. A. Williams, 1969, Bedforms on an inter-tidal estuarine sandbank, *Proc. geol. Soc.*, **1662**, 71–72.
Meade, R. H., 1969, Landward transport of bottom sediments in estuaries of the Atlantic Coastal Plain, *J. sedim. Petrol.*, **39**, 222–234.
Meade, R. H., 1972, Transport and deposition of sediments in estuaries, *Mem. geol. Soc. Am.*, **133**, 91–120.
Migniot, C., 1968, Étude de propriétés physiques de differents sédiments très fin et de leur comportement sous des actions hydrodynamiques, *Houille Blanche*, **7**, 591–620.
Miller, C. R., and R. F. Piest, 1970, Chapter IV. Sediment sources and sediment yields, *J. Hydraul. Div., Proc. Am. Soc. civ. Engrs.*, **96**, 1283–1329.
Mishra, S. K., 1969, Heavy mineral studies of the Firth of Tay region, *J. geol. Soc. India*, **5**, 37–49.
Müller, A., A. Gyr and T. Dracos, 1971, Interaction of rotating elements of the

boundary layer with grains of a bed, a contribution to the problem of the threshold of sediment transportation, *J. Hydraul. Res.*, **9**, 373–411.

Müller, G., and U. Förstner, 1968, General relationships between suspended sediment concentration and water discharge in the Alpenrhein and some other rivers, *Nature, Lond.*, **217**, 244–245.

NEDECO, 1959, *River studies and recommendations on improvement of Niger and Benue*, North Holland, Amsterdam, 1000 pp.

Nelson, B. W., 1960, Recent sediment studies in 1960, *Virg. Polytechnic J. Inst. Mineral Industries*, **7**, 1–4.

Odd, N. V. M., and M. W. Owen, 1972, A two layer model of mud transport in the Thames Estuary, *Proc. Instn civ. Engrs*, Suppl. 7517, 175–205.

Oertel, G. F., 1972, Sediment transport of estuary entrance shoals and the formation of swash platforms, *J. sedim. Petrol.*, **42**, 857–863.

Partheniades, E., 1962, A study of erosion and deposition of cohesive soils in salt water, Ph.D. thesis, Univ. of California.

Partheniades, E., 1965, Erosion and deposition of cohesive soils, *J. Hydraul. Div., Proc. Am. Soc. civ. Engrs.*, **91**, 105–139.

Partheniades, E., 1972, Results of recent investigations on erosion and deposition of cohesive sediments, in H. Shen (Ed.), *Sedimentation*, Fort Collins, Colorado, USA, 20, pp. 1–24.

Partheniades, E., 1973, Recent investigations in stratified flows related to estuarial hydraulics, in B. W. Nelson (Ed.), *Environmental framework of coastal plain estuaries, Mem. geol. Soc. Am.*, **133**, 29–70.

Passega, R., 1957, Texture as characteristic of clastic deposition, *Bull. Am. Assoc. Petrol. Geol.*, **41**, 1952–1984.

Passega, R., 1964, Grain size representations by CM patterns as a geological tool, *J. sedim. Petrol.*, **34**, 830–847.

Postma, H., 1967, Sediment transport and sedimentation in the estuarine environment, in G. H. Lauff (Ed.), *Estuaries*, Publ. Am. Assoc. Advmt. Sci., **83**, 158–179.

Pritchard, D. W., 1955, Estuarine circulation patterns, *Proc. Am. Soc. civ. Engrs.*, **81**, 717.

Putnam, J. A., and J. W. Johnson, 1949, The dissipation of wave energy by bottom friction, *Trans. Am. geophys. Un.*, **30**, 67–74.

Ramsay, D. M., 1968, Geology, in S. J. Jones (Ed.), *Dundee and District*, Advmt. Sci., Br. Ass., David Winter, Dundee, pp. 107–124.

Riedel, H. P., J. W. Kamphuis and A. Brebner, 1972, Measurement of bed shear stress under waves, *Proc. 13th Conf. cst. Engng., Vancouver*, Vol. 1, 587–604.

Routh, C. D., 1970, The movement of a tidal wave in the Tay estuary, B.Sc. thesis, University of Dundee.

Schlichting, H., 1968, *Boundary layer theory*, 6th ed., McGraw-Hill, New York, 747 pp.

Schocklitsch, A., 1926, *Die Geschiebebewegung an Flussen und an Stauwerben*, Springer, Vienna.

Schubauer, G. B., and C. M. Tchen, 1961, *Turbulent Flow*, Princeton University Press, Princeton, New Jersey, 123 pp.

Schubel, J. R., 1968, Suspended sediment of the northern Chesapeake Bay, *Johns Hopkins Univ., Chesapeake Bay Inst. Tech. Rept.*, *35* (Ref. 68-2), 264 pp.

Schubel, J. R., (Convenor), 1971, *The Estuarine Environment*, American Geological Institute, Short Course Lecture Notes, Washington, D.C.

Scruton, P. C., 1956, Oceanography of Mississippi delta sedimentary environments, *Bull. Am. Assoc. Petrol. Geol.*, **40**, 2864–2952.

Shields, A., 1936, Anwendung der Ahnlichkeits Mechanik und der Turbulenzforschung auf die Geschiebe Bewegung, *Preuss. Vers. Anst. WassBau Schiffbau, Berlin*, **26**, 5–24.

Siefert, W., 1972, Shallow-water wave characteristics, *Proc. 13th Conf., cst Engng., Vancouver, Canada*, Vol. 1, pp. 329–347.

Simons, D. B., and E. V. Richardson, 1961, Forms of bed roughness in alluvial channels, *J. Hydraul. Div. Am. Soc. civ. Engrs.*, **87**, 87–105.

Skempton, A. W., 1970, The consolidation of clays by gravitational compaction, *Q. J. geol. Soc. Lond.*, **125**, 373–408.

Sternberg, R. W., 1967, Measurements of sediment movement and ripple migration in a shallow marine environment, *Mar. Geol.*, **5**, 195–205.

Sternberg, R. W., 1971, Measurement of incipient motion of sediment particles in the marine environment, *Mar. Geol.*, **10**, 113–119.

Sternberg, R. W., 1972, Predicting initial motion and bedload transport of sediment particles in the shallow marine environment, in D. J. P. Swift, D. B. Duane and O. H. Pilkey (Eds.), *Shelf sediment transport*, Dowden, Hutchison and Ross, Stroudsburg, Pa., pp. 61–72.

Stevenson, R. E., and K. O. Emery, 1958, Marshlands at Newport Bay, California, Allan Hancock Publ. 20, 110 pp. University of Southern Calif. Press, Los Angeles.

Sundborg, Å., 1956, The River Klarälven. A study of fluvial processes, *Geogr. Annlr.*, **38**, 127–316.

Sundborg, Å., 1958, A method for estimating the sedimentation of suspended material, *C.R. and Repts., Toronto, Int. Assoc. Sci. Hydrology*, **43**, 249–259.

Sverdrup, H. V., and Munk, W. H., 1947, Wind, sea and swell theory relations for forecasting, *Tech. Rept. U.S. Hydrographic Office, No. 1. Publ. 601.*

Teleki, P. G., 1972, Wave boundary layers and their relation to sediment transport, in D. J. P. Swift, D. B. Duane and O. H. Pilkey (Eds.), *Shelf Sediment Transport*, Dowden, Hutchison and Ross, Stroudsburg, Pa., pp. 21–59.

Terwindt, J. H. J., 1970, Observation on submerged sand ripples with heights ranging from 30 to 200 cm occurring in tidal channels of S.W. Netherlands, *Geologie Mijnb.*, **49**, 489–501.

Tucker, M. E., 1973, The sedimentary environments of tropical African estuaries: Freetown Peninsula, Sierra Leone, *Geologie Mijnb.*, **52**, 203–215.

Van Straaten, L. M. J. U., 1960, Transport and composition of sediments, in *Symposium Ems-Estuarium, Nordsee, Verh. K. ned. Geol. Mijnb. Genoot.*, **19**, 279–292.

Van Straaten, L. M. J. U., 1953, Megaripples in the Dutch Wadden Sea and in the Basin of Arcachon (France), *Geol. en Mijnbouw*, **15**, 1–11.

Verger, F., 1968, *Marais et wadden du littoral français*, Biscaye Frères, Bordeaux, 541 pp.

West, J. R., 1973, On the salinity of the Tay estuary, Ph.D. thesis (unpublished), University of Dundee.

Williams, D. J. A., and J. R. West, (in press), Salinity distribution in the Tay estuary, *Proc. R. Soc. Edinb., B.*

Yalin, M. S., Geometrical properties of sand waves, *J. Hydraul. Div., Proc. Am. Soc. civ. Engrs.*, **90**, 105–120.

Yalin, M. S., 1972, *Mechanics of sediment transport*, Pergamon Press, Braunschweig, 290 pp.

Znamenskaya, N. S., 1965, The use of the laws of sediment dune movement in computing channel deformation, *Soviet Hydrol.* (Am. geophys. Univ. abstr.), No. 2.

CHAPTER TEN

Dynamic Structures in the Holocene Chenier Plain Setting of Essex, England

J. T. GREENSMITH and E. V. TUCKER

Abstract

Present deposition in the intertidal and supratidal zones of the study area is partly a legacy of Pleistocene events. Along the Blackwater and Colne estuaries sea level terrace gravels of Pleistocene age are being reworked into asymmetric mobile banks, cheniers (marsh beach ridges) and spits exhibiting landward movement and lateral extension. Growth of the gravel structures is a partial function of the increased availability of sandy gravel resulting from the erosion of a veneer of Holocene marsh and upper tidal flat deposits. There is evidence for several episodes of increased and decreased availability. The structures are sometimes surmounted by mobile shell debris. Between the Blackwater and Thames estuaries cheniers composed of shell and sand are developing at the presently retreating marsh edges, the Sales Point chenier being a type-example. Recent changes in the form of this chenier are discussed. All the modern cheniers and older analogues farther inland help to identify the late Holocene chenier plain and allow a sequential picture of chenier evolution to be deduced from juvenile to old age stages. The prograding plain, though similar in many respects to those of Louisiana, New Zealand and Queensland, differs fundamentally in having become established in an area of relative tectonic instability.

Introduction

The Outer Thames Estuary, which includes the submerged river valleys of the Colne, Blackwater, Roach and Crouch, is an area of considerable geological complexity. Pleistocene times witnessed considerable fluctuations in sea level, the advance of at least one major ice-sheet onto the northern periphery (Bristow and Cox, 1973) and changing patterns of drainage with outlets to the east and southeast. There is evidence in the area that sea level fell by at least 34 m during the late-Pleistocene and indications from farther afield that a fall of at least 100 m occurred (Guilcher, 1969; Greensmith and Tucker, 1973a). During this, and earlier Pleistocene phases of sea level change, considerable incision by rivers occurred so that there now exists beneath the Outer Thames Estuary a network of buried channels (D'Olier and Maddrell, 1970; Greensmith and Tucker, 1971; D'Olier, 1972) (see Figure 10.1). At other times of relative stillstand or rising

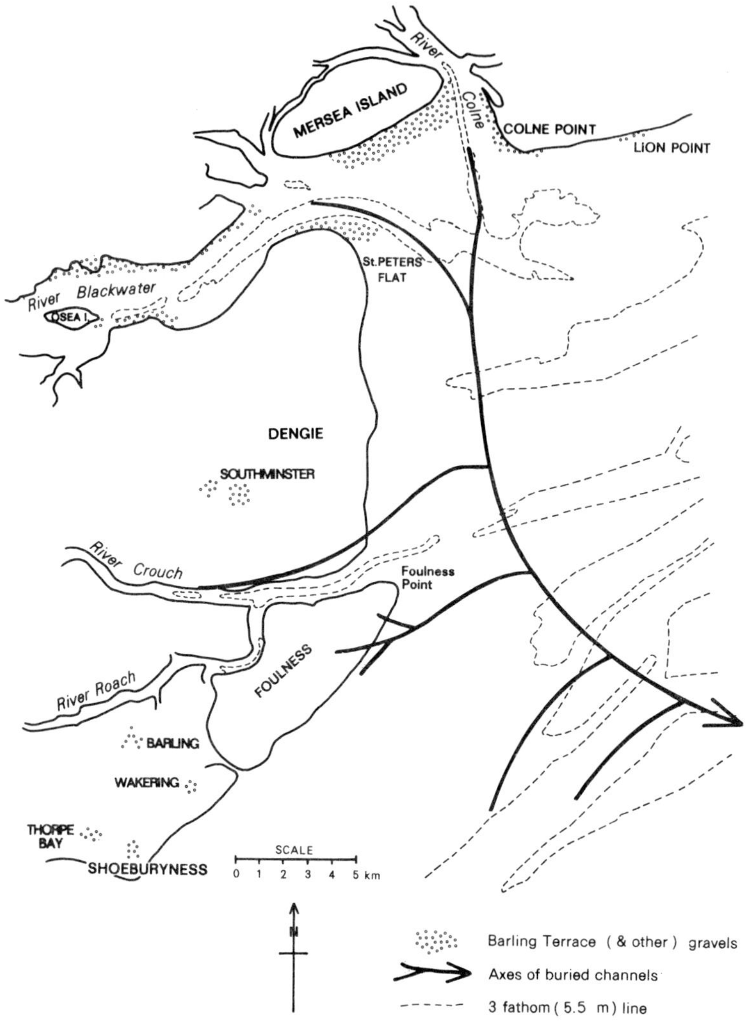

Figure 10.1. Known distribution of gravels adjacent to present mean sea level and position of major Pleistocene buried channels (partly after D'Olier and Maddrell, 1970).

sea level, aggradation created fluviatile sandy gravel spreads in the valleys, now well recognized as terraces along the flanks of the River Thames, but less well known immediately to the north.

The purpose of this paper is to describe a number of present-day mobile sedimentary structures in the intertidal and supratidal zones of the Outer Estuary and to discuss their origin in terms of sediment supply and sediment movement. These findings will then be briefly related to certain similar, older structures which constitute a significant part of the Holocene Essex chenier plain, originally reported by the writers (Greensmith and Tucker, 1966 and 1969).

Gravel availability

Of all the known *in situ* Pleistocene terrace gravel deposits associated with the Thames drainage system, the ones which are most relevant to this study are restricted to a range of a few metres around present mean sea level. Their precise age and correlation are controversial yet immaterial (e.g. Oakley and Leakey, 1937). At Barling, terrace gravels at about 8 m above O.D. (Ordnance Datum, Newlyn, which is approximately mean sea level) have a Middle Acheulian archaeological assemblage and cool temperate fauna, and have been tentatively assigned to the end of the Great Interglacial (Hoxnian) by Gruhn and Bryan (1972). It is therefore possible that some of the lower level gravels straddling zero metres O.D. are Last Interglacial (Ipswichian-Eemian-Sangamon) or younger (Cornwall, 1971, personal communication). Some of these are exposed inland, as, for example, at Shoeburyness, while others are recorded in boreholes, as at Southminster Hall, and in temporary exposures in reclaimed ground of the Dengie peninsula and near Goldhanger. Bristow (in Ambrose, 1973) refers the gravels at the last locality to the First Terrace of the River Blackwater, but by far the most extensive exposures occur in the intertidal zone of the northern part of the area, particularly good spreads being visible on the Mersea Flats and along the flanks of the Colne and Blackwater estuaries. Some of the most closely studied used to occur on the foreshore at Lion Point, near Clacton-on-Sea, but the site is no longer accessible (Warren, 1923 and 1924). At Lion Point and East Mersea the gravels have yielded abundant remains of extinct temperate forest or parkland mammalian fauna and freshwater shells (Cornwall, 1958).

In all instances the deposits have a polymodal size distribution and consist of angular to well-rounded pebbles, dominantly flint, set in a matrix of coarse sand and silt, whilst clay-size particles may be admixed on the tidal flats. The pebbles are up to 10 cm in maximum diameter, the larger diameters being most abundant in the Blackwater–Colne area. It is intriguing that these terrace gravels, so well preserved in the northern part of the area, are not well preserved in the corresponding zone farther south along the Crouch and Roach estuaries. This anomaly probably reflects late-Pleistocene lowerings of base-level and valley widening downstream from nickpoints located at about zero metres and −23 m O.D. (Greensmith and Tucker, 1971, p. 315). Erosion of the zero metre terrace gravels proceeded farther along the Crouch and Roach valleys in late-Pleistocene times than it did along the Blackwater and Colne valleys.

In cliff exposures, and inland, the irregular upper surface of the gravels is commonly mantled by a light brown, generally structureless silt layer of Pleistocene age, which may be soliflucted material, but which also resembles *in situ* brickearth and loess. A gravel with such a veneer is clearly exposed in a seacliff near East Mersea (Greensmith and coworkers, 1973, p. 15). The gravel alone passes seawards without break into a gravel pavement extending virtually over the full width of the intertidal zone and extending for several kilometres along it (Figure 10.2). In 1958 these intertidal gravels were poorly exposed but, since then, extensive tracts have been uncovered by erosion of a layer of Holocene clayey silts at least 1·5 m thick (Davis, 1964; Greensmith and Tucker, 1973a and 1973b). The present processes affecting the stripping at Mersea are operative throughout the Outer Thames area. Evidence for the efficacy of these processes need only be summarized here because full details appear in Greensmith and Tucker (1965, 1967, 1969, 1973a and 1973b). It includes:

(1) Extensive salt marsh cliffing, not only where the marshes face the open flats but also in protected areas to the rear of cheniers and spits.
(2) Diminution of the area of salt marshes as a consequence of (1).
(3) Extensive development of ebb-scour channels and mud-mound topography in the upper tidal flat zone, especially between the Blackwater and Crouch estuaries. These structures are formed on the site of original salt marshes.
(4) Re-exposure of ancient *in situ* marsh deposits in the intertidal zone. These deposits include peat seams with various radiocarbon dates from 4959 ± 65 B.P. to 118 ± 48 B.P. The erosion of a peat seam dated at 173 ± 60 B.P. at East Mersea is exposing an 'occupation surface' of Romano-British age (*c*. 1960–*c*. 1750 B.P.) and possibly even Neolithic–early Bronze age (*c*. 4500 B.P.). This 'surface', or non-sequence, can be recognized intermittently throughout the area.
(5) Exposure of tidal flat deposits of earlier cycles of deposition.

The effect of the present re-exposure of gravels is to allow the material to become mobilized and capable of being utilized in the formation and further evolution of a range of sedimentary structures in the tidal zones. This material is additional to that being supplied to the zones from the retreat of unprotected seacliffs at Osea Island, Mersea Island and intermittently as far north as Walton-on-Naze.

The present condition along the Essex coast is one of marine transgression and the patterns of sedimentation reflect this fundamental fact. Moreover, there is evidence in the local Holocene marine succession for several earlier transgressive episodes with intervening regressions (Greensmith and Tucker, 1973a). Some of the later transgressive episodes, such as the one which may have culminated between 1434 ± 110 B.P. and 1265 ± 100 B.P., probably had a marked effect by increasing shell, sand and pebble availability in the intertidal zone.

Lower and middle tidal flat structures

Gravel banks

The main physical expression of the reworking of terrace deposits in the low-water and mid-water tidal zones along the Blackwater and Colne estuaries and at Mersea Island are asymmetric banks (Figures 10.2 and 10.3). Their location appears to be predominantly controlled by the original position and extent of the

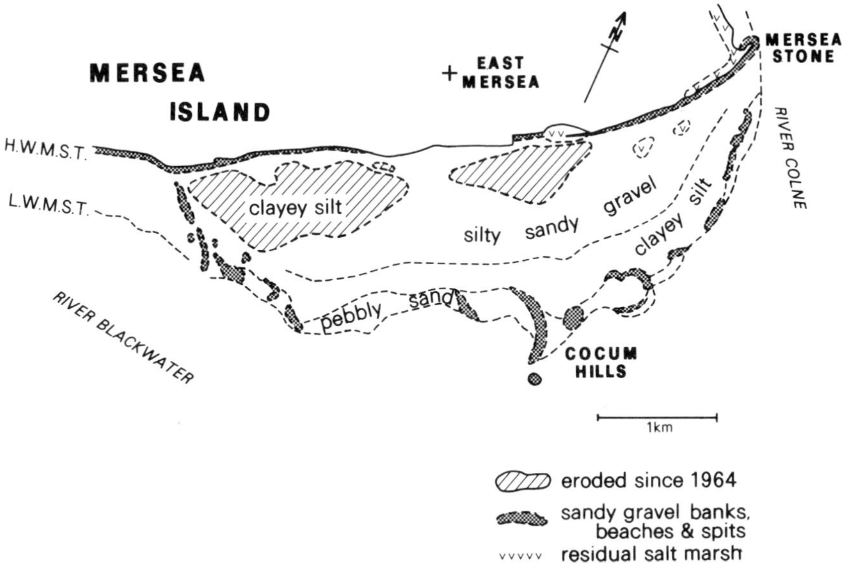

Figure 10.2. Lithofacies map of the eastern end of the Mersea Flats (modified after Davis, 1964).

Pleistocene terrace and associated channels. Some of the banks trend roughly parallel to present low water mark, others extend obliquely or almost at right angles across the zones. They vary in width up to 75 m, stand proud of the adjacent surface by heights up to 1·5 m, and reach up to 500 m in length. The larger bodies are normally compounded of several originally independent banks which have coalesced by progressive extension, possibly as a result of littoral drift. The more asymmetric structures have a smoothly curved crestline and have steep leading sides which attain angles near to 30° and gentle trailing sides sloping at about 5°. A crude internal cross-lamination parallels the leading slopes. In the mildly asymmetric structures significant internal structure is not readily discernible.

The translation of these bodies across the flats is very slow and marked movement only occurs during prolonged storm conditions. The amount of movement can be gauged from the fact that certain banks paralleling low water mark at the northeastern end of the Mersea flats and initiated since 1964 have moved only about 10 m shorewards over the following decade.

The position of present channels clearly affects the disposition and progressive elongation by wave action of the banks. As the flood water velocity does not exceed 85 cm/s, a speed incapable of moving particles larger than sand and small pebbles 10 mm in diameter, it is likely that the necessary energy requirements for movement of the larger pebbles are met by turbulence induced by storm waves moving along the channels and across the flats (Talbot, 1967).

Recurvature of the banks sometimes occurs at both ends, being most pronounced adjacent to shallow channels presently draining the flats. Along the

Figure 10.3. Landward face of mobile gravel banks on the East Mersea Flats showing, in the foreground, transitory discordant relationships with ridges and runnels in the back-bank clayey silts. In the middle distance the ridges and runnels show a more normal sub-parallelism to the leading edge of the banks. The banks were about 1 m high in 1973.

flanks of these channels the banks extend for several tens of metres towards the upper tidal flats. This mode of extension, however, does not satisfactorily account for the origin of strings of banks crossing the tidal zones, e.g. Cocum Hills (Figure 10.2). Control on the position of these banks is conjectural but, by comparison with modern channel margin processes and effects, it may relate to the position of old fluvial drainage channels (? late-Pleistocene) associated with the terraces. During the subsequent Holocene marine incursions along the channels, the banks may have been initiated by wave action at the feather-edge of the residual gravels. Whatever the origin, these banks are still in a condition of being moulded by present tidal and wave activity.

Langhorne (1973) has described asymmetric structures similar to the banks from depths between 12 m and 20 m in the Outer Thames Estuary. He suggests that they are stable features of relict origin, located along the margins of a partially buried channel and its tributaries and developed by increased turbulent flow at the break in slope on the seabed. He does not indicate, however, whether he considers that the features were initiated in an intertidal zone during an early lower sea level stage of, or prior to, the Flandrian transgression.

There is little evidence that any of the present intertidal gravel banks reflect human interference, except possibly at Tollesbury pier on the Blackwater estuary. But this possibility, however remote, always warrants consideration.

Shell banks

In addition to pebbles and sand, the banks described above are gathering grounds for a wide range of drifted subtidal and intertidal shells, though they rarely accumulate to any great thickness (< 5 cm). These shell materials are mobile and under the influence of wave action are moved horizontally over the leading edge of the gravel banks and onto the adjacent silty muds of the flats, where they adhere and progressively accumulate as shell banks to thicknesses of 2 m or more. There is little evidence for the shell accumulations moving any significant distance away from and ahead of the less mobile adjacent gravel banks. In this respect, restraint may be aided by the development within the shell structures of vertically packed layers of broken shell which are relatively resistant to movement (Greensmith and Tucker, 1968). Such a shell structure formed at the mouth of the Blackwater on its southern side between 1953 and 1970 (Greensmith and Tucker, 1969, p. 406). Gravel sheets were intermittently exposed on the western and northwestern sides of the asymmetric bank. Since 1971 the bank has degraded considerably with such a marked loss of shell that by 1973 it had diminished in height from about 1·5 m to 0·5 m above the level of the adjacent flats. The loss of shell is caused by many factors, including progressive comminution and dispersal of the material by wave action and a local decrease in shell supply which may reflect poor spat-fall in preceding seasons. A period of gales can be destructive, but the greatest changes are achieved during rare storm surges, such as on 31 January–1 February 1953. The Sales Point shell bank was swept almost clear of shells on that occasion. Conversely, it is important to appreciate that mass-mortality of epifaunal (surface-living) and infaunal (burrowing) organisms affects the availability of shells. Following the severe winter of 1962–63, when the Blackwater and Crouch estuaries were frozen over and occupied by ice-floes, the growth in height of some offshore shell banks doubled in 2–3 years.

At Foulness Point on the southern side of the Crouch estuary is a shell bank which, in contrast to the Sales Point structure, has shown considerable mobility. Since 1946 it has moved bodily at an average rate of 12 m per year (Greensmith and Tucker, 1969, Figure 6) over a silt and sand substrate, gravel being totally absent. During its progress in a west-southwesterly direction the bank has been moulded into a crude horseshoe-shape with a protected centre occupied by soft clayey silts. This soft sediment becomes re-exposed on the seaward side of the migrating structure and is quickly eroded.

The Foulness Point bank is one of the few known to have been affected by human activity within recent decades. However, aerial photographic evidence indicates that the removal of moderate quantities of shell has only had a marginal effect on its evolution and mobility.

Ridges and runnels

Under normal meteorological and tidal circumstances the partly protected area behind offshore gravel and shell banks, as at Foulness Point, Sales Point and Mersea, sees the accumulation of a slurry-like clayey silt and fine sand. The sediment occurs in the form of flat, broad, mildly asymmetric ridges and ebb tide erosional runnels with long axes roughly paralleling the adjacent banks, except where recent bank movement has taken place. When that has happened there is a marked discordance between the alignment of the structures for a considerable period (several months to more than a year) before the ridges and runnels adjust to the new situation and re-align themselves. The moulds stand generally not more than 0·5 m high, are up to 3 m broad and are 10–30 m long (Figure 10.3). They carry a characteristic living fauna including *Macoma* and *Tellina*, and in certain sandier runnels, small communities of *Cerastoderma* (*Cardium*) *edule*.

The effect of extended storm turbulence on the ridges is not very marked. There is no evidence of reorientation or destruction, but it is probable that some of the fine sediment is dispersed. Northwesterly gales at West Mersea during the winter of 1961–62 caused clayey silt to be eroded on the flats and redeposited on oyster layers, resulting in a significant mortality (Davis, 1962).

It should be emphasized that these low tidal flat ridges formed of unlithified sediment are not technically the 'mud-mounds' described by the authors in 1965. 'Mud-mounds' *sensu stricto* are low amplitude (*c.* 25 cm), narrow (*c.* 0·5 m) and elongated (*c.* 10 m) structures formed of partly lithified deposits of salt marsh and upper tidal flat origin, which are undergoing active erosion; they are commonly aligned in groups at right angles to a retreating marsh front.

Upper and supratidal flat structures

Cheniers (marsh beach ridges) and barrier spits

The most obvious constructional forms at the upper extremities of tidal action are cheniers (marsh beach ridges) and barrier spits and there are several situated along the open coast as well as along the flanks of the Blackwater and Colne estuaries. Following Price (1955) cheniers are defined here as shallow-based, perched ridges resting on clay along a marshy, seaward facing tidal shore. In Essex they consist of pebbles, sand and shells in very variable proportions. Along the Blackwater and Colne, sand and pebbles are the dominant constituent, whereas southwards from Sales Point to the Inner Thames Estuary sand and shell are dominant. Generally, they run parallel to, or at small angles to, the marsh edges.

In most instances the longshore drift of the constituent materials is determinable by direct observation; regionally it virtually boxes the compass (Figure 10.4). Movement is to the west from Clacton to Colne Point, then there is a marked swing to the north up the Colne estuary. Mersea Island shows both

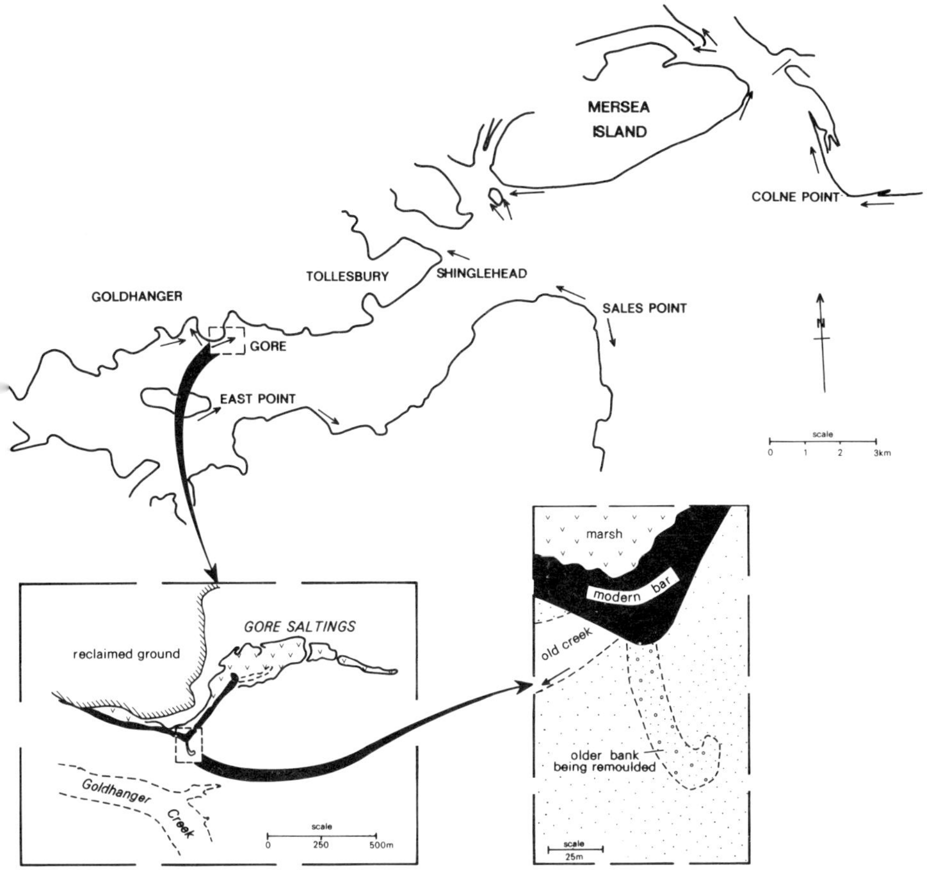

Figure 10.4. Direction of longshore drift deduced from cheniers located at high water mark. The Gore Marsh chenier appears to show two periods of growth. The modern structure transects an old infilled channel which originally drained an area to the rear (i.e. north) of the chenier. The channel margin gravel bank is now being displaced further eastwards.

eastwards and westwards movement at Mersea Stone (Figure 10.2) and westwards at West Mersea. The Mersea Stone chenier is a rather unusual structure in extending in spit-like form across the full width of the intertidal zone as far as the edge of the main deep channel into the Colne, then tightly doubling back on itself on its northern side. On that side it is not only elongating westwards back towards the seawall but also encroaching southwards by washover action.

At various points near Gore Marshes and Goldhanger there is movement to the east, northeast and northwest. At the eastern end of Osea Island movement is to the east-northeast, whereas near Sales Point it is both westwards up the Blackwater estuary and southwards down the open Dengie coastline.

The controls over the position of the structures and their movement are numerous and include the physiography of the site, the supply of materials, tide and wave patterns, and wind directions. The importance of fetch is implied in most instances. For example, the southern limb of the Mersea Stone structure shows the effect of wave energy being imparted at an angle to the beach face by prevailing winds blowing from the west-southwest sector. Wave refraction at the eastern extremity of the structure, where it abuts the deep channel, probably accounts in part for the remarkable recurvature of the northern limb. However, the recurvature is such that the influence of strong winds blowing from the north-northeast sector cannot be entirely discounted.

The chenier–barrier spit structures at Colne Point are most intriguing (Figure 10.5). Field observations and aerial photographs demonstrate the presence of at least two older, degraded chenier–barrier spit complexes preserved within the marsh sediments behind the present-day structure. The innermost and oldest rises about 0·5 m above the level of the flanking marshes and carries a gorse-

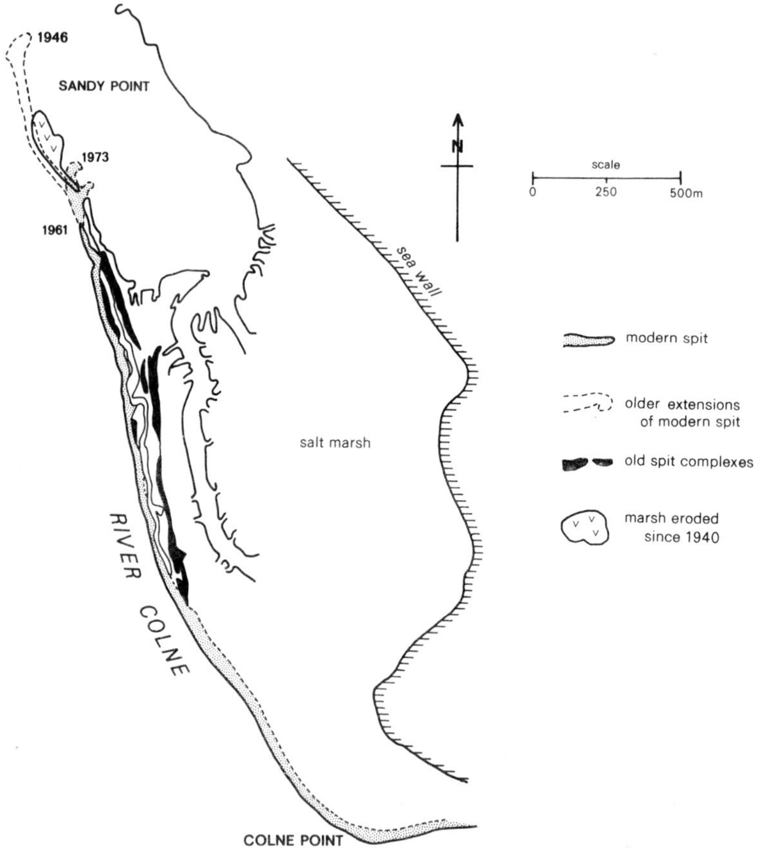

Figure 10.5. Colne Point chenier complexes, the oldest being to the east. The modern sandy gravel structure is actively extending northwards and simultaneously moving bodily eastwards, leading to a reworking of adjacent marsh deposits and older complexes.

Figure 10.6. View looking northwards showing the convergence of the modern and the oldest sand and gravel cheniers near Colne Point. The modern structure is progressively overriding and 'cannibalizing' the gorse-covered old complex.

dominated vegetation. It extends for just over 1 km in an approximate north–south direction from a position about 1 km northwest of Colne Point proper (Figure 10.6). At various places along its length it is marked by washover deltas and bifurcation, the latter being most pronounced at its northern end. The structure also has a narrow central swale running along its length in which contemporary or subsequent clayey silts were thinly deposited.

Some 20–70 m to the west of the oldest complex, and separated from it by a zone of salt marsh deposits, are the remnants of a second chenier. Very little of it is now preserved because the modern chenier is progressively overriding it eastwards and reworking the gravel constituents into its own structure—a process which can be referred to as 'cannibalism'.

The age of the two old cheniers is a matter of pure speculation at present. It is possible that the older originated between about 1550 and 1200 B.P. because there is evidence from the Dengie peninsula and the Low Countries that an important marine transgression occurred at that time (Tavernier and Moormann, 1954; Pons and coworkers, 1963; Greensmith and Tucker, 1973a). The date of the second chenier complex probably falls between 1200 and 250 B.P. and may eventually be narrowed down to minor transgressive episodes falling between 1000 and 700 B.P. (Bennema, 1954; Tavernier and Moormann, 1954) but conclusive evidence in Essex remains to be found. In any event, an ample and persistent supply of sand and gravel would have been required for a considerable length of time, and it might be anticipated that this condition would be best met during extended phases of marsh and cliff erosion towards Clacton, plus enhanced exposure of terrace gravels at sea level.

The modern chenier has a history which may extend over the last 250 years, evidence from Mersea broadly suggesting that the latest episode of marine transgression in the area was initiated at about that time. The structure is undergoing marked changes over the northern 700 m at present, but these probably reflect human interference in the last few decades more than the longer term consequences of marsh and cliff erosion. The termination of the northern spit was 400 m longer in the early 1940s and 160 m shorter in 1959–61 than it is at present. Gravel extraction adjacent to high water mark at Colne Point commenced in 1947 but ceased in 1962, so restoring the uninterrupted movement of material alongshore (Robinson, 1953). At the same time, northwards growth has been enhanced by an injection of fresh material reworked from the older chenier complexes and tidal flats. What may be one of the older complexes is exposed near Colne Point, at high water mark, at the foot of a cliff about 1 m high and extending for several tens of metres laterally (Figure 10.7). The succession exposed in the retreating cliff face is:

modern complex	45 cm	sandy gravel with *Ostrea*, *Cerastoderma* (*Cardium*) and *Crepidula*
	——— sharp contact ———	
	5 cm	sandy peat with *Crepidula* on surface
old complex ?		grading down into medium
	30 cm	sand with rootlets
	(>20) cm	sandy gravel with *Ostrea* and *Littorina littorea*

Crepidula fornicata (slipper limpet), introduced accidentally into the area from North America in 1870–80, appears to be absent from the lower sandy gravel (Davis, 1966).

Although the modern chenier is actively extending northwards as a spit, and thereby increasing the area of protection to its rear, the present indications are that the marsh edges to the rear are retreating in some places and static in others. Erosion seems to be at least in balance with, and probably is outpacing, sedimentation. This situation is duplicated behind the gravel chenier at Gore Marshes (Figure 10.4).

The onshore advance of these Colne Point cheniers is commonly such that the marsh re-emerges at their trailing edge, where it immediately becomes susceptible to active erosion. Elsewhere along the open coast and for several kilometres up the estuaries similar erosion has produced marsh edge cliffs up to 2 m in height. In 1973 tidal silts, interbedded at two distinct levels with *in situ* marsh silts and peat seams, were exposed in a cliff at East Mersea (Figure 10.8). These sediments imply a series of marsh front recessions and marsh front advances, respectively. Radiocarbon dates from the peats of 173 ± 60 B.P. and 118 ± 48 B.P. indicate oscillations of the marsh edge probably spanning the last 250 years. Evidence, for much older oscillations also occurs in the area (Greensmith and Tucker,

Figure 10.7. Present-day cannibalism at Colne Point, exposing in the lower half of the small cliff the sandy peaty top layers of an older gravel chenier.

Figure 10.8. Salt marsh retreat at East Mersea with residual sea-stacks. The cliff faces are about 1·5 m high and consist of intertidal clayey silts carrying two peat seams and a distinct peat ball horizon.

1973a and 1973b). The fact that marsh fronts change in position, either locally or regionally, probably has had important repercussions on the supply of sand, pebbles and shell to prograding structures at high water mark and may have played a significant role in the evolution of multiple chenier complexes as at Colne Point.

At Sales Point the progress of a chenier (Figure 10.9) across the marsh surface has been accomplished in conjunction with the development of marsh edge cliffs and erosional mud-mounds, *sensu stricto*, the latter occurring in groups which extend for several hundreds of metres seawards. This chenier is one of several presently forming along the open coast between the Blackwater and Inner Thames estuaries and can be taken as representative of the sand and shell types forming in Essex. In compositional terms it differs considerably from those at Colne Point, Mersea and Gore Marsh in consisting of a much higher proportion of sand and shell with few pebbles. On average, sand forms 47% and shells 51%, the rest being pebbles, clayey silt and plant debris (Greensmith and Tucker, 1966 and 1969). Internally there is a crude alternate layering of coarse shell, comminuted shell, and sand with no tendency towards overall coarsening-upwards, as in the case with the Louisiana cheniers (Byrne and coworkers, 1959). Moreover, in contrast to the Louisiana structures, the sediments of the backslope are characteristically coarse, being formed predominantly of large shell fragments.

The evolution of the Sales Point chenier is of considerable interest. Since 1947 certain parts have migrated landwards across the marsh surface at a rate of 8 m per year. The main change at present involves southwards elongation which has amounted to 80 m since 1965. This has been accomplished by unification of what then were isolated sand and shell sheets swept onto the marsh surface, a linkage resulting from longshore drift during high spring tides. Simultaneously, some 250 m of the northern landlocked end of the chenier has become more stabilized than in 1965 with increased growth of vegetation on its uppermost surface. There also has been enhanced summer growth of filamentous algae on the immediately adjacent flats with the consequent effect of producing dense mats of drifted decaying algal material which adhere to the trailing edge of the chenier. Although capable of being removed by storm waves these mats in general effectively prevent, during the summer seasons at least, active redistribution of the sand and shell. Growth at the southern end of the chenier, therefore, predominantly results from the reworking of material from the middle unprotected reaches of the structure, plus the acquisition of materials from the adjacent flats. This acquisition can involve the 'cannibalizing' of older shell sheets and pockets incorporated in the upper tidal flat sediments.

Except where it is actively elongating at the southern end, the chenier exhibits an asymmetric cross-section with the steepest slope on the landward side (Greensmith and Tucker, 1969, Figure 2, Section 4). This shape is characteristic and contrasts with the cheniers of Louisiana, New Zealand and Queensland where the steepest flank is on the seaward side (Byrne and coworkers, 1959; Schofield, 1960; Cook and Polach, 1973). The comparatively greater mobility of the Essex structures appears to be the main reason for the difference in

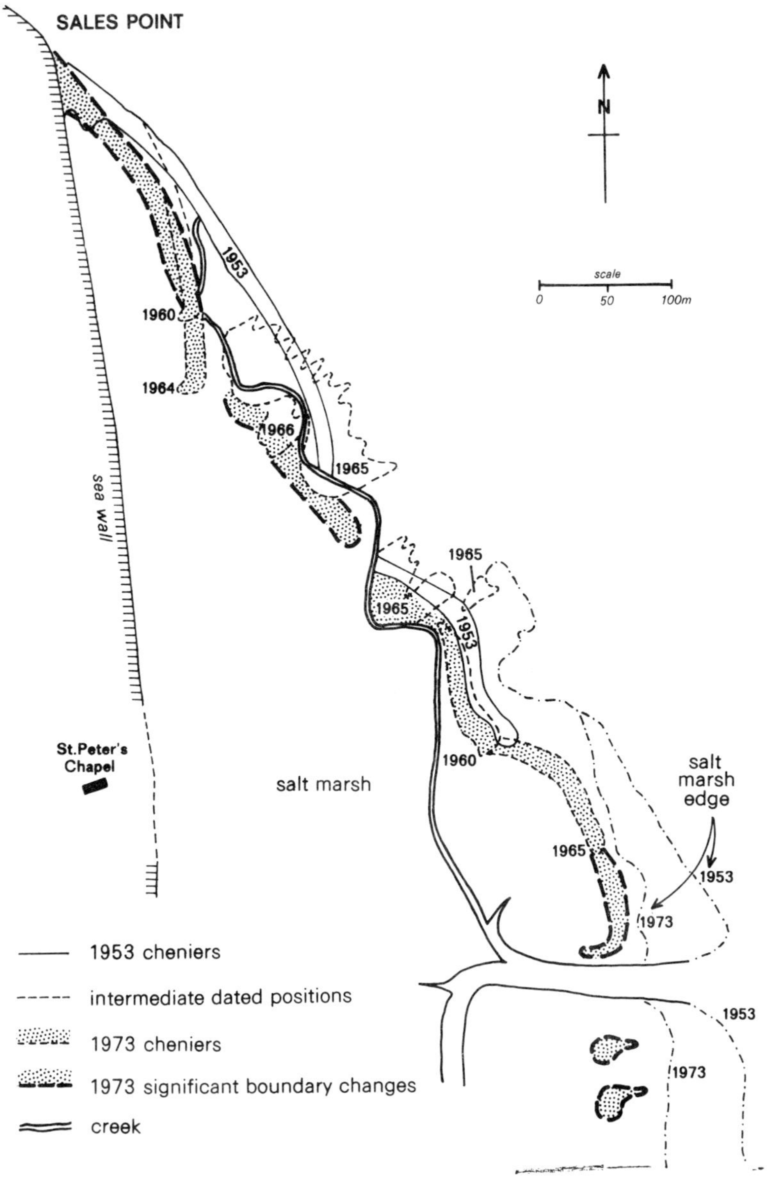

Figure 10.9. The evolution of the Sales Point chenier 1952–73. In 1946 the northern end was isolated completely from the seawall and located a further 80–250 m seawards. Also since 1946 the salt marsh front adjacent to the southern part has retreated *c.* 150 m.

cross-section. In turn, this mobility may also account for the fact that the cheniers, in general, rest directly on salt marsh organic silts and clays, not directly upon shallow intertidal and subtidal deposits as is the case in Louisiana. There is no evidence for the seaward growth of any of the cheniers.

The Essex chenier plain

The evidence presented by the cheniers developed along the Essex coast, allied to that of contemporary and widespread erosion, confirms that the coastal region is passing through a significant phase of marine transgression. This is

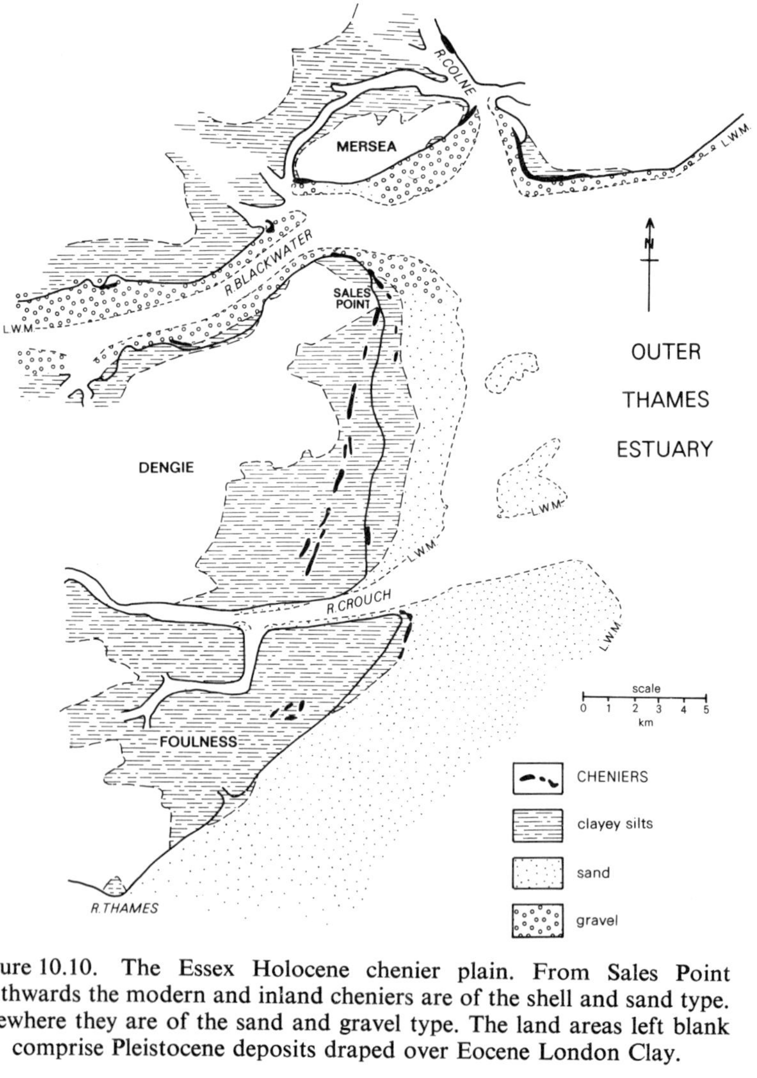

Figure 10.10. The Essex Holocene chenier plain. From Sales Point southwards the modern and inland cheniers are of the shell and sand type. Elsewhere they are of the sand and gravel type. The land areas left blank comprise Pleistocene deposits draped over Eocene London Clay.

one of several such phases (separated by regressive phases) comprising what is generally referred to as the Flandrian transgression. One important phase, possibly spanning the interval 1434 ± 110 B.P. to 1265 ± 100 B.P., is indicated by the outcrop of a linear zone occupied by cheniers and running through the low-lying, marshy, partially reclaimed ground at the eastern end of the Dengie peninsula and on Foulness Island (Figure 10.10). The zone, essentially a relict shoreline, is 0·1 to 1 km wide and within it the individual sand and shell cheniers are up to 3 m thick, 10–250 m wide and several hundreds of metres long. Flint pebbles are incorporated in small amounts at some localities. The structures are similar in most respects to those forming presently on the nearby salt marshes, even to the extent of having mud-mound structures on their original seaward flanks.

None of the inland cheniers has a marked surface relief because of the fact that materials have been removed during intensive farming activities and the probability that many of the structures were subject to late-stage degradation into sheet-like forms (cf. southwestern Louisiana and New Zealand).

Although the history of this low-lying ground is complicated by reclamation since medieval times and seawall construction since the 17th century (Grieve, 1959), it is clear that the area has evolved over at least the last 2000 years into a chenier plain—a progradational landform analogous to, but much smaller than, that of the type region of southwestern Louisiana (Gould and McFarlan, 1959). However, the Essex cheniers represent intervals, not of coastal stability as in Louisiana, but of coastal instability and extensive erosion. Furthermore, it is probable that these intervals reflect in part the fundamental tectonic instability of the southern North Sea region. Table 10.1 illustrates the differences, and similarities, between the Essex chenier plain and others where tectonism has probably played a lesser role.

Table 10.1. A comparison of cheniers and chenier plains

	Essex	Louisiana	New Zealand	Queensland
Chenier constitution	shell, sand and gravel	shell and sand	shell, sand and gravel	shell, sand and gravel
Chenier width	35 m average	200 m average	100 m average	up to 50 m
Chenier thickness	2 m average	2 m average	2·5 m average	2 m average
Chenier length (max.)	3 km	48 km	2·5 km	5 km
Relationship of chenier to marsh deposits	mainly superimposed	mainly marginal	mainly marginal	partly marginal, partly superimposed
Width of chenier plain (max.)	10 km	24 km	2·5 km	5 km
Age of chenier plain	at least 2000 years	*c*. 3000 years	*c*. 4000 years	*c*. 5000 years
Relict coastlines	1 (possibly 2)	7	13?	6–7
Tectonic framework	unstable	stable	moderately stable	stable

In evolutionary terms the events which have collectively produced the Essex chenier plain can be expressed by sequential stages as follows:

(1) Marine transgression resulting in the erosion of the upper tidal flats and salt marshes, together with the initiation of mud-mound relief.
(2) Release of living and dead shells, sand particles and pebbies present within and on the surface of the tidal flats; accumulation of these materials as offshore banks and in scour depressions adjacent to marsh edges.
(3) Juvenile chenier stage, as evidenced by the displacement of shells, sand and pebbles by tidal and wave action onto the marsh surface. Formation of irregular flat and thin patches, and sheets of shell, sand and pebbles.
(4) Juvenile chenier stage continues with the merging of patches and sheets by longshore drift during spring high tides or storm tides to form low continuous ridges which extend onto the tidal flats as spits.
(5) Mature chenier stage displays progressive growth, extension and landward movement of cheniers, and increase in complexity of structures with branching and deflation during extreme storm conditions. Also, there is temporary division by materials being displaced into major marsh creeks, modification of creek drainage paths on marshes, increases in scouring capacity of tidal water in certain creeks leading to increased rates of creek lateral migration, and erosion of re-emergent marsh surfaces on the seaward side of cheniers.
(6) Old age chenier stage accompanied by marine regression and progressive deposition of clay and silt on seaward and landward sides of cheniers, isolation of cheniers from sea by salt marsh extension, stabilization and obscuration by vegetation, increase in stability of creek drainage paths to rear of structures and possible human interference if in reclaimed ground. At present this stage is represented on the Essex coast at Colne Point, but more predominantly inland on Dengie and Foulness.

Conclusions

An understanding of Holocene coastal deposition in Essex can only be achieved by paying appropriate regard to the earlier Quaternary history of the region. The working and reworking of intermittently exposed relict terrace sandy gravel deposits of Pleistocene age during the Flandrian transgression is now expressed by a wide range of mobile sedimentary structures within the intertidal and supra-tidal zones. In broad terms, the present (and past) changing patterns of sedimentation in the zones appear to reflect relatively minor episodes of transgression and regression superimposed onto the main Flandrian transgression. Hence, the detailed analysis of many of the structures indicates a very complex genesis and evolution, which can be related partly to these episodes and partly to the increased or decreased availability of constructive materials. The most important consequence of these episodic events, during approximately the last 2000 years, has been the development of a prograding chenier plain.

Acknowledgements

The authors gratefully acknowledge the assistance given by various officers of the Institute of Geological Sciences and Professor Shotton in determining the radiocarbon age of certain peat layers and shells. The Essex River Authority gave valuable information on gravel working at Colne Point. Mrs. J. Fyffe and Mr. B. Samuels of Queen Mary College assisted with the illustrations.

References

Ambrose, J. D., 1973, The sand and gravel resources of the country around Layer Breton and Tolleshunt D'Arcy, Essex, *I.G.S. Report 73/8, Assessment of British Sand and Gravel Resources, No. 7*, 34 pp.

Bennema, J., 1954, Holocene movements of land and sea-level in the coastal area of the Netherlands, *Geologie Mijnb.*, **16**, 254–262.

Bristow, C. R., and F. C. Cox, 1973, The Gipping Till: a reappraisal of East Anglian glacial stratigraphy, *J. geol. Soc. Lond.*, **129**, 1–19.

Byrne, J. V., D. O. S. Leroy and C. M. Riley, 1959, The chenier plain and its stratigraphy, southwestern Louisiana, *Trans. Gulf-Cst Ass. geol. Socs.*, **9**, 237–260.

Cook, P. J., and H. A. Polach, 1973, A chenier sequence at Broad Sound, Queensland and evidence against a Holocene high sea level, *Mar. Geol.*, **14**, 253–268.

Cornwall, I. W., 1958, *Soils for the Archaeologist*, Phoenix House, London, pp. 204–208.

Davis, D. S., 1962, Some observations on the movement of sediments on the shore at Bradwell, Essex, *Central Electricity Res. Lab. Note RD/L/N39/62*, 2 pp.

Davis, D. S., 1964, The physical and biological features of the Mersea Flats, *Central Electricity Res. Lab. Note RD/L/N131/64*, 14 pp.

Davis, D. S., 1966, Studies on the American slipper limpet (*Crepidula fornicata*) in the Blackwater estuary: 1960–1965, *Central Electricity Res. Lab. Report RD/L/R1349*, 13 pp.

D'Olier, B., 1972, Subsidence and sea-level rise in the Thames Estuary, *Phil. Trans. R. Soc. A*, **272**, 121–130.

D'Olier, B., and R. J. Maddrell, 1970, Buried channels of the Thames Estuary, *Nature, Lond.*, **226**, 347–348.

Greensmith, J. T., and E. V. Tucker, 1965, Salt marsh erosion in Essex, *Nature, Lond.*, **206**, 606–607.

Greensmith, J. T., and E. V. Tucker, 1967, Morphology and evolution of inshore shell ridges and mud-mounds on modern intertidal flats, near Bradwell, Essex, *Proc. geol. Ass.*, **77**, 329–346.

Greensmith, J. T., and E. V. Tucker, 1968, Imbricate structure in Essex offshore shell banks, *Nature, Lond.*, **220**, 1115–1116.

Greensmith, J. T., and E. V. Tucker, 1969, The origin of Holocene shell deposits in the chenier plain facies of Essex (Great Britain), *Mar. Geol.*, **7**, 403–425.

Greensmith, J. T., and E. V. Tucker, 1971, The effects of late-Pleistocene and Holocene sea-level changes in the vicinity of the River Crouch, east Essex, *Proc. geol. Ass.*, **82**, 301–322.

Greensmith, J. T., and E. V. Tucker, 1973a, Holocene transgressions and regressions on the Essex coast, Outer Thames Estuary, *Geologie Mijnb.*, **52**, 193–203.

Greensmith, J. T., and E. V. Tucker, 1973b, Peat balls in late-Holocene sediments of Essex, England, *J. sedim. Pet.*, **43**, 894–897.

Greensmith, J. T., R. G. Blezard, C. R. Bristow, R. Markham and E. V. Tucker, 1973, The Estuarine Region of Suffolk and Essex, *Geol. Ass. Guide, No. 12*, 41 pp.

Gould, H. R., and E. McFarlan, 1959, Geologic history of the chenier plain, southwestern Louisiana, *Trans. Gulf-Cst Ass. geol. Socs.*, **9**, 261–272.
Grieve, H., 1959, *The Great Tide*, County Council of Essex, Chelmsford, 883 pp.
Gruhn, R., and A. L. Bryan, 1972, *A contribution to Pleistocene chronology in southeast Essex*, private publication, Dept. Anthropology, University of Alberta, Edmonton.
Guilcher, A., 1969, Pleistocene and Holocene sea level changes, *Earth Sci. Rev.*, **5**, 69–97.
Langhorne, D. N., 1973, A sandwave field in the Outer Thames Estuary, Great Britain, *Mar. Geol.*, **14**, 129–143.
Oakley, K. P., and M. Leakey, 1937, Report on excavations at Jaywick Sands, Essex, *Proc. prehist. Soc.*, **10**, 217–260.
Pons, L. J., S. Jelgersma, A. J. Wiggers and J. D. De Jong, 1963, Evolution of the Netherlands coastal area during the Holocene, *Verh. K. med. Geol. Mijnb. Genoot.*, **21**, 197–208.
Price, W. A., 1955, Environment and formation of the chenier plain, *Quaternaria*, **2**, 75–86.
Robinson, A. H. W., 1953, The changing coastline of Essex, *Essex Nat.* **29**, 1–15.
Schofield, J. C., 1960, Sea level fluctuations during the last 4000 years as recorded by a chenier plain, Firth of Thames, New Zealand, *N.Z. J. Geol. Geophys.*, **3**, 467–485.
Talbot, J. W., 1967, The hydrography of the estuary of the River Blackwater, *Minist. Agriculture Fish and Food, Fishery Invest.*, H.M.S.O., London, 25, 92 pp.
Tavernier, R., and F. Moormann, 1954, Les changements du niveau de la mer dans la plaine maritime flamande pendant l'holocene, *Geologie Mijnb.*, **16**, 201–206.
Warren, S. H., 1923, The *Elephas antiquus* Bed of Clacton-on-Sea (Essex), *Q. Jl. geol. Soc. Lond.*, **79**, 606–634.
Warren, S. H., 1924, The Elephant Bed of Clacton-on-Sea, *Essex Nat.*, **21**, 32–40.

Discussion

Dr. A. H. W. Robinson, Department of Geography, University of Leicester. Are the marine transgressions a result of local subsidence or a rising sea level?

The causes of the marine transgressions and regressions are problematic and it is clear that any conclusions, however tentatively expressed, are likely to be controversial. The southern part of the North Sea has long been recognized as a region of subsidence through Quaternary times. So it would be anticipated that during periods of more active subsidence marsh erosion and transgression would occur. Conversely, during periods of crustal stability marsh extension and regression would ensue. Evidence is present within the Essex Holocene succession which points to possible local variation in subsidence rates and certainly to different rates of subsidence between Essex and the Low Countries. In part this probably reflects the influence of the local basement structure. But there are certain dated levels of transgression on both sides of the southern North Sea and elsewhere in the world which are so similar in age that the possibility of eustatic changes in sea level has to be considered. If there have been eustatic oscillations they are essentially minor in quality and appear to be superimposed on the universally recognized and general Holocene rise in sea level.

Dr. W. R. Parker, Institute of Oceanographic Sciences, Taunton, Somerset. Could you suggest why the gravel, interpreted as being related to old salt marsh creeks, has accumulated on the flanks of the creeks and not on the bed of the creeks?

The displacement of gravel adjacent to ancient channels (*not* old salt marsh creeks) crossing the present tidal flats appears to involve a Pleistocene phase(s) of fluviatile channel formation followed by a Holocene phase(s) of marine incursion along the same channels. During the Pleistocene phase(s) there is evidence in the area for gravel removal into channel axes. During the later Holocene phase(s) wave-induced turbulence at the edge of the channels seems to have been adequate to displace the residual gravel laterally and mould it into levee-like bodies. These marginal bodies have been displaced subsequently and, as at present, are being remoulded by wave and tidal activity.

F. J. T. Kestner, Hydraulics Research Station, Wallingford. I do not agree that the salt marsh edges described mean active erosion. Rather, they are the limits of salt marsh advance.

It is most unlikely that the present position of the marsh edge, markedly cliffed over an extensive area in the Outer Thames Estuary, indicates the maximum progradation of the salt marshes in historically recent times. The cumulative evidence from ancient maps, Ordnance Survey records, archaeological remains, successive aerial photographic surveys and prolonged field observations all indicate that a period of longstanding and widespread erosion (the latest of a number of Holocene cycles of erosion) accounts for the present disposition and increasingly reduced area of the Essex salt marshes.

Dr. G. Evans, Department of Geology, Imperial College, London. Is not the development of a marsh cliff within many of the Essex estuaries a result of unusually violent wave action produced artificially by bow waves of various coastal craft? The latter appear to produce unusually violent wave action in the most sheltered conditions.

It has to be appreciated that the evidence for erosion is not derived solely from observations on cliff retreat in the estuaries and along the open coast, but includes that deduced from mud-mound formation *sensu stricto* (see text) over wide tracts of the present flats and also the presence of features such as peat balls preserved within the tidal flat sediments and known to be of an age predating that of powered craft.

Dr. A. P. Carr, Institute of Oceanographic Sciences, Taunton, Somerset. Are the small changes in salt marsh level associated with eustatic changes in sea level? Could the changes be an indication of changing wave energy or different rates of compaction? At Caerleverock, in the Solway Firth, there are abrupt changes in salt marsh height which appear to be related to local mass slumping and creek distribution. These factors might usefully be taken into consideration.

Regarding the stability of shell and shingle structures some appear to be highly mobile, as at Foulness Point. Commercial extraction of shell from the Foulness structures appears to have taken place to the order of between 100 and 500 tonnes annually, but to have been largely replaced by natural transport of material from elsewhere. This seems to imply that extraction would have been likely on the banks further to the north also.

I would refer you to an earlier answer regarding eustatic sea level changes. In the paper transgressions and regressions simply refer to advances and retreats of the sea margin respectively. In this area it would be unwise to isolate any one factor and deduce that as being the sole cause of the transgressions and regressions. Long- and short-term cyclical climatological changes could have affected wind and wave patterns, but so could changing coastal morphology. Then, these changes have to be set into a context of regional subsidence, local differential subsidence and compaction and even, possibly, minor eustatic changes. The undermining of the widespread marsh cliffs, which are basically formed as a consequence of variations in the influence of the above controls, commonly leads to slumping, sliding and the generation of cliff-foot mud–pebble conglomerates.

It should be strongly emphasized that the shell material stranded in bulk on and adjacent to the various gravel structures does clearly show mobility. The Foulness offshore shell bank, though not gravel-based, is very mobile (see text, and Greensmith and Tucker (1969)).

Dr. J. R Hails, Institute of Oceanographic Sciences, Taunton, Somerset. Could the shell and shingle structures that you relate to minor oscillations of sea level also be attributed, perhaps entirely, to variations in the distribution and dissipation of wave energy? Channel migration and ultimate elimination can be associated with the growth and decay of salt marshes, and thus reflect the complexity of migrating shell and shingle structures.

I would repeat my answers to Dr. Robinson and Dr. Carr regarding the first question. On the second point, creek migration occurs irrespective of whether the marshes are retreating or advancing, but it is our experience that creek patterns are modified most during transgressions, when cheniers are at their most mobile. During these episodes, creeks are progressively infilled and ultimately eliminated, often at the temporary expense of the continuity of the advancing mobile structures.

CHAPTER ELEVEN

The Transportation and Deposition of Suspended Sediment over the Intertidal Flats of the Wash

G. EVANS and M. B. COLLINS

Abstract

The broad conclusions which have been reached from this study are listed below. These can, of course, by the nature of the study, only be strictly applied to the immediate area which has been investigated; however, they do have some broader implications.

(1) The tidal waters do not advance and retreat over the intertidal flat normal to the shoreline. There is a marked clockwise rotation of these waters throughout the tidal cycle.
(2) Resolution of the tidal velocities into onshore–offshore and alongshore components indicates that the latter is of equal magnitude to the former and that it is relatively consistent in strength throughout a tidal cycle.
(3) There is a velocity gradient, which decreases from low to high water mark, normal to the shoreline.
(4) Levels of suspended sediment concentration are generally less than 200 p.p.m., but concentrations of up to 1200 p.p.m. have been measured.
(5) There is a gradient in the concentration of suspended sediment from surface to bottom of the water column, and from high to low water mark.
(6) The suspended sediment consists mainly of inorganic material and has a highly variable granulometric composition with sand percentages rising to 96% by weight of the total suspended matter.
(7) Both the pattern of currents and the variation in the levels of suspended sediment concentration appear to be highly unpredictable; this makes the calculation of overall sediment fluxes a difficult exercise.
(8) A study of the flux of sediment for one tide yields useful data on the magnitude of the sediment transport. Such calculations show that with levels of concentration of 100 p.p.m. approximately 456 tonnes pass over a transect across the intertidal flats in a direction oblique (22°) to the shoreline. A more accurate calculation based on the adoption of the varying average suspended sediment concentration throughout a tidal cycle shows that approximately 540 tonnes pass over this same transect in a similar direction.
(9) Fluvial sediment supply appears to be inadequate to account for the known

volume of suspended sediment concentration present in the area and the known rate of deposition since early Romano-British times.

(10) It is impossible, with the available data, to relate the levels of concentration of suspended sediment to any particular individual hydrodynamic or meteorological process. Each sampling occasion is a unique event due to a particular combination of the effects of various processes (supply and lag effect). Generally the controlling factors in the supply of sediment appear to be external to the immediate area and it is the availability of material for transportation by tidal currents onto the intertidal flats of the Wash, rather than the strength of the local tidal currents, which is of prime importance.

Introduction

The Wash is a large embayment on the east coast of England with a width of approximately 20 km and length of about 30 km. The average spring and neap tidal ranges are 6·5 m and 3·5 m respectively. At high water spring tides approximately $73{\cdot}2 \times 10^3$ hectares of the embayment are covered. At low water spring tides roughly half of this area dries out to expose a broad marginal intertidal zone and a series of large offshore intertidal banks which are joined to the marginal area in the southern parts of the embayment. This embayment has a maximum depth greater than 40 m at its entrance but, over most of its area, depths are generally less than 10 m.

The Wash is bordered by an artificial embankment for most of its length. This rises 3–4 m above the inner parts of the intertidal zone, except in the extreme east where a small area of Cretaceous rocks form a cliff and a gravel ridge runs along the inner margin of the intertidal zone for approximately 13 km; however, even this is backed by an artificial embankment. To the north and east of the Wash, on the more exposed coastlines of East Lincolnshire and North Norfolk, beaches and dunes (sometimes enclosing marshes) characterize the coastline. A series of rivers flow into the Wash and drain a large area of central eastern England ($1{\cdot}3 \times 10^6$ hectares) between the Trent drainage basin to the north and the Thames to the south. These rivers have controlled sluice outlets whilst the configuration of their estuaries has been drastically modified by man so that they now flow between confining training walls. Considerable difficulties have arisen in the past in keeping the channels stable and deep enough to allow shipping to enter the small, tidal, Fenland ports of Boston, Spalding, Wisbech and Kings Lynn (Darby, 1940).

Although higher ground, which reaches the coast to form the cliffs of Hunstanton, is fairly close to the shoreline on the east of the Wash, elsewhere this embayment is bordered by a flat plain—the Fenland—which covers an area of some $0{\cdot}3 \times 10^6$ hectares. This plain is mostly just above high water spring tides but in some places, owing to drainage and wastage of peat, it may be as much as 3·4 m below mean sea level. The flatness of its surface is broken only by isolated hills of Pleistocene sand and gravel, or till, and Jurassic clays, some man-made mounds and the embanked present-day or old abandoned river courses.

This Fenland plain has originated by sediment filling part of a depression

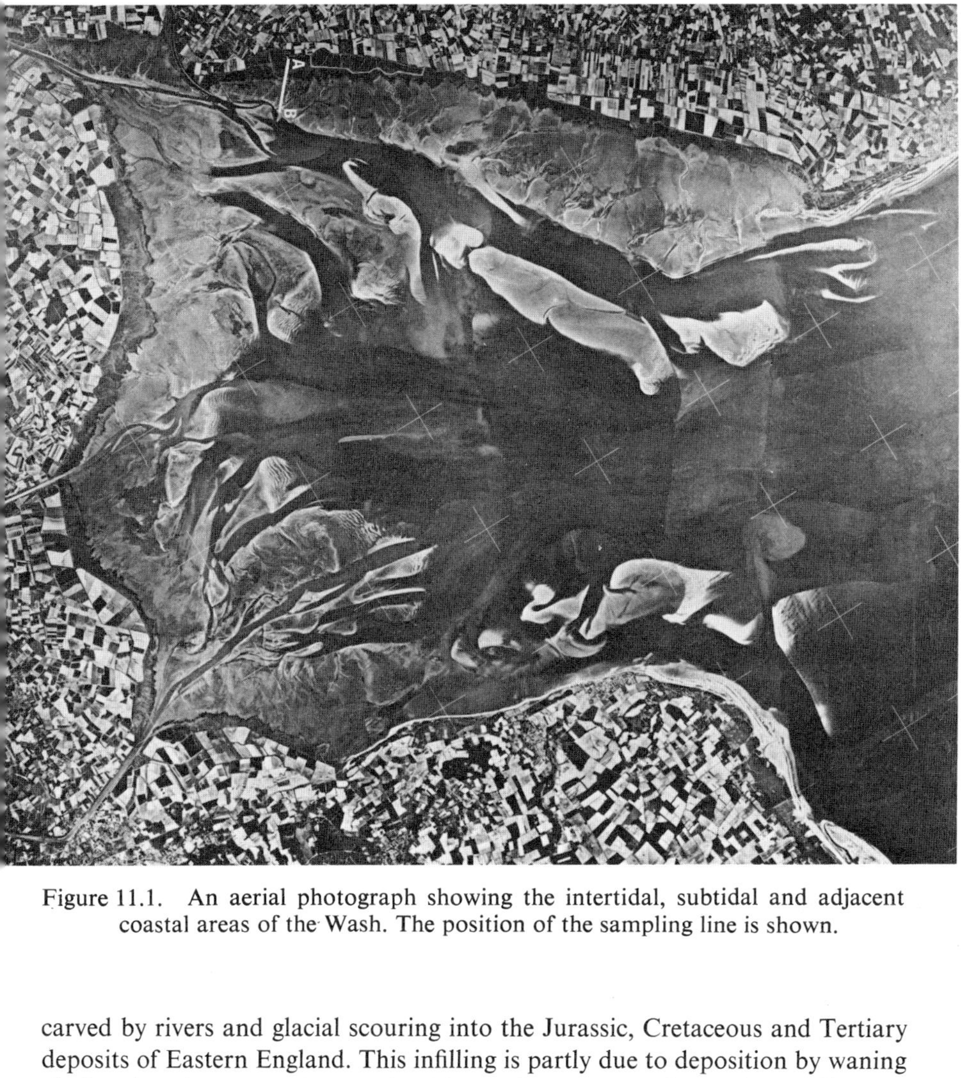

Figure 11.1. An aerial photograph showing the intertidal, subtidal and adjacent coastal areas of the Wash. The position of the sampling line is shown.

carved by rivers and glacial scouring into the Jurassic, Cretaceous and Tertiary deposits of Eastern England. This infilling is partly due to deposition by waning glaciers but, in many places, Flandrian sediment unconformably overlies pre-Pleistocene deposits. The greater part of the infill is composed of Flandrian sediments which have been deposited during the rise of the sea from its low levels of the last glaciation: these consist of buried forest beds, peats, lake marls, lagoonal and marsh silty clays, silty sands, sandy silts, silty clays and clayey silts of intertidal flat and subtidal origin, as well as fluvial channel sands and gravels. The last episode of this sedimentation was the development of the belt of intertidal flats around the shoreline of most of the Wash approximately 2000 B.P. (Godwin, 1940; Godwin and Clifford, 1938; Willis, 1961). This environment has persisted until the present and the growth and development of these intertidal flats has produced, by coastal progradation, a wide belt of so called 'siltlands' which appear to be continuing to develop today. Man has found it profitable to reclaim and cultivate these coastal siltlands and, since Roman times, large areas have been reclaimed with the greatest development of the reclamation taking place since the mid-17th century (Darby, 1940); approximately $1{\cdot}2 \times 10^3$ hectares have been reclaimed since that date.

Many debates have arisen because of the conflicting interests of the farmer, the hunter, the fishermen and the merchants of this area. Generally, it appears that the reclamation has aided and substantially increased the rate of outgrowth of the intertidal flats and the seaward progradation of the coastline (Hydraulics Research Station, 1953; Inglis and Kestner, 1958a and 1958b; Kestner, 1962). Drainage and reclamation of the inner Fenland has led to peat wastage, to a lowering of the surface level and to considerable problems in attempting to drain the water into the sea.

Long discussions have continued on the origin of the sediment which make up the siltlands. Local river engineers, over the centuries, have found that siltation has only been a problem at the river mouths, and then always on the seaward side of the sluices; furthermore they have found that river waters never seem to carry large quantities of material so that even when they are impounded there is hardly ever any build-up of the bed. It has therefore been assumed firstly, that the most of the sediment originates in the adjacent sea and coastal areas and, secondly, that it has been carried shoreward and deposited from the tidal waters which constantly flood and ebb in the estuaries and over the adjacent intertidal flats (Clark, 1959). There has been no definite evidence that this is so; however, the balance of the evidence seems to indicate it to be true.

Geological studies of the sediments seem, without being conclusive, to support such a contention. The sediments of the intertidal flats contain remains of marine organisms: foraminiferids, fragments of molluscs, polyzoa, etc., which fairly clearly have originated from the adjacent seafloor (Macfadyen, 1938). The sand fraction of the siltlands sediments contains a suite of heavy minerals which are clearly related to the sediments of the adjacent seafloor and coastal exposures outside the Wash (Chang, 1971). They are slightly different in composition from the sands found in the sediments of the beaches and dunes of the exposed coastline of East Lincolnshire and North Norfolk, but this difference is thought to result not from difference in source but through sorting according to grain size and density of the constituent minerals (Chang, 1971). Furthermore, these minerals appear to indicate that the sediments are derived from the reworking of the Hunstanton-Hassle boulder clay which is found cropping out on the adjacent coasts and seafloor and which also underlies part of the outer Fenland. These sediments do not relate to those of the glacial sediments found over much of the catchment area of the rivers flowing into the Wash; and as the latter only cross the Hunstanton-Hessle glacial deposits in their lower courses they are unlikely to derive much of their sediment load from the latter. The study of the composition of the clays has not been helpful in assigning a source for this material because it has a relative uniform composition in most of the potential source areas (Shaw, 1971 and 1973).

The present study

Various studies have been carried out on the rate of sedimentation on the present intertidal flats of the area under consideration and the effects of man-made embankments on this process. However, little work, apart from studies by Kestner (1961, 1963), has been done on the actual amount of sediment in suspension in the waters on the intertidal flats, or in the adjacent Wash. Also, although some very interesting investigations have been carried out on the problem of transportation and deposition of sediment on intertidal flats (Postma, 1954 and 1967; van Straaten and Kuenen, 1957 and 1958; Kestner, 1961; Anderson, 1973) these are comparatively few in number from which to draw conclusions.

In an attempt to understand the transport and deposition of suspended sediment over and on the intertidal flats of the Wash, a programme of work was initiated to:

(a) determine the levels of suspended sediment concentration over the intertidal zone;
(b) calculate the onshore–offshore balance (or resultant flux) of suspended sediment by combining the result of suspended sediment concentration with tidal current data. It was also hoped that this would be correlated with accretion studies;
(c) determine the importance of the various factors which influence the concentrations of sediment in the tidal waters;
(d) examine the possibility of long-term prediction of sediment transport and deposition;
(e) consider the potential source areas and their present and past importance on intertidal flat sedimentation of the Wash.

In order to achieve these aims a series of experiments were carried out which included:

(i) the measurement of tidal current velocities across a section, considered to be typical of the intertidal flats of the Wash, including measurements from top to bottom of the water column for both neaps and spring tides and
(ii) the collection of the material in suspension in the flooding and ebbing tidal waters for determination of total suspended sediment concentration.

These measurements were made against a background of determination of height of tidal waters and observed local hydraulic and meteorological conditions.

The section across the intertidal flats was near Frieston Shore on the western side of the Wash. Previous work by one of the writers (Evans, 1960 and 1965) and current studies by C. Amos (unpublished) have provided considerable background data for this area while the Frieston Shore section is particularly accessible. The location and the general topography of the line is given in Figure 11.2. It shows the usual pattern of salt marshes and mud and sand flats which are so characteristic of the marginal intertidal zone of the Wash. However,

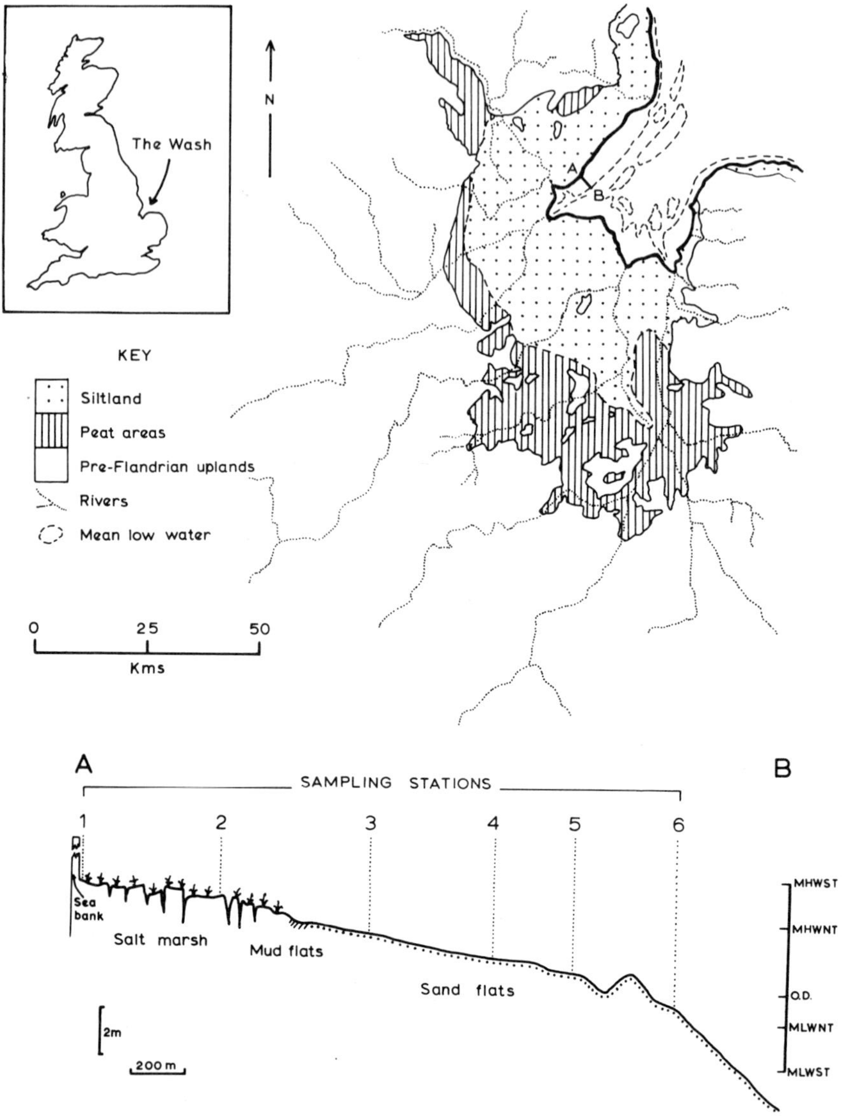

Figure 11.2. Location map of the area which has been studied and the profile across the intertidal flat of the Wash at Frieston Shore.

the lower mud flat, characteristic of other areas, does not occur on this section. Repeated measurements and surface sediment sampling along the line have shown (e.g. Table 11.1) that the intertidal flat in the immediate vicinity is slowly accreting over most of its length, particularly in its upper part, but with some local erosion resulting from creek migration. It is eroding in its lower part owing to a lateral displacement of the adjacent, deep offshore channel. The granulometric composition of the surface sediment of the intertidal flats has remained fairly constant along the line during the period of investigation, except for a slight coarsening near stations 4 and 5 resulting from erosion or local scour (Amos, personal communication).

Unfortunately, because of the problems of organizing sufficient manpower and of the financial limitation imposed on the investigation, it was not possible to collect information on water velocity throughout a tidal cycle simultaneously with the suspended sediment sampling. Thus only in a few limited cases can actual field accounts of the two variables be directly correlated. Also, the measurement of accretion/erosion of the intertidal flat surface along the transect were only initiated halfway through the investigation. Complementary work has been carried out on the changing bed configuration along the transect, the frequency of particular bedforms, rate of bed load transport and the relative effects of waves and currents on the sediment movement (see discussion by Amos and Collins at the end of this chapter).

The techniques used in the study

Water velocity information was collected throughout a range of neap and spring tidal cycles. The frequency of observation, in relation to tide height and the suspended sediment sampling positions across the intertidal zone, is listed in Table 11.1. Water velocities were measured at various levels within the water column using current meters suspended from small boats. On some occasions (see Table 11.1) it was possible to collect information synoptically; on others, observations had to be limited to an individual station located on, or adjacent to, the sampling line. The distribution of current meter positions over the intertidal zone has provided sufficient water velocity information to examine the general pattern of water movement over the area which has been studied.

The material in suspension in the water mass passing over the intertidal zone was sampled using a tower-type sampling system (Figure 11.3). The equipment was developed specifically for use in the investigation and enabled the collection of virtually synoptic samples at predetermined positions across the intertidal zone.

Each sampling tower consisted of an aluminium Dexion framework to which sample containers and sampling control mechanisms were attached. The latter devices enabled the instant of sampling to be controlled either in relation to the state of the tide or at a particular time. Four such towers were constructed, ranging in height from 1·6 to 4·3 m. Five sampling stations were selected across the intertidal zone (see Figure 11.2) and the positioning of each of the towers was arranged in accordance with the particular tidal range being sampled. The towers were maintained in position at each of the stations by being attached to fixing bolts let into reinforced concrete bases embedded in the bed; supplementary support was provided by splayed guy ropes which were anchored some distance away from the towers.

The original design of the sampling systems was based on the following five

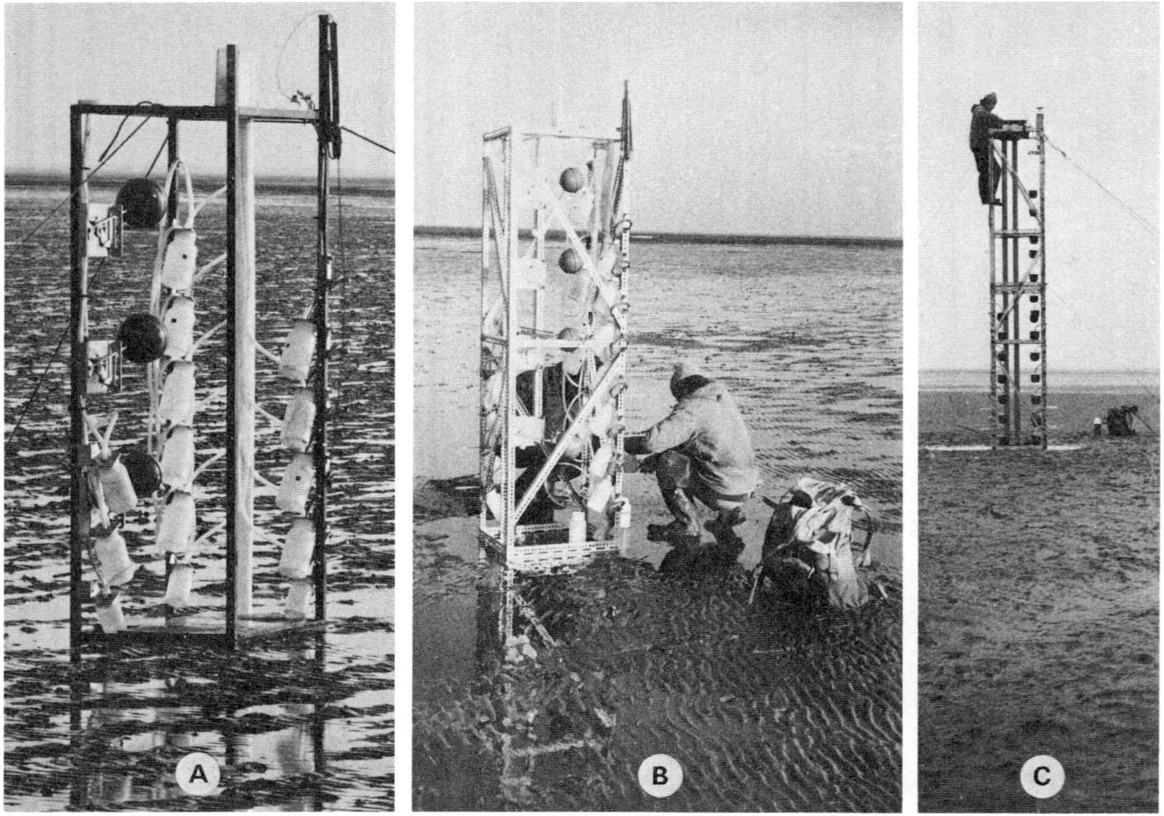

Figure 11.3. The tower systems used for the sampling of suspended sediment concentration in this study. (a) Sampling Tower 1 (1·6 m in height) between stations 2 and 3. (b) Sampling Tower 2 (2·2 m in height) at station 3. (c) Sampling Tower 4 (4·3 m in height) at station 5.

considerations: weight and manoeuvrability; strength to resist tidal, wind and wave forces; ability to resist corrosion; easily replaceable components; low cost.

The ultimate design adopted for the sampling system will be fully described elsewhere, but it is sufficient to comment here that each of the original requirements was adequately and successfully satisfied. In particular, the towers withstood exposure to a wide range of meteorological and hydrodynamic conditions (including wind strengths up to Beaufort Scale force 8) and no serious corrosion problems developed in the equipment throughout the duration of the investigation.

Individual components of the tower framework were examined hydrodynamically in a flume prior to their use in the field. It was verified that no excessive turbulence was created by the presence of the components in water flowing at velocities equivalent to those experienced in the field. In addition, the efficiency of the sampling control mechanisms was tested by lowering and raising each of the towers in an observation tank. The control mechanisms incorporated into the design included a modification to the single-stage suspended sediment sampling device (as described by the United States Subcommittee on Sedimentation (1963)); a simple alarm clock and coiled spring release mechanism devised by Carruthers (1968) and a unique electro-mechanical sampling control system. The electro-mechanical system worked on the principle of sensing the rise and fall of the water level by means of a damped floating tube and transferring and transposing this information to the energy output of a solenoid: the induced movement in the solenoid caused the rubber inlet tube of a sample container to be released.

By using this type of sensing and release mechanism, it was possible to collect samples of the water/sediment mixture at predetermined water levels during a tidal cycle. Throughout this investigation, samples were collected at mid-depth positions in the water column on the incoming and outgoing tide, whilst generally a complete spectrum of samples was taken from top to bottom of the water column at high water. A limited number of spectrum samples were also collected during rising or falling water stages. The information collected from these samples was sufficient to verify that mid-depth samples on the rising and falling period of the tidal cycle were representative in concentration of the overall distribution of suspended material in the water column; on each occasion that spectrum samples were collected, the abstracted mid-depth concentration was reasonably close to the average concentration of material in the water column at the time of sampling (see Evans and Collins, 1971).

Sampling was carried out at approximately two-monthly intervals: the frequency of sampling, in relation to date and tide height, is shown in Table 11.1. On each sampling occasion the towers were fixed in position on the intertidal zone for periods of up to a week. Normally six consecutive tides were sampled during each visit to the field. This necessitated visiting the sampling positions during the day and the night, in order to retrieve samples from an immediately preceding tide and to prime the equipment for a subsequent tide.

The concentration of suspended material in each of the samples was determined

Table 11.1. The availability of suspended sediment and water velocity data used in this paper.

Date of observations	Tidal height in m O.D.	Field observation reference	Stations 1 *(3·59)	2 (3·13)	2–3 (2·38)	3 (1·93)	3–4 (1·75)	4 (1·21)	4–5 (0·95)	5 (0·68)	Beacon (−0·31)
17 Oct. 71	2·92	Test				I		I		I	
14 Dec. 71	2·00	A						I		I	
15 Dec. 71	2·03	B			I	I		I		I	
15 Dec. 71	3·03	C			I	I		I		I	
16 Dec. 71	2·92	D			I	I		I		I	
16 Dec. 71	2·87	E			I	I		I		I	
17 Dec. 71	2·87	F			I	I		I		I	
15 Feb. 72	4·24	G		I							
16 Feb. 72	4·05	H	I	I		I					
17 Feb. 72	3·60	I		I		I					
17 Feb. 72	4·20	J	I	I		I					
18 Feb. 72	4·30	K	I	I		I		I			
19 Feb. 72	3·70	L		I		I		I			
13 Apr. 72	3·70	M								X	
13 Apr. 72	4·30	N	I	I		I		I		X	
14 Apr. 72	$4{\cdot}15^{p}$	O	I	I		I		I		X	
14 Apr. 72	$4{\cdot}57^{p}$	P	I	I		I		I		X	
15 Apr. 72	$4{\cdot}24^{p}$	Q	I	I		I		I			
15 Apr. 72	$4{\cdot}60^{p}$	R	I	I		I		I			
17 Jun. 72	3·06	S				I		I		I	
28 Jun. 72	3·59	T				I		I		I	
28 Jun. 72	3·20	U				I		I		I	
29 Jun. 72	3·75	V		I		I		I		I	
29 Jun. 72	3·42	W				I		I		I	
13 Aug. 72 → 24 Aug. 72	1·36 } 4·16	includes X–e									} 0^{c}

16 Aug. 72	2·79					0^1					0
17 Aug. 72	1·67	X				I	I	X^2		I	0
18 Aug. 72	1·63	Y					I	I		I	
18 Aug. 72	1·05	Z					I	I		X^3	0
19 Aug. 72	0·88	a								I	
19 Aug. 72	0·67	b								I	
20 Aug. 72	0·91	c								I	
20 Aug. 72	0·91	d								I	
21 Aug. 72	2·09	–						0	0	0	0
23 Aug. 72	3·19	–				0	0		0	0	
24 Aug. 72	3·50	e			I	I		I		I	
25 Aug. 72	3·80	f		I	X	X		X		0	
26 Aug. 72	3·68	g		X	X	I		I			
27 Aug. 72	3·59	h	X	I	I	I		I			
30 Nov. 72	2·35	i				I	I	I		I	
30 Nov. 72	2·47	j				I	I	I		I	
01 Dec. 72	2·50	k				I	I	I		I	
01 Dec. 72	1·75	l								I	
02 Dec. 72	2·82	m				I	I	I			
26 Jun. 73	2·58	–								0	
27 Jun. 73	2·81	–								0	
28 Jun. 73	3·43	–				0					
29 Jun. 73	3·09	–						0			
29 Jun. 73	3·46	–						0			
30 Jun. 73	3·58	–				0					
01 Jul. 73	3·91	–						0			
02 Jul. 73	3·95	–		0							

Key: p = Predicted tidal height, no actual measurement. * = Level of sampling position in metres above O.D. I = Suspended sediment sampling only. X = Suspended sediment sampling and current measurement. 0 = Current measurement only. 0^c = Continuous current measurement only. 0^1 = Current measurement only; 100 m N.E. of station 3. X^2 = Suspended sediment sampling and current measurement; 200 m S.W. of station 4. X^3 = Suspended sediment sampling and current measurement; 200 m S.W. of station 5.

in the laboratory by filtration through Oxoid membrane filter papers; although this method of sediment abstraction arbitrarily delimits the minimum size of material considered to be transported in suspension at 0·45 micron, it is possible that finer material could have been retained owing to clogging of the filter papers.

A slightly different analytical technique was employed on a selected number of samples in order to examine the texture of the suspended material. The samples were first filtered through a 63-micron nylon mesh sieve prior to filtration through the membrane filters. In this way it was possible to determine the relative percentages of sand to silt and clay in the samples. However, it should be realized that the water-sediment samples were so small (approximately 500 ml) that it was hardly adequate to carry out a grain-size analysis.

Tidal data (i.e. observed time and height of high water) were collected for each of the suspended sediment sampling periods by setting up an automatic recording tide gauge on a refuge beacon at the seaward end of the sampling line. In general, the time of high water at Freiston Shore was found to occur 10 to 20 minutes earlier than at the dock sill in Boston, but to be greatly affected by localized meteorological conditions; similarly, there was an inconsistent relationship between the observed and predicted height of the tide, presumably controlled by meteorological conditions in the Wash and adjacent North Sea.

During the sampling, observations were made on local meteorological and hydrodynamical conditions: estimates of wind speed and direction and the height, wavelength, and periods of waves were recorded as well as any other relevant phenomena.

The tidal current velocities

One of the most striking features about the current measurements is that there is a pronounced set of the current towards the northeast (i.e. in a longshore direction) in all the data which have been recorded (see Figure 11.4). Similar trends can be seen in the results obtained by the Hydraulics Research Station for measurements on the surrounding intertidal area.

This very strong longshore component was unexpected on the upper parts of the intertidal flats, although it was realized that it was probably important on the lower more seaward parts of the intertidal zone (Evans, 1965). The other factor which was most noticeable was that the oblique longshore component was, throughout the whole of the tidal cycle, directed towards the northeast (see Figure 11.3). The importance of the longshore component is illustrated even better if the various current measurements are resolved into their longshore and onshore–offshore components (Figure 11.5). Such a manipulation shows that, apart from a few results, the longshore component is of the same magnitude as the onshore–offshore component. As can be seen from Table 11.2 the maximum onshore–offshore velocities decrease landwards and the mean onshore and offshore velocities also show this trend but not so well. It can be seen that the maximum longshore velocity component decreases as one approaches the shore but the mean does not show this trend so clearly. These results therefore indicate the type of gradient suggested by the grain sizes of the sediment of the intertidal flats which decreases from low to high water mark.

The mean velocity (average of bottom, mid-depth and surface) measured at each station at various states of the tide in relation to high water has been plotted for tides of various heights, in order to see if there is a relationship between tidal height and tidal velocity. The averages of the actual recorded measurement have been plotted and no consideration has been given to the direction of motion. It is important to establish such a relationship if any generalizations are to be made on the likely current velocity conditions to be found at any particular part of the intertidal flat. As can be seen from Figure 11.6 the data show a reasonable relationship with an increase in water velocity as the height of the tide increases but there is a broad scatter of the data and the relationship between tide height and current velocity at various stages of the tide is not as convincing as perhaps would be expected.

The nature of the suspended sediment

A preliminary examination of the suspended sediment, collected in the sampling containers, shows it to be composed mainly of inorganic material. The breakdown of this sediment into sand, silt and clay fractions by washing through a nylon sieve shows that the sand content is very variable in concentration, reaching up to 96% by weight of the total suspended load. Although the data are variable there is an increase of the percentage of sand through the water column towards the bed, but there is no unambiguous decrease in sand content from low to high water mark. The highest percentage of sand, although not the highest total suspended sediment concentration, is found in samples collected during the summer (e.g. Station 2, tide K, 18 February 72, range in suspended sediment concentration 290–634 p.p.m., range in sand content 3–15%; at station 5, tide S, 27 June 72, range in suspended sediment concentration 23–103 p.p.m., range in sand content 33–68%). If the concentration of sand is approximately the same throughout the year, this variation may be merely a reflection of the higher concentrations of silt and clay during the winter.

The concentration of the suspended sediment

The techniques and procedures described in an earlier section seem to give meaningful representative results. Comparison of the mid-depth samples with a mean of the whole spectrum of samples from top to bottom of the water column endorses the validity of measuring the latter to determine the amount of material in suspension. However, it should be fully realized that this is only material in suspension and perhaps a much greater amount of material is moving along in the water column close to the bed.

Most of the values of concentration of suspended matter in the waters above the intertidal flats of the Wash are low, of the order 200 p.p.m. Generally, the few unusually high values of suspended sediment concentration have been obtained from the sampling containers nearest the bed, and it is assumed that

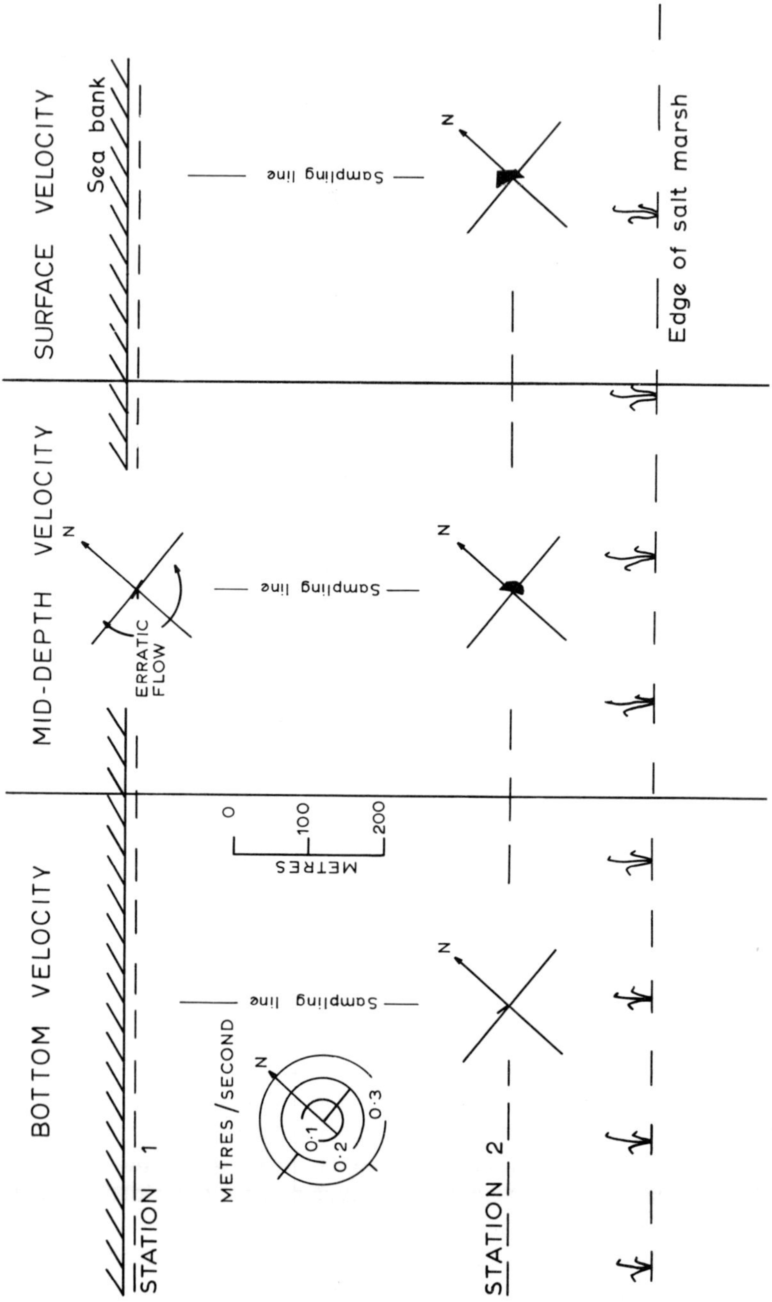

SURFACE VELOCITY
MID-DEPTH VELOCITY
BOTTOM VELOCITY
Sea bank
Sampling line
Sampling line
Sampling line
Edge of salt marsh
N
N
N
N
N
ERRATIC FLOW
METRES
0
100
200
METRES/SECOND
0·1
0·2
0·3
STATION 1
STATION 2

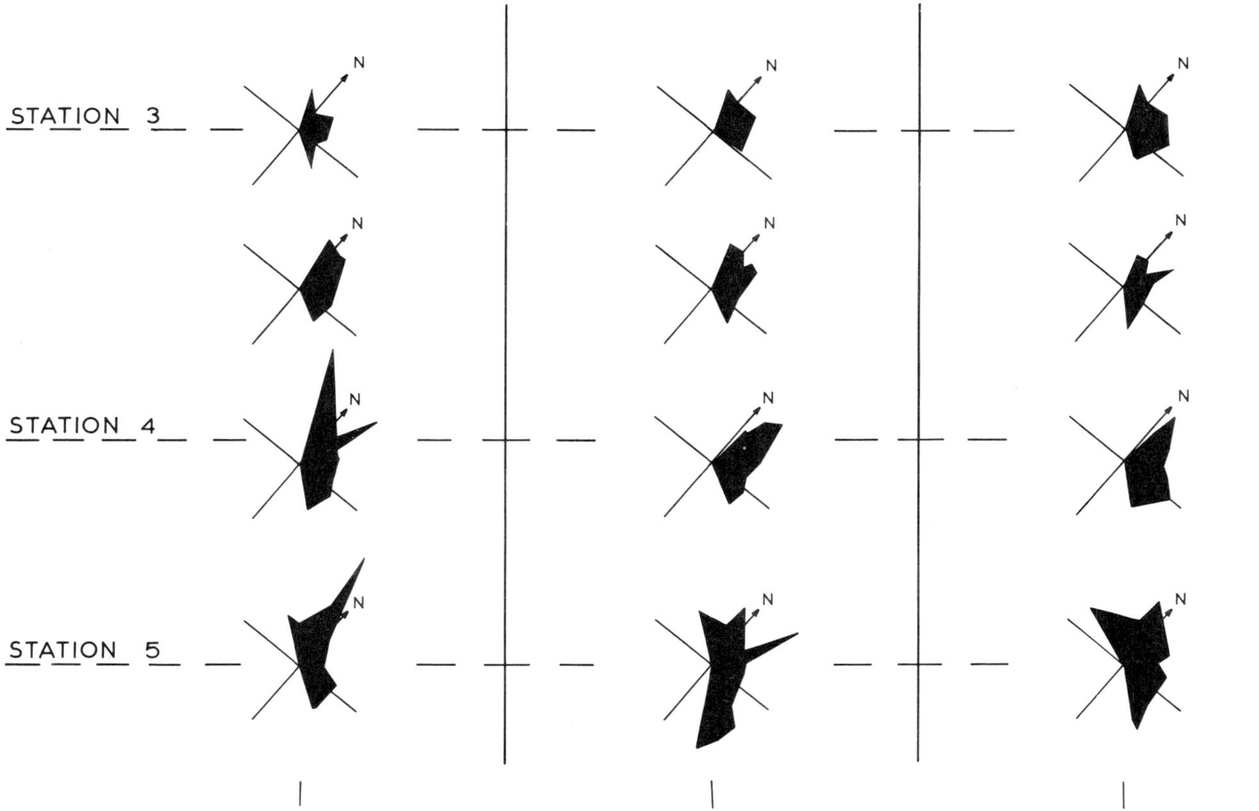

Figure 11.4. The results of measurement of tidal current velocities during a selected tide along the sampling transect.

LONGSHORE COMPONENT

ONSHORE / OFFSHORE COMPONENT

STATION 3

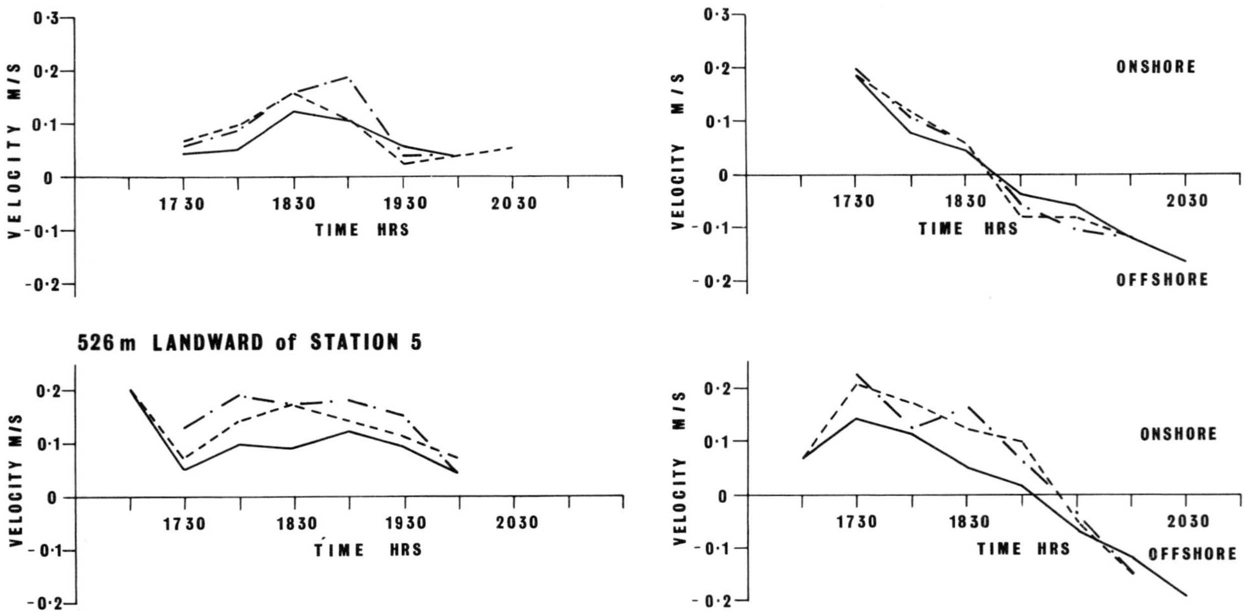

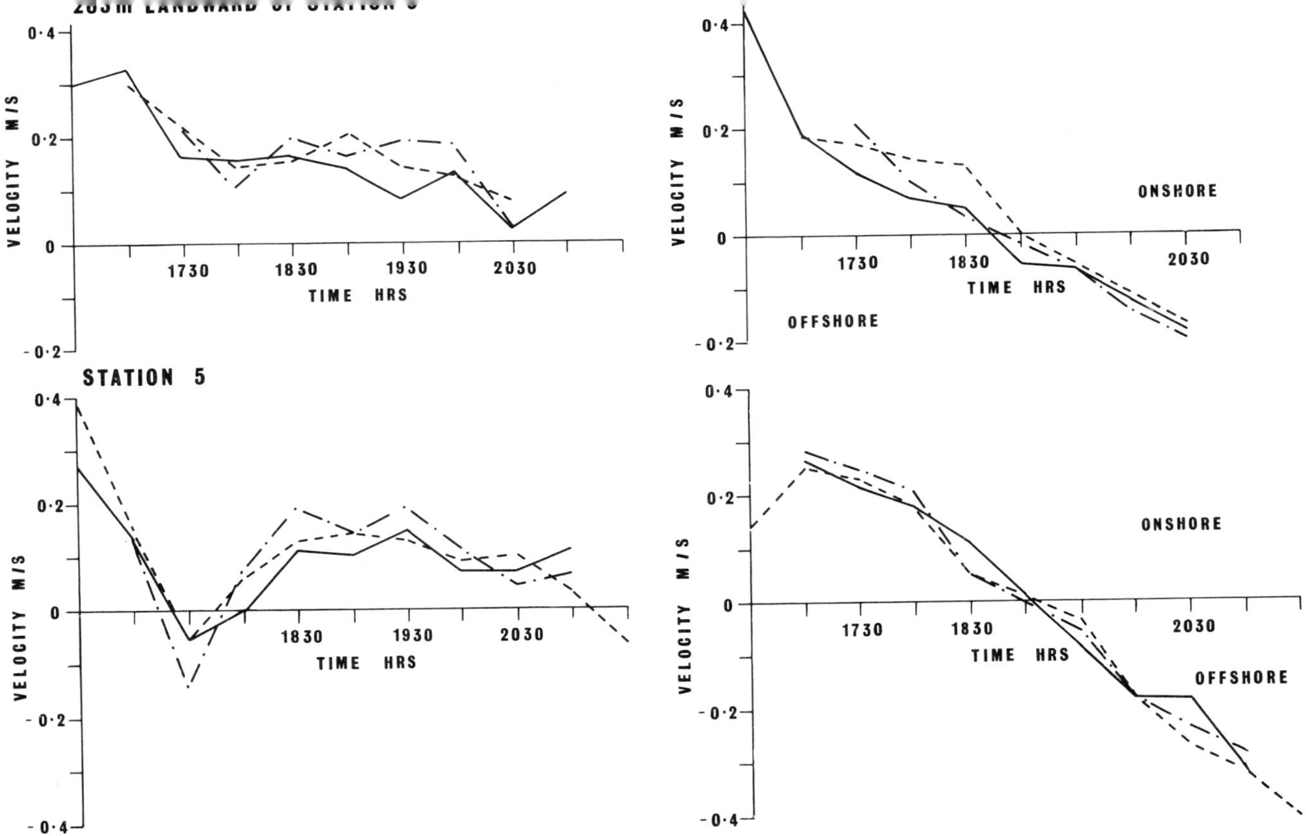

Figure 11.5. The results of resolution of the tidal current velocities into onshore–offshore and alongshore components during a selected tide.

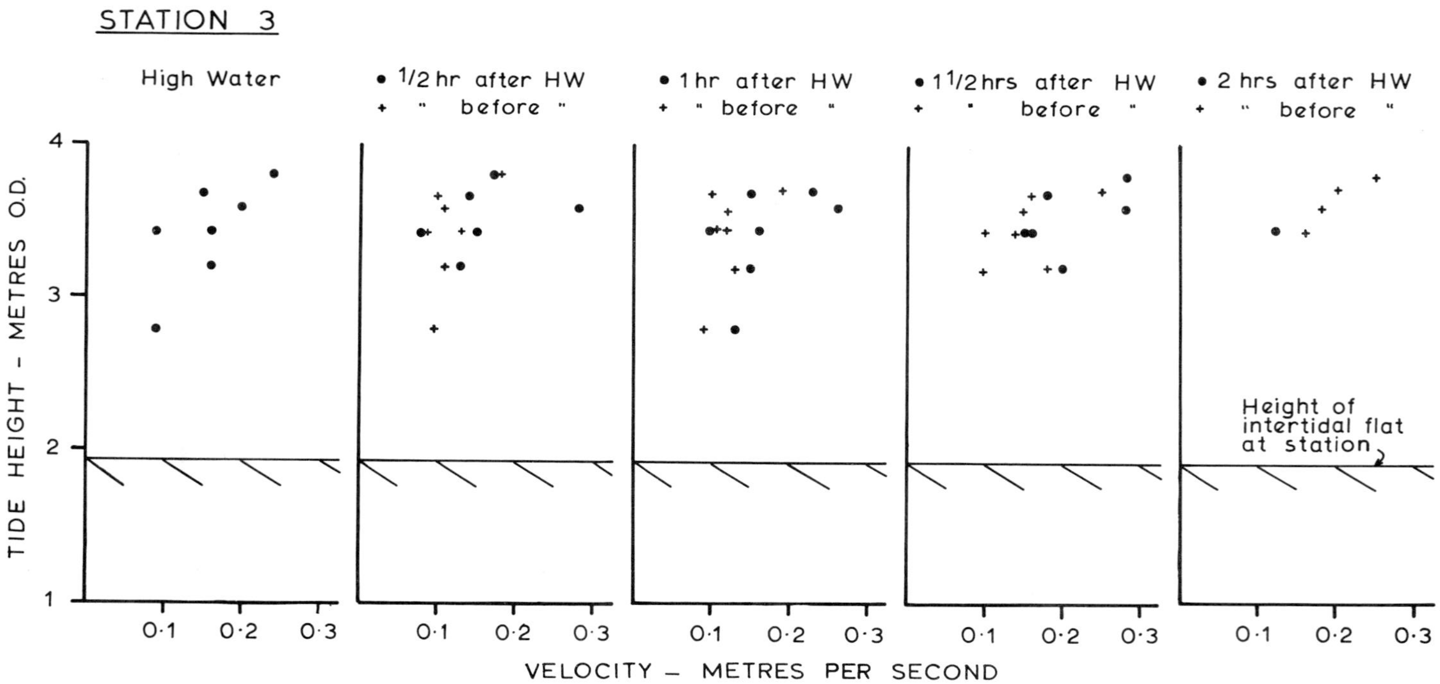

Figure 11.6. The relation between the velocity of the tidal currents and the height of the tidal waters, at a particular station.

these were a result either of increased local scour, possibly produced artificially around the base of the towers by severe current action, or of increased wave activity at shallow water depths.

It was possible, on several occasions, to trigger off the sampling towers so that a synoptic picture of suspended sediment concentration was obtained at various stages of the tide. When this was done, the results for both neap and spring tides showed that there was an increase in suspended sediment concentration towards the base of the water column and also a general decrease in suspended sediment concentration from low to high water mark at any one stage of the tide. However, in spite of the presence of these broad patterns, localized high values occur, even high in the water column, which may be merely sampling errors or possibly may indicate plumes of water with a high concentration of suspended sediment, because of rapidly fluctuating turbulence in the water column.

The total number of determinations of suspended sediment concentration, excluding some dubiously high values, have been presented in Table 11.3 with the various maximum, mean and minimum values shown. An examination of all these results show that there is a general decrease in the maximum values of suspended sediment concentration from low to high water mark (although this may be merely the effect of the greater frequency of sampling at the outer stations); there is not a clear trend in the mean value of suspended sediment concentration (which range between 77 p.p.m. and 190 p.p.m.). The maximum concentration values for stations 1, 2, 3 and 4 all relate to a particular tide sampled during February 1972 when, unfortunately, it was impossible to sample at station 5 because of insufficient sampling equipment (tide reference K). This emphasizes the problem of attempting to generalize from data which are unfortunately not just incomplete but are incomplete during a period of exceptionally high concentrations. The minimum concentration values range from 2 p.p.m. to 12 p.p.m. but these values are obviously of little significance.

A typical series of results from the sampling towers for a tidal cycle is shown in Figure 11.7 which illustrates the relationship of the availability of samples

Table 11.2. The magnitude of the velocities of the tidal currents across the intertidal flats of the Wash at Frieston Shore (see Figure 11.2 for location of stations). Tide of 25 Aug. 72.

	Velocity in metres per second				
Station	Maximum onshore–offshore	Maximum longshore	Mean onshore	Mean offshore	Mean longshore
2*	0·10	0·06	0·04	0·04	0·03
3	0·28	0·26	0·14	0·19	0·16
4	0·28	0·35	0·17	0·14	0·16
5	0·39	0·44	0·27	0·17	0·16

* Tide of 26 Aug. 72.

Table 11.3. The distribution of the suspended sediment concentrations across the intertidal flats of the Wash at Frieston Shore (see Figure 11.2 for location of stations).

Station	Number of observations	Concentration (p.p.m.)		
		Maximum	Mean	Minimum
1	48	754	101	12
2	86	658	141	6
2–3	59	736	77	6
3	268	1830	137	6
3–4	30	1042	190	10
4	278	2190	136	4
5	304	820	123	2

and the passage of the tidal water column past the sampling stations on the intertidal flat. In addition, the values of suspended sediment concentrations at mid-depth positions on a rising and falling tide, and throughout the whole water column at high tide, are shown. Values of suspended sediment concentration in Figure 11.7 are highest during the first phase of the flood and the last phase of the ebb, when the current is flowing most strongly. Their lowest values tend to occur at the high water stage.

As the concentration of the suspended matter shows great variations (e.g. at station 3, from approximately 40 p.p.m. to 350 p.p.m. throughout the tidal cycle, and from 40 p.p.m. to 100 p.p.m. throughout the water column at high water) a problem arises in the choice of a representative value for a particular tide. This problem becomes very apparent when an attempt is made to see if the suspended sediment concentration shows any relationship to tidal height or any other variable. One of the most useful parameters which suggests itself is that of mean concentration of suspended matter for each station (see Halliwell and O'Connor, (1966). When this parameter is plotted against tidal height for some of the stations (see Figure 11.8) there is clearly no evident relationship (cf. the results from the Mersey presented by Halliwell and O'Dell (1969)).

Another factor which needs consideration is the relationship of the tide samples to the phase of that particular tide (i.e. springs to neaps or vice versa). It is possible that the measurements of suspended sediment concentration made at any one time would be influenced by the conditions which preceded it (for example, a period of very turbulent conditions, produced by either high wave activity or by stronger current activity). Assuming such to be the case, it would be expected that there would be a much greater concentration of suspended sediment during a particular tide height which succeeded such a period than during a tide of similar height immediately preceding it and which had possibly followed a period of calm conditions. However, the data available from this investigation do not show any distinct relationship between the suspended sediment concentration and the phase of the tide as was found by the British

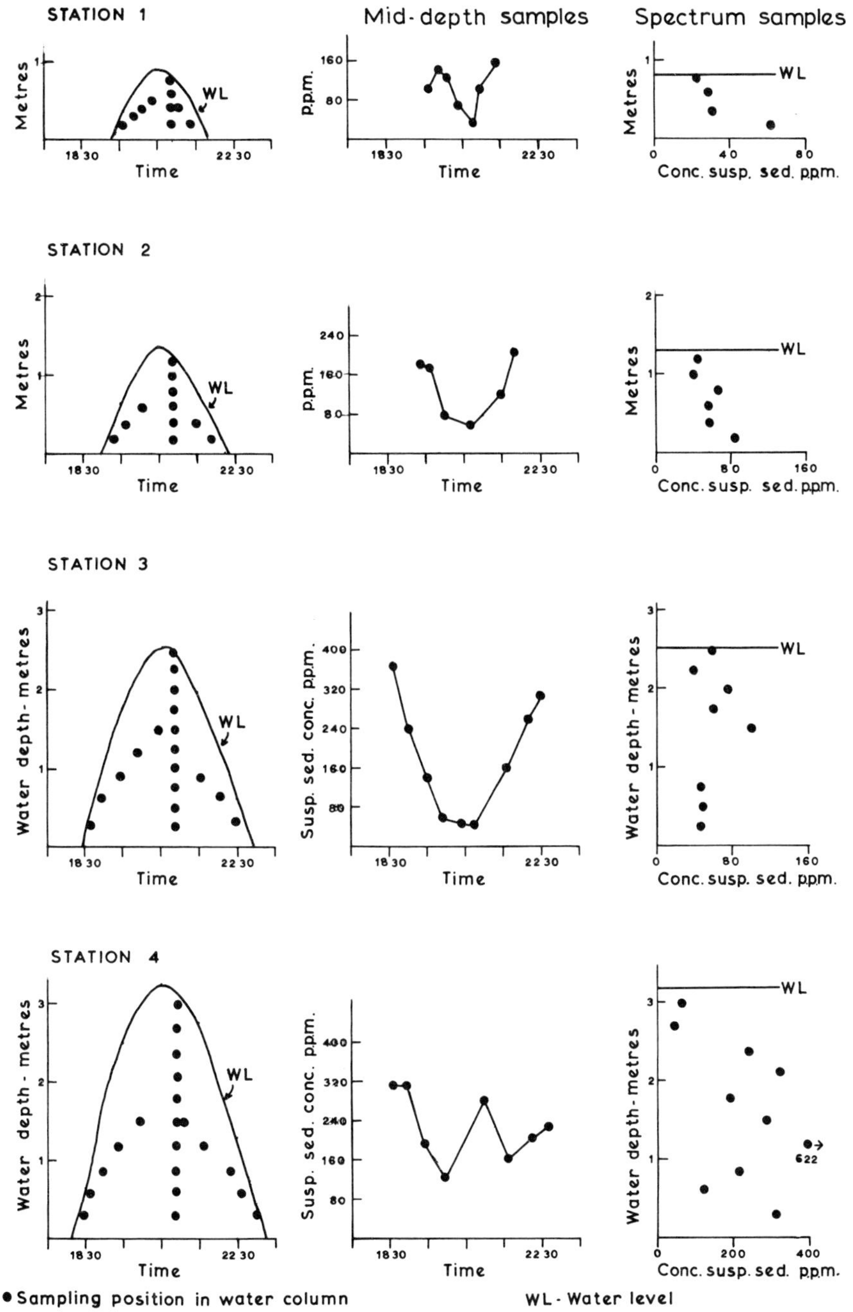

Figure 11.7. A typical record of the results of sampling the suspended sediment for a selected tide (15 April 1972, tide reference R).

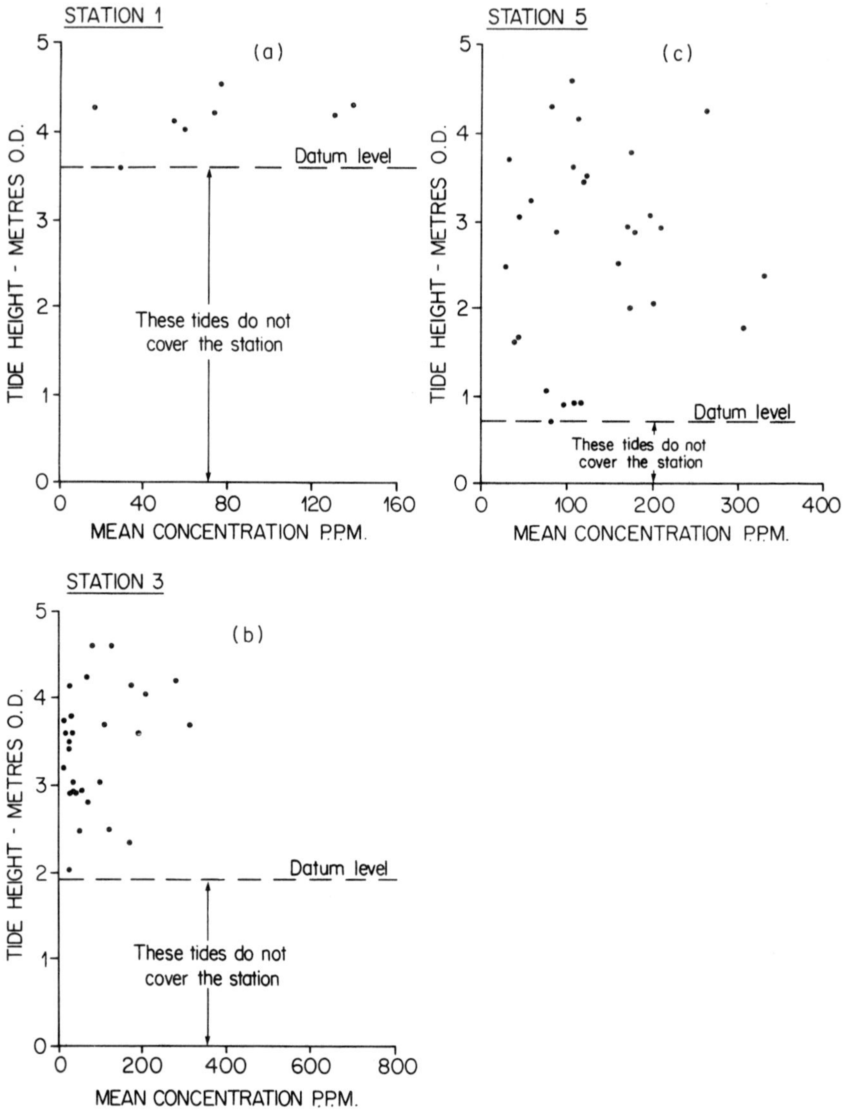

Figure 11.8. The relationship between the mean suspended sediment concentration and the height of the high water for stations 1, 3 and 5.

Transport Docks Board (1970) to be generally, but not always, true for the Humber.

The mean concentration of suspended sediment has been plotted, for the higher tides (i.e. those that reach 3·0 m above O.D.), sequentially throughout the period investigated (Figure 11.9). These results show quite a large variation in suspended sediment concentration. The reasons for these variations in values are not known.

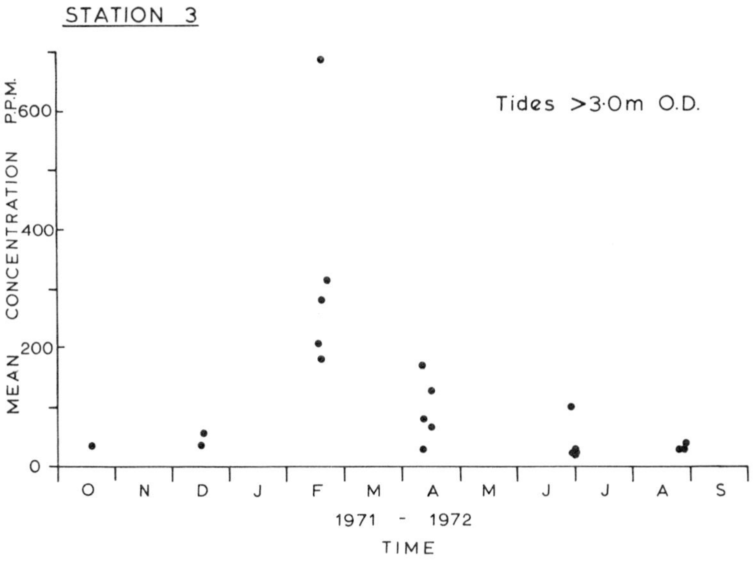

Figure 11.9. The average concentration of suspended sediment for tides whose high water level exceeds 3·0 m O.D. at station 3.

However, it is of interest to compare the suspended sediment concentrations of tide with high water reaching 4·3 m O.D. during the February period with that found in a tide of similar height sampled during April. In the former tide the mean suspended sediment concentration at stations 1, 2, 3 and 4 were respectively 139, 403, 688, 878 p.p.m. whilst for the latter they were 17, 22, 30, 65 p.p.m.

These differences are possibly related to the markedly different meteorological conditions which occurred on the two occasions: there were very intense and sustained winds over the February period contrasting with relatively calm conditions in the April period. Whether the differences in suspended sediment concentration at the various sampling periods (Table 11.3) is due to a seasonal pattern or variation in the meteorological variables, or merely a haphazard series of events is, with the limited data available, impossible to ascertain. There may well have been many other occasions when the tidal waters had high concentrations of suspended sediment, even over the period of the investigation, but which were missed by the actual sampling occasions.

It is also possible that the high February value could have been the result of essentially local turbulent conditions—such as a period of unusually intense wave action—but no such unusual effects were observed at the time in the field. The writers are inclined to the view that it is the conditions outside the immediate sampling area which are influencing, and perhaps controlling, the levels of suspended sediment concentrations in the waters covering the intertidal flats of the Wash, and that it is the availability of material for transportation by tidal currents onto the intertidal flats of the Wash, rather than the strength of the local tidal currents, which is of prime importance.

The implications of these differences in suspended sediment concentration are that certain optimum hydrodynamic and/or meteorological conditions in the Wash, North Sea and adjacent coastline can create high suspended sediment concentrations. Intense wave action or very strong tidal currents (see Johnson and Stride, 1969) would lead to the expenditure of very high amounts of energy on the seafloor and cause the entrainment and transport of unusually large volumes of sediment. However, numerous other factors need to be considered in attempting to understand these differences.

The effects of sediment discharge into the Wash via rivers are thought to be minimal. However, it is possible that periods of high-water discharge from the Humber would increase the level of the suspended sediment in the North Sea. Similarly, a period of high waves, coinciding with high tides, would yield a large amount of sediment by erosion of the boulder clay and glacial outwash sands of the cliffs of adjacent Norfolk, Lincolnshire and Yorkshire.

Each of these events could obviously be important and each could ultimately lead to the build-up of suspended matter which could possibly be carried into the Wash. However, there would be a lag between the event causing the initial entrainment of sediment into the waters of the North Sea and the transportation of that sediment onto the intertidal flat. Coincidence of these events would obviously culminate in exceptionally high suspended sediment concentrations over the intertidal flats or, conversely, if the various events were completely out of phase this would lead to the obscuring of the relationship between cause and effect. This clearly indicates the problems of relating cause and effect without a large and very costly investigation.

In the previous arguments it has been assumed that all the sediment in suspension has been brought to the intertidal flats from or via the adjacent subtidal areas of the Wash. Generally it is thought that this is true for the major part of the material, as the intertidal flats of the Wash have grown, albeit intermittently, seawards over the last 2000 years. However, it has been argued elsewhere that it is redistribution of the material on the intertidal flat which is responsible for much of this accretion on the higher intertidal flats (Kestner (in Inglis and Kestner 1958b)). The exact proportions of material which are derived locally from the intertidal flat, as contrasted to that which is brought onto the flats, is difficult to ascertain. It is further complicated by the fact that material eroded from the intertidal flats during the falling tide will, or may, be brought back onto them during the next tide.

Obviously, periods of very high tides and hence higher tidal velocities are likely to be occasions of maximum erosion. Similarly periods of high wave activity can be important. Heavy rainfall may well produce a slurried surface which will lead to the surface sediments being more easily eroded, just as, on the other hand, periods of very hot dry weather coinciding with neap tides leads to a dessication of the surface muds and a widespread peeling off of their surface layers as the tides begin to rise to higher levels. Similarly, local biological effects tend to be seasonal and hence the variations in density of the sediment-binding algae will influence the potential erodibility of the surface.

The flux of suspended sediment across the intertidal zone

As has been described in the foregoing paragraphs the study of the tidal currents over the sampling line revealed a distinct persistent oblique longshore component to the northeast; this obviously produced difficulties in attempting to calculate a balance of water and sediment moving on and off the intertidal flats. Although it is possible to resolve the data into onshore–offshore components it is considered more useful to attempt to calculate, using a simple approach, the resultant movement of water and sediment over the intertidal flats to obtain some estimation of the magnitude of the transportation. For this exercise a tide, for which fairly complete current data are available, has been chosen. It was possible to calculate the resultant movement of water per metre width past a particular point during the tidal cycle by using the half-hourly interval water velocity measurements, and knowing the depth of water over the measuring site at each station during that interval. Similar calculations were made for all stations across the intertidal flats except for station 1 where current data were unavailable. It should also be noted here that the data of station 2 are also of dubious value as this station is close to a creek drainage system. However, the omission of station 1 and the limited consideration of the data from station 2 has little effect on the overall result as the magnitude of water and sediment movements past these stations is very small when compared to that passing the stations to seaward.

Such calculations indicate that during this tide 163 m^3 moved to 026° past station 2; 2680 m^3 of water moved to 039° past station 3; 3500 m^3 of water moved to 015° past station 4; and 6340 m^3 moved to 021° past station 5. If it is now assumed that each of the stations are representative of particular sections of the transect across the intertidal flat it is possible to convert these figures into a total water movement across the sampling line during the particular tidal cycle being considered (for this calculation station 5 has been assumed to be representative of the outer 250 m; stations 4 and 3 to be representative of the central portion, which is 1000 m across; and station 2 is representative of the 750 m of the inner intertidal zone). The resultant of these motions yields a figure of $4{\cdot}56 \times 10^6$ m^3 moving to 022° across the line during the tidal cycle. It may be argued that the results obtained from station 2 should really be ignored in such calculations because they are so obviously affected by the presence of a creek; however, the magnitude of the resultant water volumes passing across the inner intertidal zone (calculated using station 2) are of the order of magnitude of ten times less than those of each of the other stations to seaward; the inclusion or exclusion of data from this station therefore has little effect on the overall pattern, but it has been included for completeness.

To convert these water volumes into sediment fluxes it is necessary to use a sediment concentration factor. If, for example, 100 p.p.m. or 200 p.p.m. are taken as sensible average concentrations (see earlier section) this yields 456 tonnes or 912 tonnes moving across the sampling line towards 022° respectively. However, it is possible from the available data to get a better estimate than this since a large amount of data is available on the suspended sediment concentration for various states of the tide at each station. It is therefore possible to obtain the most likely concentration for successive half-hour intervals through the tidal cycle and then to combine these with water volumes passing through a 1 m section during that particular period. Summation of such data yields 540 tonnes moving across the sampling line towards N 22° E during the particular tidal cycles being considered.

The range of values of sediment concentration found in the water over the intertidal flats, along the sampling line at Frieston Shore at the west of the Wash, may be compared with the large amount of offshore data recently collected by the Hydraulics Research Station in connection with the Wash Feasibility Study (Hydraulics Research Station, 1972). These show fairly low (generally <100 p.p.m.) maximum suspended sediment concentrations for the greater part of the offshore area of the Wash and only show higher values in the inner parts adjacent to, and in, the immediate offshore areas near the outfalls of the Fenland rivers. Also, for any offshore station, where suspended sediment information is available for both spring and neap tides the latter are usually lower, particularly in the areas adjacent to the river outfalls where they are significantly so.

Some general considerations

The measurements of suspended matter in the waters over the intertidal flats of the Wash near Frieston Shore allow some general conclusions to be drawn. A 100 p.p.m. is a fairly average value for concentration of such suspended matter. If it is assumed, as has been argued in the previous section, that the suspended sediment of the waters above the intertidal flats has been largely—or almost completely—derived from offshore, then during a neap tide a minimum value of $0{\cdot}3 \times 10^5$ tonnes of sediment is brought onto the intertidal zone of the Wash during a neap tide and a minimum value of $0{\cdot}12 \times 10^6$ tonnes during a spring tide. This means that in a normal year approximately $5{\cdot}0 \times 10^7$ tonnes of sediment is brought onto the intertidal flats of the Wash by the advancing tidal waters.

The volume of material supplied by the rivers has been estimated by Collins (1972) to be $0{\cdot}15 \times 10^6$ tonnes and by the Crown Estate Commissioners Report (1969) to be $0{\cdot}2 \times 10^6$ tonnes; it would appear that approximately 330 years supply of river material would be needed to build up the concentrations which are found in the waters which cover the intertidal zone. This is, of course, ignoring the volume of sediment present in the permanently submarine parts of the Wash and also assumes no loss of material by sedimentation. The latter assumption is known to be untrue as extensive sedimentation has occurred around the margins of the Wash. Obviously, only a small proportion of the sediment present in the water above the intertidal zone actually settles on the surface.

It is possible to take the argument a little further and consider the volume of material which has been deposited around the borders of the Wash since the initial development of the Romano-British siltlands (assumed here to have been deposited in the last 2000 years). In such a calculation, it is impossible to consider the submarine parts of the Wash as there are not sufficient data to make even a broad estimation at the volumes of material involved (however, it is hoped that such a situation will be corrected by the studies currently being made by the Institute of Geological Sciences).

Even for the bordering Fenland the data are very inadequate, although Godwin (1940), Godwin and Clifford (1938) and Willis (1961) have given a good general stratigraphic section across the Fenland. However, the present writers have assumed that over the last 2000 years at least 6·5 m of sediment (i.e. the average thickness of the intertidal zone) has been deposited over the area covered by the Romano-British siltlands. It has been assumed that intertidal flat sedimentation has prograded the coast from the inner limits of the Romano-British siltlands to the present shoreline. This rough approximation is undoubtedly too great in some areas, such as the inner edges of the siltlands, but is undoubtedly much too small in the outer parts of the reclaimed lands, particularly in the old estuaries of the Nene and Ouse where approximately 20 m of material has been deposited. Assuming, that the various inaccuracies tend to cancel one another out, it may be calculated that the average deposition of coastal sedimentation over the last 2000 years is 8×10^6 tonnes/year. And, as has been already stated, the river supply is only $0{\cdot}15$ to $0{\cdot}20 \times 10^6$ tonnes per year so that it would take approximately 50 years of river supply to produce enough material to maintain the calculated level of coastal accretion. It is known that the rivers supply very little sand to the Wash and that their load is composed mainly of silt and clay (Collins, 1972). Also, it is known that the average intertidal zone section, from high to low water mark, consists dominantly of sand and only contains 20% silt and clay. So even if it were assumed that the rivers supplied all the silt and clay present in the Romano-British siltlands it would need approximately 10 years river supply to maintain the silt and clay contribution to the estimated rate of accretion of sediment.

It may be argued, of course, that the Fenland rivers once transported much more sediment to the sea and that their contribution has been substantially reduced by man's engineering works in the area. However, this seems unlikely, as generally the deforestation and more intensive agriculture, with an increase in proportion of arable land, has almost certainly led to an increase of sediment run-off in the past 2000 years (Langbein and Schumm, 1958; Schumm, 1969). Also, it may be noted that even when river water is trapped in large channels, such as the New Bedford River, and the suspended load is given an opportunity to settle, very little accretion occurs (Kestner, 1961). Thus, it would appear, from the above arguments, that the volumetric studies of the suspended sediment in the waters of the Wash and of that locked in the coastal siltlands indicate that fluvial supplies are unlikely to be adequate to explain the vast volumes of material which have been deposited and that a marine source seems to be an absolute necessity.

This argument, of course, merely restates and substantiates the opinion of engineers working in the area for a century or so (Wheeler, 1875–76; Clark, 1959) as well as geologists and other people who have studied the problem.

The state of equilibrium of the Wash

A further point should be made concerning the apparent state of equilibrium (O'Brien, 1931, 1969) of the Wash (see Figure 11.10). The entrance of this embayment is floored by boulder clay and Cretaceous rocks; it is bordered on the southeast by a rocky cliff, consisting of Cretaceous sediments, which is receding only slowly; and it is bordered on the northwest by the prograded beach-dune complex of Gibraltar Point. It is unlikely that the cross-sectional area of the entrance of the Wash has decreased much in size over the last 2000 years by any other process than the progradation of Gibraltar Point; and it is possible that once the rise of sea level since Romano-British times is taken into account that the cross-sectional area has in fact increased.

However, the tidal prism or average volume of water which flows in and out of the Wash in an average tide has been decreased by extensive coastal accretion and even furthermore by man's reclamation and the subsequent exclusion of the tidal waters from wide areas. The expected consequences of such a process would be to decrease the cross-sectional area by progradation of the adjacent shorelines or build-up of the floor of the entrance; alternatively the floor of the

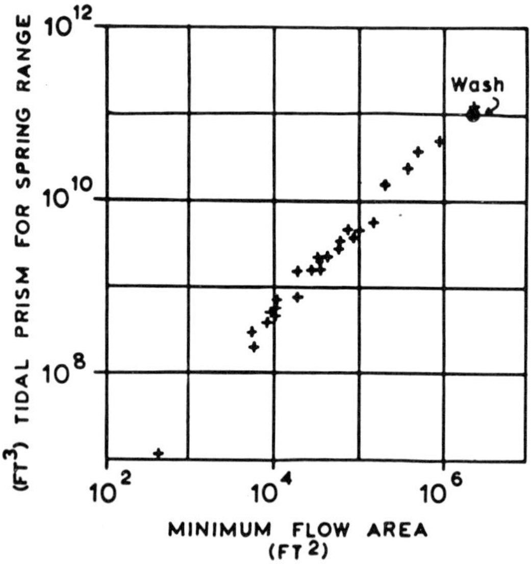

Figure 11.10. The relationship of the volume of the tidal prism for spring range to the minimum cross-sectional area of the entrance for some natural inlets in the USA (O'Brien, 1969), with the Wash also plotted.

main embayment itself could have been deepened thus maintaining the volume of the tidal prism and allowing the cross-sectional area of the entrance to remain constant. It is interesting to note, in view of this fact, that studies by the Hydraulics Research Station indicate that the Wash has apparently deepened since the earliest hydrographic surveys of 1828 (Hydraulics Research Station, 1953). It is unlikely that such deepening can continue since once the boulder clay or underlying solid rock formations are reached erosion will be minimal and thus the volume of the tidal prism cannot be increased in this way. This may not of course have been true earlier, as intertidal flat deposits and glacial outwash sands, etc. may have originally cloaked much of the floor of the Wash and these sediments would have been fairly easily removed by the strong tidal currents so characteristic of the area.

Conclusion

The results of a series of experiments have been described. These were initially planned in an attempt to see if it were possible to calculate the flux of suspended material carried by the advancing and retreating tidal waters over a typical cross-section of an intertidal flat in the Wash embayment. Various problems, largely financial, led to the measurements of the tidal currents and the concentrations of suspended matter on different occasions; the unpredictability, particularly of the sediment, has led to problems in attempting to fuse the data into a single calculation. However, these measurements have highlighted the problem of attempting to predict the likely volumes of material moving on and off the intertidal flats, and the complex problem of attempting long-term forecasts. They have further illustrated how, in such a study, the boundaries of the system are relatively infinite: external forces and factors, often far away and apparently divorced from the area under study, probably control the concentration of suspended sediment in a nearshore embayment like the Wash. The lag effect, which only allows the particular effects to be recognized days, weeks or months later must surely complicate the problems even further. This study has again stressed the problems of carrying out and understanding the detailed changes in a small coastal area without an abundance of background data from the neighbouring seas and coasts. Also, it has indicated how important it is to carry out repeated measurements day after day throughout the seasons and not merely on isolated occasions. Engineering predictions based, as they often are, on such short-term measurements may be disastrously wrong.

The circulation of water on the intertidal zone has been shown to be more complicated, in the area studied, than was originally envisaged. This may partly be attributed to an unfortunate choice of area (and perhaps it is more simple on the south coast of the Wash), or the pattern which has been found may be a result of a large-scale clockwise circulation of water in the Wash. However, the intertidal flat circulation may be very complicated locally, possibly controlled by the creek patterns and perhaps consisting of a series of cells, rather like those set up in a breaking wave system on beaches with rip current channels (Shepard and Inman, 1950). The original balance studies became difficult, in view of the strong lateral motion of the advancing and retreating tide, and nothing more than a typical flux of material for some specific occasions could be attempted.

Large-scale considerations, using admittedly poorly controlled geological data, river data collected by one of the writers and average suspended sediment concentrations, appear to indicate that the fluvial supply of sediment is relatively unimportant in the construction of the coastal Romano-British siltlands and that these would never have originated without an abundant supply from the adjacent seafloor or possibly from the coasts of East Lincolnshire and North Norfolk.

It must be admitted, that these conclusions are based on general considerations and it is possible that the rates of accretion of the coastal belt have been greater in the past than the present and hence perhaps river supply may proportionally be more important in modern times. Similarly, it is possible that during the deepening of the Wash, the marine supply of sediment may have been more important than at present or in the future; there is, however, no proof that the latter is the case and it seems likely that there is an almost inexhaustible supply of sediment present on the adjacent seafloor.

Acknowledgements

Many individuals have contributed to the work reported in this paper. The authors wish to record their thanks to: Mr. C. Harness for supplying accommodation, for his hospitality and for considerable local help; Mr. A. Kuhn for his help in the field and for his most generous hospitality; Mr. A. Heath and other members of the Lincolnshire River Authority for considerable local help and the loan of equipment; to Messrs. S. Phethaan, M. Gill, P. Bush, D. Shelton and C. Amos for help in the field measurements; Messrs. C. Amos and P. Jones for invaluable help in the calculation and the preparation of the papers; Mr. F. J. T. Kestner for many useful discussions; Miss Mary Pugh for drafting the figures. Finally, the writers would like to record their thanks to the Natural Environment Research Council for supporting the studies which led to this project and to Hunting Surveys and the Water Resources Board for permission to reproduce the aerial photograph of the Wash.

References

Amos, C., unpublished Ph.D. thesis, London.

Anderson, F. E., 1973, Observations of some sedimentary processes acting on a tidal flat, *Mar. Geol.*, **14**, 101–116.

British Transport Docks Board, 1970, Silt movement in the Humber estuary, *Research Station Report No. R221.*

Carruthers, J. N., 1968, The Pooh-Bah automatic float, *Cahiers Ocean.*, **20**, 13–17.

Chang, S. C., 1971, A study of the heavy minerals of the coastal sediments of the Wash and adjacent areas, M.Phil. thesis, London.

Clark, H. W., 1959, In discussion of Inglis and Kestner (1958b), *Proc. Instn. civ. Engrs*, **13**, 393–407.

Collins, M. B., 1972, The Wash—sediment contribution from freshwater sources, *Interim Report* to Natural Environment Research Council, U.K.

Crown Estate Commissioners Report, 1969, *Wash and Reclamation. Technical and Economical Aspects of Reclamation at the Wingland Grontmij n.v., The Netherlands.*

Darby, H. C., 1940, *The drainage of the Fens*, Cambridge University Press, London.

Evans, G., 1960, Recent sedimentation in the Wash, Ph.D. thesis, London.

Evans, G., 1965, Intertidal flat sediments and their environment of deposition in the Wash, *Q. Jl. geol. Soc. Lond.*, **121**, 209–245.

Evans, G., and M. B. Collins, 1971, Transportation and deposition of fine-grained sediments on the intertidal flats—Wash, England, *First Interim Report*, Natural Environment Research Council, U.K.

Godwin, H., 1940, Studies of post-glacial history of British vegetation. III, Fenland pollen diagrams. IV, Post-glacial changes of relative land- and sea-level in the English Fenland. *Phil. Trans. R. Soc. B*, **230**, 239–303.

Godwin, H., and M. H. Clifford, 1938, Studies of the post-glacial history of British vegetation. I, Origin and stratigraphy of Fenland deposits near Woodwalton, Hunts. II, Origin and stratigraphy of deposits in southern Fenland, *Phil. Trans. R. Soc. B*, **229**, 323–406.

Halliwell, A. R., and B. A. O'Connor, 1966, Suspended sediment in a tidal estuary, *Proc. 10th Conf. cst. Engng., Tokyo*, Vol. 1, pp. 687–706.

Halliwell, A. R., and M. O'Dell, 1969, Differences in silt patterns across an estuary, *Dock Harb. Auth.*, **50**, 125–129.

Hydraulics Research Station, 1953, *Annual report.*

Hydraulics Research Station, 1972, Wash storage scheme, *Field Studies Report, Ex. 601.*

Inglis, C. C., and F. J. T. Kestner, 1958a, The long-term effects of training walls, reclamation and dredging on estuaries, *Proc. Instn. civ. Engrs*, **9**, 193–216.

Inglis, C. C., and F. J. T. Kestner, 1958b, Changes in the Wash as affected by training walls and reclamation works, *Proc. Instn. civ. Engrs*, **11**, 435–466.

Johnson, M. A., and A. H. Stride, 1969, Geological significance of North Sea sand transport rates, *Nature Lond.*, **224**, 1016–1017.

Kestner, F. J. T., 1961, Short-term changes in the distribution of fine sediments in estuaries, *Proc. Inst. civ. Engrs*, **19**, 185–208.

Kestner, F. J. T., 1962, The old coastline of the Wash, *Geogr. J.*, **128**, 457–478.

Kestner, F. J. T., 1963, The supply and circulation of silt in the Wash, *Proc. Int. Ass. Hydraul Res. Congress, London*, 231–238.

Langbein, W. B., and S. A. Schumm, 1958, Yield of sediment in relation to mean annual precipitation, *Trans. Am. geophys. Un.*, **39**, 1076–1084.

Macfadyen, W. A., 1938, Post-glacial Foraminifera from the English Fenland, *Geol. Mag.*, **75**, 409–417.

O'Brien, M. P., 1931, Estuary tidal prism related to entrance areas, *Civ. Engng.*, **1**, 738.

O'Brien, M. P., 1969, Dynamics of tidal inlets, *Lagunas Costeras, Un Simposio. Meim. Simp. Intern. Lagunas Costeras UNAM-UNESCO, 1967, Mexico*, 397–406.

Postma, H., 1954, Hydrography of the Dutch Wadden Sea, *Archs néerl. Zool.*, **10**, 405–511.

Postma, H., 1957, Size frequency distribution of sands in the Dutch Wadden Sea, *Archs néerl. Zool.*, **12**, 319–349.

Schumm, S. A., 1969, Geomorphic implications of climatic changes, in R. J. Chorley (Ed.), *Introduction to Fluvial Processes*, Methuen, London, pp. 201–203.

Shaw, H. F., 1971, The clay mineralogy of Recent sediments in the Wash, E. England, Ph.D. thesis, London.
Shaw, H. F., 1973, Clay mineralogy of Quaternary sediments in the Wash Embayment, eastern England, *Mar. Geol.*, **14**, 29–45.
Shepard, F. P., and D. L. Inman, 1950, Nearshore water circulation related to bottom topography and wave refraction, *Trans. Am. geophys. Un.*, **31**, 196–212.
Straaten, L. M. J. U. van, and Ph. H. Kuenen, 1957, Accumulation of fine grained sediments in the Dutch Wadden Sea, *Geologie Mijnb.*, **19**, 329–354.
Straaten, L. M. J. U. van, and Ph. H. Kuenen, 1958, Tidal action as a cause of clay accumulation, *J. sedim. Petrol.*, **28**, 406–413.
United States Subcommittee on Sedimentation, 1963, A study of methods used in measurement and analysis of sediment loads in streams, in *Determination of fluvial sediment discharge, Report 14*, Hydraulic Lab., St. Anthony's Falls.
Wheeler, W. H., 1875–76, Fascine work at the outfalls of the Fen rivers and reclamation of the foreshore, *Minut. Proc. Instn. civ. Engrs.*, **46**, 61.
Willis, E. H., 1961, Marine transgressions in the English Fenland, *Ann. New York Acad. Sci.*, **95**, 368–376.

Discussion

Written contribution from M. B. Collins and C. L. Amos

The effect of waves on an intertidal sand flat of the Wash

In addition to the information on suspended sediment presented in the paper, a limited number of wave observations are available for stations 3, 4 and 5 on the transect used by Evans and Collins. Observations were limited to a period of rela-

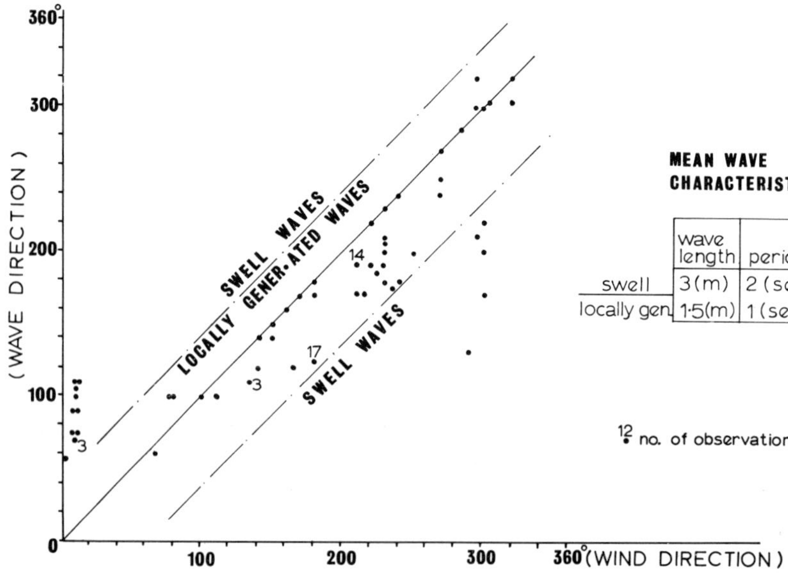

	wave length	period	height
swell	3(m)	2 (sec)	7(cm)
locally gen.	1·5(m)	1 (sec)	9(cm)

Figure 11.11. Relationship between wind direction and direction of wave approach.

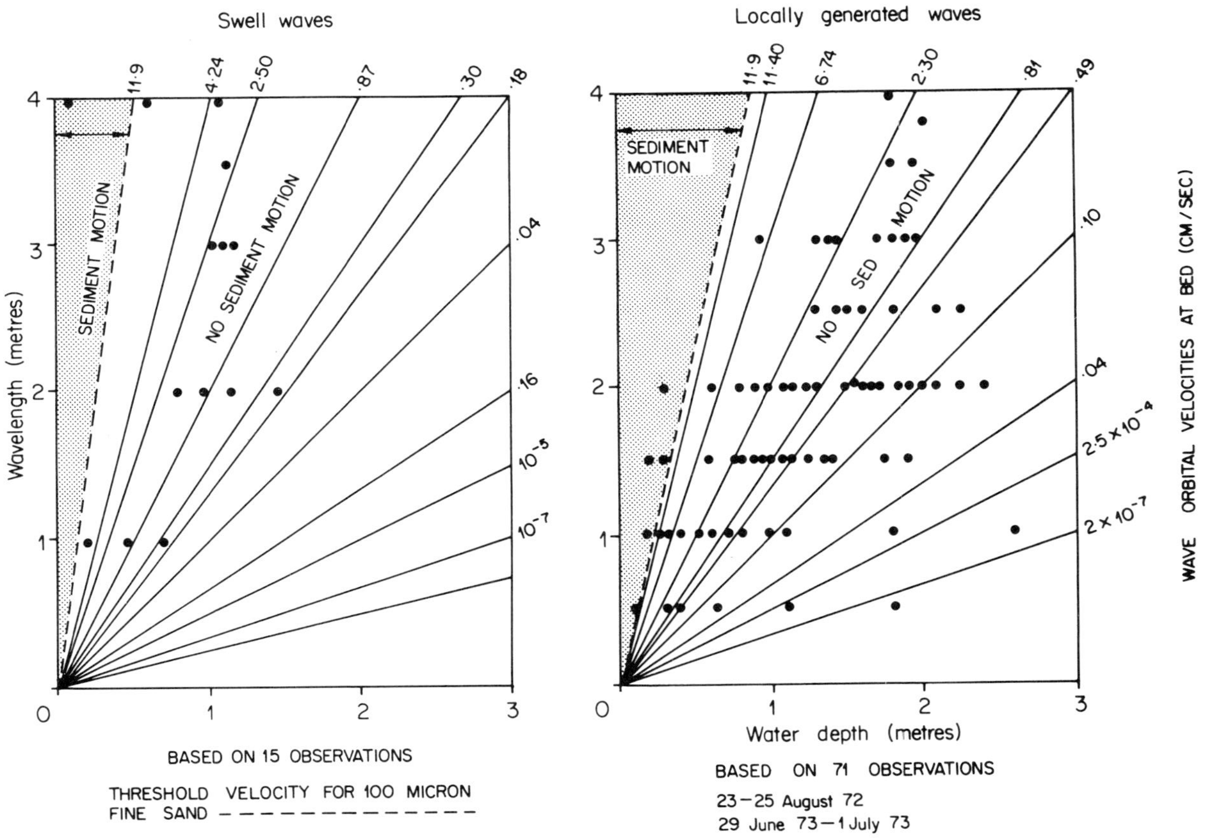

Figure 11.12. Orbital velocity diagrams.

tively calm summer conditions but, nevertheless, an assessment of the influence of these waves on the sediment transport processes was considered justified.

Under normal wave conditions across the intertidal flats, observations have shown that there was only a swell zone and no breaker zone. Thus the energy of waves is not imparted to a specific area of the intertidal zone, but is dissipated across the entire width of the sand flat.

The wave characteristics, based on the classification of progressive surface waves (Defant, 1961) were indicative of deep and intermediate water waves. In attempting to simplify the effects of these waves on sediment movement, it was found possible to further subdivide the waves into two groups according to their direction of approach and morphology (see Figure 11.11). The directions of approach of the locally generated waves have a strong positive correlation with wind direction, whereas swell waves do not follow this relationship.

The orbital velocity of the two types of waves, at the bed, were deduced in accordance with the small amplitude Airy theory (Defant, 1961):

$$U_{\max} = \frac{AN}{\sinh Kd} = \frac{A}{T} \times \frac{2\pi}{\sinh 2\pi d/\lambda}$$

A graphical mean A/T value for each wave type was used (4·3 cm/s for locally generated waves; 1·6 cm/s for swell waves). The relationship between the computed $U_{\max}$ value for each wave type and the threshold velocity for the bed material of the sand flats (100 microns, mean grain size) indicates (see Figure 11.12) the importance of waves in the sediment transport process. Based on data presented by Inman (1957), the threshold velocity of the bed material was 11·9 cm/s (i.e. 7 times the critical friction velocity determined by Sternberg (1972)). It can be seen that 7% of the observed swell waves and 8% of the observed locally generated waves induced orbital velocities sufficient to cause movement of the bed material.

Although these results appear to indicate that the influence of waves is relatively small, it is evident that the combined influence of waves and tidally induced currents is the important and realistic entrainment and transport within the intertidal zone of the Wash. This combined influence will be discussed elsewhere by the writers; however, it can be generally stated that only over approximately 40% of a tidal cycle (i.e. during the first phase flood and final stage ebb) is the tidally induced current sufficient to cause sediment motion independently.

References

Defant, A., 1961, *Physical Oceanography*, Pergamon Press, London.

Inman, D. L., 1957, Wave generated ripples in nearshore sands, *Beach Erosion Board Tech. Mem. No. 100.*

Sternberg, R. W., 1972, Predicting initial motion and bed load transport of sediment particles in the shallow marine environment, in D. Swift, D. Duane and O. Pilkey (Eds.), *Shelf Sediment Transport*, Dowden, Hutchinson and Ross, Stroudsburg, Pa.

Index

Index of Principal Place Names